CAMBRIDGE LIBRARY COLLECTION

Books of enduring scholarly value

Earth Sciences

In the nineteenth century, geology emerged as a distinct academic discipline. It pointed the way towards the theory of evolution, as scientists including Gideon Mantell, Adam Sedgwick, Charles Lyell and Roderick Murchison began to use the evidence of minerals, rock formations and fossils to demonstrate that the earth was older by millions of years than the conventional, Bible-based wisdom had supposed. They argued convincingly that the climate, flora and fauna of the distant past could be deduced from geological evidence. Volcanic activity, the formation of mountains, and the action of glaciers and rivers, tides and ocean currents also became better understood. This series includes landmark publications by pioneers of the modern earth sciences, who advanced the scientific understanding of our planet and the processes by which it is constantly re-shaped.

The Life and Letters of Hugh Miller

This biography, edited by the writer and critic Peter Bayne (1830–1896), was published in 1844. Miller (1802–1856), a Scottish geologist, palaeontologist and evangelical Christian, is best known for his geological arguments for the existence of God. Miller sought to demonstrate the accuracy of the biblical creation story by demonstrating that the seven days of creation correspond to seven geological periods. Volume 2 covers Miller's developing intellectual life and religious ideas; his publications; his marriage and the birth and loss of children; and his own tragic death after a long history of mental illness. The work is a key source for the life and thought of this fascinating nineteenth-century man whose life was marked by a passionate commitment to both science and religion and the attempt to reconcile the two. It will invigorate and entertain the modern-day reader.

The Life and Letters of Hugh Miller

VOLUME 2

PETER BAYNE

CAMBRIDGE UNIVERSITY PRESS

Cambridge, New York, Melbourne, Madrid, Cape Town,
Singapore, São Paolo, Delhi, Tokyo, Mexico City

Published in the United States of America by Cambridge University Press, New York

www.cambridge.org
Information on this title: www.cambridge.org/9781108072380

This edition first published 1871
This digitally printed version 2011

ISBN 978-1-108-07238-0 Paperback

Vincent Brooks Day & Son lith

Æt. 50.

THE LIFE AND LETTERS OF HUGH MILLER

By PETER BAYNE, M.A.

TWO VOLUMES.—II.

The Sutors of Cromarty from "Uncle Sandy's" Garden.

STRAHAN & CO., PUBLISHERS
56, LUDGATE HILL, LONDON
1871

CONTENTS OF VOL. II.

BOOK IV.—THE BANK ACCOUNTANT.

BOOK V.—EDITOR.

BOOK VI.—MAN OF SCIENCE.

BOOK IV.

THE BANK ACCOUNTANT.

Would you understand modern society? Study it in a Bank parlour.

CHAPTER I.

OFFER OF A SITUATION—RESOLVES TO ACCEPT IT—SAILS FOR EDINBURGH—INTERCOURSE WITH SIR THOMAS .D LAUDER—LINLITHGOW.

MANY as are the happy circumstances which we have noted in Hugh Miller's life, it is to be remembered that, at the age of thirty-two, he still finds himself a stone-mason; and that he is ardently attached to a lady, whom he has inflexibly resolved not to marry while he continues to earn his bread by the labour of his hands. The scheme of emigration to America, almost insuperable as were his objections to it, begins to be again entertained. 'My mother,' says Mrs Miller, 'had at length agreed, if nothing suitable turned up, to give us three hundred pounds of mine, of which she had the life-interest; and with this sum we were to face the great wilderness.' Such is Hugh's outlook towards the end of 1834; the final decision on the question of emigration being, I suppose, deferred until the volume of Traditions, of which we have heard so much, and which is now getting actually into the printer's hands, shall have seen the light.

One morning he sits down, by invitation, to breakfast with Mr Robert Ross, just appointed agent of the Commercial Bank in Cromarty. Mr Ross is a warm friend of Miller's, and has asked him to his house on this occasion to have some talk on a matter of business. Mr Ross mentions that he will want an accountant, that the young man who had been thought of for the situation cannot find security, and that his guest may have the place if he will. Hugh is taken by surprise, and, with his usual diffidence, commences to make excuse. 'I know nothing,' he says, 'of business, and very little of figures; there is not a person in the country worse qualified for the office.' Mr Ross understands his man, and persists. 'Say, however, that you accept, and I shall become responsible for the rest.' Hugh reflects for a few moments. 'I thought of the matter; I remembered that no man was ever born an accountant; and that the practice and perseverance, which do so much for others, might do a little for me. The appointment, too, came to me so unthought of, so unsolicited, and there seemed to be so much of the providential in it, that I deemed it duty not to decline.' This last was no mere conventional phrase on the lips of Miller. His religion, quiet and unobtrusive as it was, had impressed itself upon all his habits of thought and life. It had become the one thing essential to his happiness, that he should feel a Divine hand leading him. As usual in the changes of his life, he regarded the alteration in his circumstances with calmness and equanimity, deliberately glad to behold the prospect of life in Scotland with the woman he loved opening before him; but not forgetful of the tranquil hours, so rich in delicate enjoyment of heart and mind, which he had passed, mallet in hand, on the chapel brae of Cromarty, or in sequestered country churchyards, his thoughts busy with some problem of science, or thesis of philosophy, or newly-discovered jewel of poetry, while nature prepared for him, in every changing aspect of the landscape, a fresh delight for eye and soul.

To be initiated in banking, it was necessary for him to proceed a second time to the South of Scotland. He sailed for Edinburgh, expecting to be taken into the office of the Commercial Bank there, but found, on his arrival, that he was to be stationed in the branch office at Linlithgow. He spent a few days in Edinburgh, both before going on to Linlithgow, and on his return thence; and experienced, on both occasions, great kindness from Sir T. Dick Lauder, Mr Robert Paul, manager of the bank in Edinburgh, and others.

Hugh was no sooner out of sight of Miss Fraser, than he began writing to her. He embarked after nightfall, November 27, 1834; the ship weighed anchor at dawn, and we find him, pen in hand, 'tossing on the Moray Frith, on the swell raised by the breeze of the previous night.' The breeze freshens, and he betakes himself to his berth; but the pen is not laid aside. He overhears the master and one of the passengers discussing 'a highly interesting topic,—woman.' They appear to be no enthusiasts on the subject of the sex. 'Your sweetheart, for instance,' says the master, 'was just as fine a lassie as a man could wish to meet with anywhere, and yet you were only a short while in London when she got married to another. But they are all the same.' 'Rubbish,' thinks Hugh; 'I don't believe a word of it. The last things I looked on with interest, as we swept along the Sutor, were the little grey rock and the beech tree. How much happiness, my own Lydia, have I enjoyed beside them!' But he has too much contempt for the speakers, or is too sea-sick, to rise and do battle for the fair. 'We have an eagle aboard,' he proceeds; 'a noble-looking bird, but quite as bad a sailor as myself. He is so affected by the motion of the vessel, that he has not tasted food since we left Cromarty. And yet, from his previous mode of life, he must surely have been accustomed to motion much more violent. I brought you acquainted one evening, if I recollect right, with Coleridge's rhyme of the "Ancient Mariner." It is a wonderful poem. On shore I regarded it as merely a wild and somewhat barbarous trophy of a powerful, but unregulated, imagination; at sea I can perceive that there is much of truth as well as of imagination in it. I never thoroughly understood the love-poetry of Burns, never felt that his language is, so entirely above that of all other poets, the language of the heart, until I became acquainted with you. A short trip at sea has enabled me to make a similar discovery regarding the language of Coleridge. It is thoroughly nautical; there is sea in every line of it.'

On Friday he is watching the pale sands upon the coast of Moray; on Sunday morning he is in the Frith of Forth. 'We are bearing up the Frith in gallant style, within less than a mile of the shore. Yonder is the Bass, rising like an immense tower out of the sea. How have times changed since the excellent of the earth were condemned, by the unjust and the dissolute, to wear out life on that solitary rock! My eyes fill as I gaze on it. The persecutors have gone to their place; the last vial has long since been poured out on the heads of the infatuated Stewarts, whose short-sighted policy would have rendered men faithful to their princes by making them untrue to their God. But the noble constancy of the persecuted, the high fortitude of the martyr, still live. There is a halo, my Lydia, encircling the brow of that rugged rock; and from many a solitary grave, and many a lonely battle-field, there come voices and thunderings, like those which issued of old from oehind the veil, that tell us how this world, with all its little interests, must pass away; but that for those who fight the good fight, and keep the faith, there abideth a rest that is eternal. In a few hours, if the breeze does not fail, we shall be within sight of Edinburgh. And there, my Lydia, I must enter on a new scene of life, a scene for which my previous habits have little fitted me. But I shall think of you, and bend all my mind to my untried occupation. The sun has just sunk behind the Pentlands, and the coast looks dim and blue through the twilight. One part of the line, about three miles in extent, is entirely blotted from the landscape, by the smoke of Edinburgh and Leith. And now the Isle of May light is beginning to twinkle behind, and the Inchkeith light to twinkle before. A few short hours, and our voyage shall have come to its termination. The various objects around me—the hills, the islands, the buildings—recall a hundred recollections of my residence here of ten years ago, which have lain buried in my memory ever since. There was a double row of trees beside the cottage in which I lodged, through which I used to see the Inchkeith light twinkling every evening. I remember the dark, lonely road which led past the door to Edinburgh; and how, when travelling on it in the night-time, which I did often, I used to grasp my stick every time I heard footsteps approaching. I must strive to ascertain whether my old landlady is yet alive, and what has become of my companion, John Wilson.' A little touch this last, but characteristic. Hugh, at this rather exciting moment, thinks more of old friends, however humble, than of new.

'Edinburgh, Monday.

'Our vessel got into harbour this morning, about

two o'clock, just in time to escape a tremendous storm of wind and rain from the south-west, which, had it caught us in the Frith, would have driven us back to Peterhead. I have taken lodgings at No. 14, Romilly Place, to which you must address for me. Do write me a very long letter, longer than mine by a great deal. You may see, from my first and second page, that I was within a hair's-breadth of being unable to write at all. Had I been pledged to write any one else, I would have pleaded incapacity, and firmly believed the plea a good one, but the thought of you nerved me when half-dead, and the words came. I love you ten times better when separated from you by about 200 miles than when I could draw in my chair beside you; or, I dare say, it would have been more correct to say, my affection is just what it was, but I have now semething to measure it by. Do you not think me a silly fellow? There is something inspiriting in the air of Edinburgh; I feel myself an inch taller as I walk the streets. But I am afraid I shall have something else to tell you when I write again. I shall ere then be set down to my desk, the least skilful and least confident of all clerks; and you shall hear of little else than blunders and incapacity, and of how ill-suited I am for the part I have to perform. Well, however dispirited I may be, I am sure no man was ever born an accountant, and that the practice and perseverance which do so much for others, cannot fail doing a little for me. I see many changes in Edinburgh. There are large open spaces, which were occupied, ten years ago, by lofty masses of building; and masses of building where there were then only open spaces. Some of the new statues I don't at all admire. I find few traces in them of the hand of the master. They seem as if finished by the journeymen of genius, and as pieces of mechanical skill do barely well enough. But the Promethean soul is wanting.'

This on Monday. On Thursday the ardent lover and indefatigable correspondent has again pen in hand. He has been to the Grange House, the residence of Sir Thomas Dick Lauder, and writes of his reception with enthusiasm. 'I cannot express to you the kindness with which I was received. He scolded me for taking lodgings. Why not come and live with him? And I was only forgiven on condition that, after arranging matters with the secretary of the bank, I should part with my landlady, and take up my abode at the Grange. "I have a snug room for you," he said; "breakfast shall be prepared for you to suit your office hours; and in the evening I shall have you to myself. To-morrow you must come and dine with me; I shall get Black, the bookseller, to meet with you, and meanwhile I shall write him a note that will be at once an introductory one and an invitation." He then introduced me to Lady Lauder and his daughters; showed me his library, a capacious room, shelved all round, and rich in the literature of the past and the present. "Here," said he, "Robertson, the historian, penned his last work; and here,"—opening the door of an adjoining room,—"he died." He next brought me to the leads of his house; pointing out the more striking features of the scenery; told, and told well, a number of little stories connected with it; showed me the extent of his lands,—but I want space to enumerate. We parted.'

Armed with Sir Thomas's note, he waits upon Mr Adam Black, future Lord Provost of Edinburgh and member of parliament for the city, and is civilly received. Next day he starts for Linlithgow, and seems hardly alighted there, when the indefatigable pen is again in requisition, to describe to Miss Fraser what occurred on the evening passed with Sir Thomas.

'Linlithgow, December.

'I do not know that I have anything more amusing to communicate to you, my Lydia, than what passed during the evening I spent at Sir Thomas's. But I am afraid you will find me no Boswell. I would fain be a faithful chronicler; but, in attempting to record dialogue, the words always slip away, and only the ideas remain. My invitation was for six o'clock, the fashionable hour for dinner here; but, by missing the road in the darkness, I was, unluckily, rather late. The Grange House is built in the style of two centuries ago, with a number of narrow serrated gables, that break the light into fantastic masses, by their outjets and indentation; here a pointed turret, there coped with stone, and bearing the family crest atop; yonder an antique balustrade; and, directly in front of the iron-studded door, there are two time-worn columns with a huge dragon sprawling on each. The garden is in quite the same ancient style, planted by some old-world mason, with flights of stairs, cross walls, and arches. The first thing that caught my eye on entering the lobby was a huge, carved settle of dark-coloured oak, with the bust of a mitred prelate frowning from the wainscoting over it; there were spears, too, resting against the wall, and in the antique staircase a host of old paintings of ladies, in strange, uncouth dresses, who were loved and married three centuries ago; and of their lovers and husbands, grim-looking fellows, with long beards and coats of mail. I was ushered into the parlour, a splendid apartment, as lofty as any two of our Cromarty rooms placed over each other, and more capacious than any four, with a carved oak roof, panelled sides, antique wainscot furniture, and an immense profusion of paintings. Sir Thomas and Mr Black were standing beside the fire, discussing the change in the ministry; * Lady Lauder was seated at a work table a little away. I was received by the lady very kindly, by Mr Black very politely, by Sir Thomas as if we had been friends and companions for twenty years. The political conversation was then resumed. Sir Thomas remarked that if the Duke of Wellington calculated on the soldiery,—and he could not well see what else he could calculate upon,—he trusted he had mistaken their spirit. For the army, said he, is composed of the people, and in a time of peace like the present must be imbibing their opinions. I stated to him, in proof of what he remarked, that I had crossed the ferry of Fort George last summer with a party of soldiers, and was interested to learn, from their conversation, that many of them were acquainted with the periodicals, and fond of reading. And I question, I said, whether a reading soldiery be the best for doing everything they are bid. Sir Thomas deemed the remark of some value, simple as it may seem . . . Sir Thomas showed us a highly-interesting relic of Queen Mary,—a watch, formed like a human skull, which was presented by her to that Lady Seaton whom Scott has made the heroine of his Abbot. The upper part of the skull is richly embossed with figures; there is the crucifixion, the adoration of the shepherds, and several other Scripture scenes connected with the history of our Saviour; on the sides there is a series of vignettes,—the frock without a seam, the nails, the scourge, the crown of thorns, and the spear. The workmanship is evidently French.

* The attempt made by Peel and Wellington, in the end of 1834, to take the reins out of the hands of the Whigs. It broke down in a few months.

'Sir Thomas took up a volume, presented by Sir Walter Scott to Lady Lauder, and showed us Sir Walter's holograph on the title-page. "This," said he, "I deem a valuable volume; and here is something I consider as equally so." He opened a portfolio, and showed us the original plan and elevation of Abbotsford, also a present from Sir Walter to the lady. The conversation then turned on Sir Walter. "I had some curious correspondence with him," said Sir Thomas, "shortly before his death. Contrary to the opinion he had formerly entertained, he then held with Dr Jamieson that the Celts had never inhabited the south of Scotland. I instanced several Gaelic names of places in the south—among the rest that of his own Melrose, or the barren promontory—and he seemed reconvinced; but half his mind was gone at the time. Our Gaelic names," continued Sir Thomas, "are strikingly characteristic of either the scenery of the places which they designate, or of some incident in their history, so very remote, perhaps, as to lie beyond the reach of written records. I was led, after writing my essay on the parallel roads of Glenroy, to examine appearances on the course of the Findhorn, very similar to those of the highland glens. Among the rest there is a holm on the Relugas property, round the sides of which I could trace very distinctly what seemed to have been at one time the shores of a lake; but what was my surprise when, on asking a Gaelic scholar for the etymology of the name of a field which occupies the upper part of the holm, I was informed that it was composed of two words which mean 'head of the loch.' Now, at how remote a period must not the name have been given?" I instanced some of our Cromarty names as apparently of very remote antiquity; stated that a moor in the upper part of the parish had, as shown by its cairns and its tumuli, been the scene of a battle at so early a period that history bears no recollection of the event, but that a farm in its vicinity still bears the name of *Achnagarne*, i. e. field of the carcasses; and that a rock in the sea, which Sir Thomas, in his survey of the burgh, has marked out as one of its boundaries, and on which tradition says a boat was once wrecked, is still known as *Clach Mallacha*, i. e. the stone of the curse. We had some conversation on the Celtic character. I described to Sir Thomas the form of the old Celtic head, as given us by the phrenologists, and as I have seen it in the skulls of the Inverness Museum; concluding my description by remarking that civilization seems to produce variety in the human species, somewhat in the manner that domestication produces it in some of the inferior animals. Sir Thomas seemed pleased with the thought, and illustrated it by a fact or two.

'He must have been a very busy man. He showed me his Travels in Italy in MS. They form four thick quarto volumes, elegantly bound, and illustrated with admirable crow-quill drawings. He showed me an elegant piece of penmanship, "The Lamentable Case of Sir William Dick," a thin folio, in which the old style of printing and engraving was so well imitated, that it was only from the freshness of the paper I detected it as a copy. This Sir W. Dick was one of his ancestors (Scott makes David Dean allude to him, in the Heart of Midlothian, as the godly provost), who was possessed, in the days of Charles I., of the then enormous sum of £200,000, but who lost almost all during the Commonwealth and the reign of Charles II. Sir Thomas stated to us besides that he had filled with notices of his family a roll of vellum about eighty feet in length. On Mr Black and I rising to go away, he took down his large stick, and accompanied us—by way of guard, he said—for about half a mile, and on parting kindly repeated his invitation to me of coming to the Grange the moment I returned from Linlithgow.

'I do not know that you have ever seen Sir Thomas. He is a noble-looking, elderly man, upwards of six feet in height, very erect, with bold, handsome features, and a profusion of grey hair, approaching to white, curling round his temples. His head is a very large one, with a splendid development of sentiment. Benevolence, veneration, and ideality seem all of the largest size. Love of approbation and combativeness are also amply developed. His forehead is broad and high, but the knowing organs are more powerful than the reflective ones. The contour of the whole is beautiful, and as much the reverse of commonplace as anything you ever saw.

'My trunk being too bulky for the coach, I took a berth in a canal-boat, which leaves Edinburgh at seven in the morning and reaches this place about ten. I saw little on the passage to interest me except the old castle of Niddrie, at which, as you will remember, Queen Mary passed the night after her escape from Loch Leven, and in which Scott has laid some of the scenes of the Abbot. My fellow-passengers were a Paisley shop-keeper and a Linlithgow farmer—the former a smart, shallow young man, the latter a shrewd, sagacious old fellow, with a decided cast of dry humour. On landing here I found the bank accountant, a Mr Miller, waiting my arrival. He introduced me to Mr Paterson the agent. Both of them are exceedingly civil—nay more—kind young men; but the patience of both must be sorely tried ere I can have done with them. I am one of the stupidest blockheads you ever knew, and, considering how extensive your experience in this way must have been among your pupils, that is saying something. For the first day or two I felt miserably depressed and sadly out of conceit with myself. I do not know what I would not have given to have had you beside me to speak me comfort, but I dare say you would have begun by laughing at me. My lodgings here are much too fine and too expensive, but they were taken for me by Mr Paterson at the request of Mr Paul, who intimated my coming by letter, and so I could on no account decline them. I dislike expense even for its own sake, and independent of the embarrassment which it always occasions,—especially when 'tis incurred for a man's self, for food a little more delicate, and clothes a little finer than ordinary. My disposition, too—as the Edinburgh phrenologists will, I dare say, find—leads me rather to acquire than to dissipate. . . . . . Remember, I expect a reply. You little know the exquisite pleasure which I derive from your letters.'

A letter from Miss Fraser arrives in due course, but it is with a feeling very different from pleasure that he peruses it. She has been suffering from illness, and her nervous excitement has been such, that 'this morning,' she writes, 'when your letter was delivered to me, I was almost fixed in the belief that I was suffering under a temporary derangement. God alone knows where I would have been to-day without it; it turned, in part, the current of my ideas.' Hugh writes, evidently without loss of time, though the letter has no date, and in the greatest agitation. His first sentence appears to intimate that he had a presentiment of evil so soon as his eye fell upon the letter. ''Twas no wonder, my own dearest girl, that I should have felt so unwilling to open your letter, and that I looked twice at the seal to convince myself that it was not a black one. You are unwell, my Lydia; and here am I, whose part it should be to soothe and amuse you, separated from you by more than 200 miles. Your temperament, my Lydia, is a highly nervous one; your delicate tenement is o'er-informed by spirit; 'tis a hard-working system, and the slightest addition to the moving power deranges the whole machine. But be under no apprehensions for your mind. Would that I were beside you to tell you how strangely I have sometimes felt when in a state of nervous irritability. The night before I received your letter, for instance, I had a world of foolish fancies about you; and my sense of hearing was so painfully acute, that I was disturbed by the noise of the blood circulating in the larger arteries. But all this was merely the effect of over-exertion. I had copied and calculated all day in the office, and had written for the printer till late at night; in the morning I was quite well. Try and get yourself amused, my Lydia, and do not suffer your spirits to droop. I shall tell you what my chief—indeed only—amusement is at present: I have a little paper book, which I carry about with me, and when I have a minute's leisure I take it out, and with the aid of my pen converse with you. This is the secret of my long letters; and of all my present pleasures this, with one exception, is the most pleasant. I need not tell you that the excepted one arises from your letters; but not such letters, my Lydia, as your last. I must hear that you are well before I shall have recovered the shock it has given me.

'I have never yet been in any part of the country where the surface is so broken into little hillocks; one might almost deem it an imitation, on a small scale, of the Highlands. Its geological character is highly interesting. Almost all the eminences are basaltic. In the hollows we meet with sandstone, lime, and indurated clay. The lime is rich in animal remains, all of the earliest tribes. They belong to a later era than the fossils of the Eathie *lias*. I have procured a few specimens, which I must try to bring home with me—among the rest four varieties of bivalves and two of zoophytes. I have procured, too, part of a fossil palm and the joint of a flattened reed, possibly the vegetable to which the south of Scotland owes its coal. But I have met with neither ammonites nor belemnites. In many places the basalt has overflowed the secondary strata; and in the side of an eminence, rather more than a mile from town, where there is a deep section of rock, it assumes the columnar form directly over a thick vein of lime. You have, I dare say, never seen basaltic columns. I saw them here for the first time, and wished for you, that we might examine them together. So regular are they, that old country people of the last age used to attribute their erection to the Picts. I have seen it stated in a paper by Creech, the Edinburgh bookseller, and deem the fact a highly interesting one, that on some occasion the furnace of a Leith glass-house having been suffered to cool, it was found that the glass, in passing from a fluid to a solid state, had assumed the columnar form. There is a particular form of hill which seems peculiar to basaltic countries, and of which there are various instances in this. You have seen a superb specimen in Salisbury Crags. It would seem as if, after the masses had been thrown up, they had split in the middle; and that when one half of each remained standing, the other sunk into the abyss out of which the whole had risen. Linlithgow forms, as you are aware, part of the great coal-field of Scotland, and there are pits on every side of it. The coal seems to have been formed out of vast accumulations of reeds, somewhat resembling in appearance, at least, the sugar-cane. To what a remote and misty antiquity do such appearances lead us? To a time in which the district in which I am now writing to you formed part of the delta of some immense river which drained of its waters a widely-extended continent, the place of which is now occupied by the Atlantic. Think how many ages must have elapsed before the vegetable spoils of even the largest stream could have formed the depositions of so extensive a coal measure; how many more must have passed in which new accumulations of strata settled above these to a depth of many hundred feet—settled so slowly, too, that each layer formed a plain on which plants and animals flourished and decayed. Continue the history till the immense continent was slowly worn away, and the sea beyond, enriched with the spoils of so many ages, became a scene of earthquakes and volcanoes; and then, after we have marked in imagination the retiring of the waters, and the ascent of a new continent from what had been the profounder depths of the sea—after we are lost in calculating the periods which must have elapsed ere the ascent of one plutonic eminence was followed by that of another—antiquity, as it regards the human race, has but its beginning. I find myself lost in immensity when I think of such matters; but I dare say you have quite enough of geology for one letter.

'The church of Linlithgow is a fine old building, well-nigh as entire in the present day as it was four centuries ago. In style it seems to hold a middle place between the simple Norman Gothic and the highly ornamental Gothic of the reign of Henry VII.; and there is a chastity in the design of at least the interior, which we may vainly look for in our modern imitation buildings. I never look up to a lofty stone roof without feelings of awe. Burke has said that fear is a necessary ingredient in the sublime. I do not know but that there is a lurking sensation of terror in the feeling which I experience; it does not owe its existence to the art in which, according to Thomson, "greatest seems the little builder man." I sit in the northern aisle every Sunday, beside a huge column, and directly opposite the gallery in which the spectre appeared to James IV. The clergyman is a fine, useless preacher of the Moderate party, who gives us rather ordinary matter dressed up in pretty good language. He does not pray on Sabbaths, like our north-country ministers, to be "preserved from thinking his own thoughts," and may, indeed, spare himself the trouble—he has none of his own to think.

'The palace is situated a little behind the church. It is a huge quadrangular pile, about sixty paces on each side, full of those irregularities which would not be tolerated in a modern building, but which, associated as they are with our conceptions of Scotland in the past, please more than elegance itself. There runs along the top a deeply-tusked cornice; the corners are crowned with turrets, and broken piles of building, which finely vary the outline a-top, rise high above the outer wall on either side. The carvings are sorely time-worn, and they seem to have been grotesque enough when at their best. On either side of an old gateway, which was shut up in the reign of James VI., there are two Gothic niches, surmounted by miniature cupolas that resemble the models of an architect; at the base of each there is the figure stretching forth his hands, and writhing in agony, as if crushed by the superincumbent weight. The Scottish shield, guarded by angels, is blazoned on an immense tablet above. But decay has been busy with the guardians, and with what they guard. There is a large court in the interior, whose corners are occupied by lofty towers, through each of which a staircase leads to the top. The view inside is very striking: all the sides are unlike. One of these was built in the reign of James VI.; and from the elegance and peculiar style of the architecture, I would deem it a design of Inigo Jones. The other sides are of a different character, and testify of an earlier age. The windows are square, and huge of dimensions, labelled a-top, and divided into compartments by mullions of stone. There is an uncouth profusion, too, of Gothic sculpture. In the middle of the court there are the ruins of a well, and beside it a hollow which must once have received its waters and formed a little lake; but the stream has long since failed. I passed through a wilderness of arched passages, with windows darkened by mullions of crumbling stone, and grated with wasted iron. I have seen the room in which the unfortunate Mary was born. I have seen, too, the large hall in which our Scottish parliament sometimes assembled, with the stone gallery in which the beauties of other days have listened to the long-protracted and often stormy debate. I ascended to the top of the building, and from an elevation of nearly a hundred feet looked over the surrounding country. The palace is built on a grassy eminence that projects into the lake, which extends about half a mile on either side of it, and nearly as much in front. A curtain of little hills rises from the opposite shore, and shuts in the scene towards the north. To the south we see the town, and the long line of the canal, with its multitudinous bridges; and all around there is an undulating and freely diversified country, studded with abrupt woody knolls of plutonic formation, and speckled with human dwellings. I need not tell you that, as I looked from the walls and saw so much of the antique and the venerable beneath me, and so much of the beautiful around, I wished for a companion to see all that I saw, and to feel all that I felt; nor need I say, my Lydia, what companion it was I wished for. Uncommunicated pleasure, you know, is apt to change its nature, and to become pain.

'Mr Miller, our accountant here, is the son of a dissenting clergyman, who died about four years ago. He is a smart, obliging young lad, and a thorough master of his business. He has introduced me to a few of his acquaintance here, and to his mother, a remarkably fine-looking woman, who, though her son is perhaps as old as you are, might still pass for a young lady. I have twice drunk tea with her. I have passed an evening, too, with Mr Paterson, our bank agent, a frank, obliging young man (he is five years younger than I am), of much general information, and with none of the little pride of our north-country bankers in his composition. Among the many causes of gratitude which Providence has given me, the kindness which I everywhere meet with is not one of the least. Mr Paterson tells me that McDiarmid, the editor of the *Dumfries Courier*, and the ablest of all our provincial editors, was at one time a clerk in the Commercial Bank; but he was by no means a very superior one. One of the tellers was astounded on one occasion to find his cash a thousand pounds short; and after vainly striving to discover some error in his calculations, he gave up the search, and deemed himself ruined. He lived a wretched life for about a fortnight, when he at length ascertained that poor McDiarmid, in one of his absent moods, had contrived, in carrying forward his balance, to leave a thousand pounds behind. On leaving Edinburgh for Dumfries, his desk was found stuffed full of literary scraps, unfinished stanzas, and broken sentences. I must try to avoid being made the subject of a similar story, by keeping my literature at home.

'This is the country of historical associations and historical relics. I described to you in my last the watch of Queen Mary. I have since seen two pieces of her needle-work. They are at present in the possession of my landlady, who was for many years housekeeper to a lady of quality, whose name I forget, and who at her death left her her wardrobe—the relics of Mary included. The one is an apron, the other a tippet, both of muslin, which was once white, but which now, both in colour and in fragility, resembles a spider's web. The apron is a complex piece of work—nearly as much so as the borders on which I have so often seen you engaged; the tippet is simpler. You will laugh at me when I tell you that, all unpractised in the art as I am, I am employed in making a pattern of it for you, that you may see how muslin was flowered in the sixteenth century, and bedeck yourself, should you deem it worth your imitation, in the same style of ornament with the beautiful Mary. I need not tell you I am no critic in such matters;—it strikes me, however, that the flowering of both pieces has a grotesque Gothic air, and differs as much from the needle-work of the present day, as the old castle of the sixteenth century does from the modern mansion-house. In the possession of such persons as my landlady one frequently meets with interesting relics on the last stage of their journey to oblivion. The work table on which I write is only about twenty inches square a-top; yet I am certain that top must have employed some skilful mechanic of a century ago for a full month. It is curiously inlaid with more than four hundred little pieces of coloured wood and bone, and represents a flower piece.'

Miss Fraser's next letter was brief, but gave a good account of her state of health. She had been visiting Mrs Taylor, a friend resident on the Nigg side of the Cromarty ferry, and had reaped much benefit, both in the way of health and of happiness, from 'the quiet cheerfulness of a united and happy family.' Miller evidently loses no time in replying.

'You were not mistaken, my Lydia, in thinking that your letter, short as it was, would meet with a sincere welcome. I had become extremely uneasy regarding you, for I attributed your silence to indisposition;—the mail drives up to the Post-office here by a street which fronts my window; and regularly as the hour of its arrival came,—'tis a late one,—I used to watch the approach of its two flaring lanterns, that seem two terrific eyes, in the hope of hearing from you. But evening after evening passed, and still no letter, and I began to indulge in the gloomiest forebodings. Think, then, what my feelings must have been, when, on perusing your kind, though brief epistle, I found you were well and not unhappy. I am much indebted to the kindness of Mrs Taylor; and the time may come, my lassie, when there will be no impropriety in thanking her for it.

'I am much gratified by your literary scheme. I have long ago told you that you are not one of those who can be at once indolent and happy; and I am sure you must often have felt that the remark was a just one. Your mind is highly active, and must have employment; and I know no exercise more suited to it than the one you propose. We must mutually assist and encourage each other, my Lydia, and should you be unsuccessful at first in forcing your way to the publisher's shop, you must just remember that there are few writers who have not failed in their earlier efforts. Genius itself seems hardly more indispensable to the literary aspirant than that mixture of firmness and self-reliance which, undepressed by failure or disappointment, can pursue its persevering and onward course till at length it triumphs over fortune and circumstance. At present your mind resembles a musical instrument of great compass and power, but nearly all the semi-tones are wanting. But, my own dearest Lydia, for your own sake and mine you must remember that your mind and body are unequally matched;—that though the one is strong and active, the other is comparatively fragile and easily worn out, and that your exertions must be modified to suit the capabilities of the weaker of the two.

'I was only a few days in Linlithgow when a gentleman called on Mr Paterson (the agent) to inquire for me, stating that the Principal (Baird) was at his country house, and very unwell, but desirous notwithstanding that I should call on him on the following Thursday. I then learned for the first time that the Principal's country house is not more than two miles from Linlithgow. I found the grounds in the vicinity of the house laid out into little patches, each bearing a different variety of field or garden vegetables, and altogether presenting the appearance of what is termed an experiment farm. Husbandry and gardening are two of the Principal's hobbies. The house is a little old-fashioned structure. I was shown into a low parlour;—the Principal was in bed, I was told, but was just going to try to get up. The poor Principal found himself unable to rise, and I was shown up to his room. He received me with great kindness, held my hand between both his for more than ten minutes, and overpowered me with a multitude of questions,—particularly regarding my new profession and what had led to it. "Ah," said he, when I had given him what he requested,—the history of my connection with the Bank,—"the choice of your townsman, Mr Ross, shows that you still retain your character for steadiness and probity." After sitting by his bed-side for a short time, I took my leave, afraid that he might injure himself by his efforts to entertain me; for they were evidently above his strength. It struck me, too, that there was a tone of despondency about him which mere indisposition could not have occasioned. Benevolent old man! from what I have since heard, I have too much reason to conclude that his sickness is of the heart.'

Miller proceeds to mention his having formed the acquaintance of a Mr Turpie, at whose house he was introduced to a Dr Waldie, both unknown to fame. 'Mr Turpie,' he goes on, describing an evening passed in the company of these gentlemen, 'took up a book, and showed me what he deemed a very old poem. I read a few verses, and pronounced it to be a modern imitation. The decision led to a few queries, and the queries to a sort of colloquial dissertation on old Scottish poetry, a subject with which, you know, I am pretty well acquainted. I quoted Barbour, Dunbar, Gavin Douglas, Lindsay, and a great many others. The obsolete literature of our country was quite a *terra incognita* to all the party, but they seemed interested by the glimpse I gave them of it; and Mr Turpie, when the conversation once more became general, asked me, half in simple earnest, half in the style of compliment (a question which by-the-by Mr Paterson had put to me a few days before), "Pray, Mr Miller, are there any books which you have not read?" A few evenings after, the Doctor and Mr Turpie called on me at my lodgings. The former is a metaphysician, and he had come to me apparently with the intention of discussing what may be termed the metaphysics of phrenology;—its connection, for instance, with the grand question of liberty and necessity, and the doctrines of the will. I communicated to him my ideas on the subject as clearly as I could; met his objections when they could be met; and showed him, —I should rather say strove to show him,—the boundaries of that horizon of darkness which, closing round the human intellect in this direction, renders many of them unanswerable, not because they are powerful as arguments, but because they cannot be understood. We parted very well pleased with each other. "The Doctor," said Mr Turpie to me a few days after, "can find no line long enough to measure you by; he has just met with a Dr Baird, a nephew of the Principal, who tells him that his uncle is quite enthusiastic regarding you, and deems you equal to anything." But enough of this. Never in my life before did I write anything so redolent of conceit as the last page and a half; but with you, my lassie, I know I am more than safe. Remember, too, I give you full liberty to laugh at me as much as you please.

'My own dearest Lydia, I must hear from you once yet; and to make up for the briefness of your former letters, do write me on a double sheet. Tell me much about yourself,—what you are doing, and saying, and thinking, and seeing, and feeling; on a theme so interesting you cannot be tedious.'

Miss Fraser's answer to this is dated Cromarty, 8th January, 1835. An extract from it will tend to elucidate Miller's next letter. The manse of Alness, mentioned by Miss Fraser, is beautifully situated on the north of the Cromarty Frith. It was occupied in 1835 by the late Rev. Mr Flyter and his family. Throughout the north of Scotland, where the manses had from time immemorial been centres of hospitality in their districts, none was more noted for cordial, generous, and delicate hospitality than the manse of Alness. 'I was received at Alness,' writes Miss Fraser, 'with great affection. The increase of wealth there has not blunted any of the finer emotions of the heart. I could perceive some changes; the simple manse is turned by additions and improvements into something like a mansion-house, and the glebe cultivated to resemble a gentleman's pleasure-grounds; I could perceive, too, among the inmates, something of an aristocratic turn of idea, caught from the society of the neighbouring proprietors. But in piety, and the discharge of pastoral duty, there is no change. The spirit of the Presbyterian minister, as he was in the days when the success of the gospel was all to him, is kept alive in Ross-shire in perhaps greater strength than in any other part of Scotland. The ministers of the contiguous parishes for many miles round meet every month in the house of each alternately, to inquire into the state of their parishioners, and to implore the aid of the Holy Spirit. Thus they pass a whole evening.

'We had for two evenings the society of which I am so enthusiastically fond,—that of a genuine Celt. I sang and played, and he showed a fine taste for music. I repeated some verses,—he criticized them at once with the most just conception. The conversation became general;—he showed the best sense and soundest practical observation;—his grasp was not extensive, but his ideas were all clear and well-defined, and he had evidently thought for himself. At supper the conversation again became poetical. He quoted some words of exquisite beauty and expressiveness in the Gaelic Psalms, which no English ones could render. He described the Gaelic word for "wind" in the line, "For over it the *wind* doth pass," as expressing to the imagination a breath so faint as to be almost dying, and yet sufficient in its extreme feebleness to destroy the still feebler thing over which it is passing. Again, in the same psalm, in the lines,

"Such pity as a father hath
Unto his children dear,"

he said there was something inexpressibly tender and delicate;—they convey the idea of a parent yearning to fondle what he so tenderly loves, and yet in his solicitude almost afraid lest he injure it by the touch. Oh what burning thoughts must have passed through the brain of Ossian! That a people with such genius, and with such a language, should be deemed incapable of producing such a poet! That those who have felt but for a moment the spirit of Northern poesy could doubt ever after that Ossian sung! Yes, annihilate the remains of Highland feeling and language and manners, and then tell us that the question is decided, but not till then. I am so exasperated at you that I would fain give you a—pinch.'

Hugh replies in a tone of quiet and kindly badinage. He will maintain against all comers that the poems of Ossian illustrate the genius of the Highlanders, particularly that of the gifted clan Mac Pherson. To match the Alness Highlander he brings out a specimen Celt of his own.

'Edinburgh.

'Dear me, what a red-hot Highlander you are! You

make me say things against the poor Celt I never so much as thought of, merely, I suppose, that you may have the pleasure of defending him. Who ever doubted that the poems of Ossian were the compositions of a Scotch Highlander? Truly not I, nor any one else I ever heard of, except a few Irishmen. They were written by a countryman every line of them,—bating the little bits that were borrowed from Milton and the Bible,—by a genuine countryman, who though not over endowed with honesty, equalled in genius any writer of his age. Ossian, indeed, or Oscian, as the Irish call him, was, as you know, a bog-trotter of the beautiful island, who made ballads in the days of the good St Patrick, and sold them for half-pence a piece; but who can say that of Mac Pherson?

'Since you love Highlanders so well, I fain wish I could introduce you to my cousin, George Munro. I would not fear to match him, as a specimen of what his country can produce, against your Alness Highlander or any Highlanders you ever saw. He resides with his wife and family in Stirling, and since I last wrote you I have spent a day with him. Let me describe him to you as he is both in mind and person. He is a well-built robust man of five feet eight, large-limbed, broad-shouldered, keen-eyed, and with resolution stamped on every feature. Nature has written *man* on his whole appearance in her most legible hand. But what I have to add will I am afraid give you a lower opinion of him. No one ever regarded me as particularly well built or handsome. I am, besides, fifteen years younger than my cousin, and yet through one of those tricks of resemblance so strangely occasioned by blood, I have been repeatedly addressed as Mr Munro. His mind is one of the most restless and most concentrated in its energies I ever knew. He never yet attempted anything which he did not master, and never mastered anything of which he did not tire. He was born in the Highlands of Sutherland, and bred a mason: —no one could have fewer opportunities of improvement, and yet he was not much turned of twenty ere he had added to the commoner rules of his art a knowledge of architecture, drawing, and the mathematics. The intellectual man is rarely an *athlete*, but George had a body as well as mind to educate, and after studying the mathematics, he set himself to study the art of defence, and became so skilful a pugilist that there are few professed boxers who would gain in a contest with him. He resided at this time in Glasgow. On his return home he married and took a little farm on the banks of a Highland loch, where he proposed to himself to spend his days. But he soon tired of the agricultural life,—it was too quiet and too monotonous, and quitting the farm he engaged as superintendent of some saw-mills erecting in that part of the country, and proved for some months, from his thorough though hastily acquired knowledge of the machinery, a most serviceable man to his employers. He sickened, however, at the ceaseless clatter of the wheels, and throwing up his superintendency, he again resumed the mallet. He then became a slater, and proved one of the best in the country, but the details of the art were too soon mastered to engage him long. He next applied himself to Gaelic literature, and published a translation of Bunyan's Visions, which has been commended as true to both the spirit and sense of the original. He then spent some time in fruitlessly attempting to square the circle, in studying botany, and in the composition of a metrical tale. He then taught a school, and applied to the General Assembly to be admitted on their list of teachers, but was fortunately unsuccessful. His next employment, unlike any of the others, was almost forced upon him,—he was nominated superintendent of a bridge erecting over the Forth, and acquitted himself with so much credit, that some of the neighbouring gentlemen urged him to stay in that part of the country. George consented, and became a Civil Engineer. Lord Abercrombie requested him to inspect, if he had courage enough, a copper mine in Airdrie, which had lain unwrought for many years, and which, damp, and dark, and full of water and unwholesome gases, was deemed inaccessible by all the other engineers of the country. George knew very little of copper mines, but he furnished himself with a torch, and without assistant or companion, explored the cavern to its inmost extremity, and then drew up a report which has since been successfully acted upon. Some works of an unusual and difficult character were projected last season on the river Dee. George undertook the superintendency of them, constructed a theodolite for himself, accomplished several difficult levellings which a recent survey has proved to be correct, departed from the original plan, and executed the whole in a manner which the original designer has pronounced more complete and effective. An eminent lawyer has described his reports as at once the plainest and most rational ever presented to him; but George has become master enough of his new profession to long for another; and ere I parted from him he told me that he wishes much for some employment such as that of a Gaelic teacher, which would afford him leisure to write a work on etymology.

'This is a curious portrait, but it is that of the individual, not that of the Highlander; a few strokes more, and you shall see it enveloped in tartan. Never was there man more zealous for the honour of his country: he finds more mind in her poets, and more meaning in her language, than in the language and the poets of every other put together. Ossian surpasses Homer, and nothing can be more absurd than to question the authenticity of his poems. He seems to have attached himself to him by a true Highland contract, and stands by him on all occasions in "the right and the wrong." To conclude, he has all the characteristic courage of his countrymen, and all their hospitality and warmth of heart. He accompanied me eleven miles on my way to Linlithgow, and as he shook my hand at parting, I saw the tear gather in his eye. Do not grudge him, my Lydia, the page and half which I have devoted to him; nor chide me when I tell you that I read to him the part of your letter in which you describe the Alness Highlander and the Ross-shire clergy. His remark on your style you will deem a neat one. "There are," said he, "more Mrs Grants than one."

'I saw much in my journey that interested me; never before did I pass over so large a tract of the classic ground of Scotland. Almost every stream and mountain in this district have been celebrated in song; almost every plain has been a field of battle. I stood at Bannockburn on the stone where Bruce fixed his standard, and repeated to my cousin the spirited description of Barbour. I have seen the scene of Wallace's conference with the elder Bruce; that of the battle of Shirramuir, of Stirling bridge, and of Falkirk; the tombs of Sir John the Graham, Sir John Stuart, and Sir Robert Munro; the site of the house in which James III. was assassinated; the room in Stirling Castle in which his father, James II., stabbed the Black Douglas;

the pulpit of John Knox; the Tor-wood in which Wallace so often sheltered from the English, and in which Cargill excommunicated Charles II.; the links of Forth, rendered classic by Macneil; and the distant peaks of Ben Lomond and Ben Ledi. Had I time for geological disquisition, I could tell you something curious of the valley of the Forth, and of some singular etymologies given me by my cousin, which, like the name of the holm mentioned by Sir Thomas, throw light on a very remote period. In founding the piers of the new bridge, the workmen dug through a layer, composed mostly of marine exuviæ, in which they found the skull of a wolf, with several other remains of a very early age, the productions of art. There is an eminence that rises out of the bottom of the valley, quite in the manner that Inchkeith does out of the Frith, which still bears in Gaelic the name of the Island, though now fully five miles from the sea; and a hollow that lies still farther up continues to be known as the Bay of the Anchors. But all these topics we shall discuss when we meet. Heigh-ho, my Lydia, we have missed many a happy meeting this winter!

'I saw the armoury in Stirling Castle; it contains 11,000 stand of arms, with immense sheaves of pikes, which we made, in 1803, under the dread of a French invasion, for arming the people. I saw in it, too, pikes of a much ruder fashion, which some of the people made, in 1819, for arming themselves. You are too young to remember how fierce an attitude the Radicals assumed in that year, and what fears were entertained of a general uprising,—you thought only of your doll at the time; I, on the contrary, was old enough to determine on taking a part in the convulsion, though not quite decided on which side. In Stirlingshire the Radicals broke into open rebellion ; and the arms I saw were taken from a body that had assembled on a moor near Falkirk, and which, better skilled in forming resolutions than fighting, were dispersed by a party of military on the first charge ; all except a boy, who, entrenching himself in a bog, continued firing at the soldiers until surrounded and captured. I saw his pistol and little sword. The pikes are strange, uncouth things, like broomsticks, with points, some of which resemble large nails, others butchers' knives. All that I have heard from history of popular commotions and uprisings,—of Jack Cades and Massaniellos and Jacks of Leyden,—came into my mind as I looked at them ; I saw, too, a Shirramuir Lochaber axe, and an old tilting spear. On my return to Linlithgow I was overtaken by a furious snow-storm, which, I find, from the newspapers, has been the occasion of much loss of life. Had you seen me as I entered the town,—resembling nothing earthly except, perhaps, a moving wreath, or one of those effigies of snow which children set up in the time of thaw,—I am certain you would not have known me.

'Since coming here I have made a very few accessions to my library. My cousin has given me an old edition of Locke's Essay on the Human Understanding, and I have picked up at stalls cheap copies of Goldsmith's Citizen of the World, Franklin's Essays, and Campbell's Pleasures of Hope. My own volume is getting on pretty well ; I have returned proofs for the first two hundred pages ; but I am afraid I have committed a sad blunder regarding it. Nothing could have been easier for me than to have rendered it an unbroken series of legendary stories ; I have materials at will, and find no difficulty in narration. As it is, however, it abounds in dissertation ; and holding, as it does, a middle station between works of amusement and abstract thinking, runs no small risk, I am afraid, of being neglected by the readers of both. Was it not strange that I should not have discovered this when the work was in manuscript? But it is, I believe, of almost general experience among writers, that their productions must appear in print before they can form an estimate regarding them at all approaching to correct. Lavater used to remark that his works, when in MS., appeared to him almost faultless, though no sooner had they passed through the press than he became frightened to look at them. Pope has expressed himself to nearly the same purpose. Well, the past can't be recalled, but I may trust that my fate is not staked on one throw, and that the next may be a better game.

'On Wednesday last I dined with the Principal (Baird), and have seldom spent an evening more pleasantly. He was in one of his happiest moods, and full of anecdote and remark. He seems to form a kind of connecting link between the literature of the past and of the present age; in his youth he was the friend and companion of men whose names leap to our tongues when we sum up the glories of our country,—of Burns and Robertson and Blair. Nearly fifty years ago he edited the poems of Michael Bruce, in behalf of the mother of the poet, who was then very poor and very old,—childless, and a widow. Twenty years after he was the warm friend and patron of the linguist Murray. He was the first who introduced Pringle, the poet, to the notice of the public. He lived on terms of the closest intimacy with Sir Walter Scott, and is thoroughly acquainted with Wilson. What a stride from the times of the historian of Charles V to those of the editor of *Blackwood's Magazine!* does it not sound somewhat strangely that the friend and contemporary of the amiable though ill-fated poet of Kinross, who died nearly sixty years ago, should be the warm friend of your own H— M—? I need not tell you how very interesting I found his anecdotes. He gave me notes of his conversations with Burns, and of his correspondence with Scott. One of his remarks regarding the former in connection with somebody else, I am too vain to suppress. "Burns," said he, "excelled all men I ever knew in force of genius; he leaped to his conclusions with a vigour altogether wonderful, but I do not agree with those who regard his mind as equally powerful in all its faculties. Any task that required prolonged and steady exertion was no task for him; and I have remarked that his good sense never reached the dignity of philosophy. The writer who chose so humble a theme as the 'Herring Fishery of the Moray Frith,' has, I dare say, never thought of entering the lists with Burns; nor, perhaps, could he produce such poems as 'Tam o' Shanter' and the 'Cotter's Saturday Night;' but, in tracing causes and deducing effects, Burns might just as vainly have entered the lists with him."'

CHAPTER II.

HIS FIRST PROSE BOOK—CORRESPONDENCE ON THE SUBJECT—LETTERS FROM MR CARRUTHERS AND MR R. CHAMBERS—RECEPTION AND CHARACTER OF THE BOOK—DONALD MILLER.

WHILE initiating himself, not without irksomeness, into the routine of bank business, and astonishing Mr Turpie by the extent of his reading, Miller occupied his spare moments in Linlithgow in correcting the proof-sheets of the 'Scenes and Legends in the North of Scotland.' It was his first grand effort in prose, his first clear preference of a claim to have his name inscribed in the list of English authors. The Poems by a Journeyman Mason had been printed, at his own expense, in Inverness. The Letters on the Herring Fishery had filled but a moderately-sized pamphlet. Here, at last, was an unmistakeable book, introduced to the reading world by publishing firms of the highest eminence in Edinburgh and London. We have had a glimpse of the difficulties encountered in bringing it this length, but what we have seen will not by any means represent to us their full extent, or the amount of exertion to which Miller submitted in the furtherance of his project. From his correspondence on the subject we shall take two or three additional passages, recollecting, while we read, that for at least two years he was engaged in the task of opening a way for the publication of his volume.

Here, first, is his account of the conception and plan of the book, as presented to his tried and faithful friend, Sir T. D. Lauder, so early as March, 1833: 'In making choice of my subject, I thus reasoned with myself: White's Natural History of Selborne is a most popular little book, and deservedly so, though Selborne itself be but an obscure parish somewhere in the south of England. The very local title of the work has not in the least militated against its interest. But why? Partly, it would seem, from the very pleasing manner in which it is written; partly because the natural history of even a single parish may be regarded as the natural history of the whole country in which that parish is included. And may not the germ of a similar popularity be found, if the writer do not fail in his part, in the traditional history of a Scottish village? Which of all the animals is a more interesting study than man? Or can those varieties of any of the numerous classes which we find in one district of country be more clearly identified with the varieties which we find in another, than we can identify with one another those multiform classes of the human character which, though everywhere different in their minor traits, are everywhere alike in their more important? Besides, the history of one Scottish village is in some measure the history of every one; nay, more, it may form a not unimportant portion of that of the kingdom at large. The people of Scotland, in all its several districts, have been moving forward, throughout the last century, over nearly the same ground, though certainly not at the same pace; and a faithful detail of the various changes and incidents which have occurred during their march from what they were in the past to what they are in the present, cannot surely be merely local in its interest. What does it matter that we examine but only a little part of anything, if from that part we acquire an ability to judge of the whole! The philosopher can subject but comparatively small portions of any substance to the test of experiment, but of how wide an application are the laws which he discovers in the process!

'You will perceive at a glance the conception I thus formed of my task was somewhat too high to leave me any very great chance of satisfying myself in the execution of it. I have not done so, and, indeed, could be almost sorry if I had. I have frequently met with an ingenious argument for the immortality of the soul, drawn from the dissatisfaction which it always experiences in the imperfect good of the present, and from its fondly-cherished expectations of a more complete good in the future; and so long as I am dissatisfied with what I write, and with how I think, I solace myself, on a nearly similar principle, with the hope that I shall one day write better, and think more justly. My traditional history, however, is, I trust, not a very dull one; it is a different sort of work, in some respects, from any of a merely local cast I have yet chanced to see; and I am of opinion, though I dare say I may be mistaken by that partiality which men insensibly form for any pursuit in which they have long been engaged, that a set of works of a similar character would not be quite without its use in the literature of our country. The occurrences of even common life constitute, if I may so speak, a kind of alphabet of invention—the types, rather, which genius employs in setting up her forms. She picks them out in little broken bits from those cells of the memory in which they have been stored up, and composes with them entire and very beautiful pieces of fiction. And I am convinced a set of works similar in character to my manuscript history—from each district of the kingdom—would form a complete part of this kind. Might not such a set be properly regarded as a magazine of materials for genius to work upon?'

Allan Cunningham, in whom, as a brother of the hammer and a brother of the pen, Miller took a particular interest, and whom he obliged with that sketch of Black Russel which we have seen, was applied to when the subscription scheme had been set on foot, in the hope that he might do something for the book in London.

'Cromarty, August, 1834.

'For the last few years I have devoted to the pen well-nigh all the hours I could spare from the mallet, and have produced a volume which I would fain see in print. It is traditional, and wants only genius to resemble very much some of your own. Our materials, at least, must have been collected in the same manner and from the same class—in prosecuting a wandering employment in a truly interesting country, rich with the spoils of the past—in the work-shed, and the barrack, and the cottage, from old men and old women—the solitary, fast-sinking remnants of a departed generation. But the mason of the north has no such creative powers as he of Galloway—powers that can operate on a darkened chaos of obsolete superstitions and exploded beliefs, and fashion it into a little poetical world, bright and beautiful, and busy with passion and life. Still, however, my traditions are not without their interest, though possibly they may owe little to the collector. They are redolent of Scotland and the past, and form the harvest of a field never yet subjected to any sickle except my own.

Our northern districts seem to have produced many that could invent, but none that could give their inventions much publicity; many that could *think* and *feel* poetry, but none that could *write* it: their literature is consequently an oral literature—their very history is traditional; they may be thought of as fields unreaped, as mines unopened: and must not some little interest attach to a work, however deficient as a piece of composition, that may properly be regarded as a sample of the grain—a specimen of the ore? I trust, however, that my mode of telling my stories will not be deemed very repulsive. I have had a hard and long-protracted struggle with the disadvantages attendant on an imperfect education. To you, at least, I need not say how hard and how protracted such a struggle must always prove; but I have at length, I trust, got on the upper side of them; and, if I eventually fail, it will be rather from a defect of innate vigour than from any combination of untoward circumstances pressing upon me from without.

'I publish by subscription—from the nature of the work and the obscurity of the writer, the only way open to me. But, trust me, I have no eye to pecuniary advantage; I would not give a very little literary celebrity for all the money I ever saw; besides, bad as the times are, I am master enough of the mallet to live by it. I could ill afford, however, the expense of an unlucky speculation; and as literature is not so much thought of in Cromarty as the curing of herrings, I find that, without extending my field, I cannot securely calculate on covering the expense of publication. Forgive me that I apply to you. I am a pilgrim, passing slowly and heavily along the path which leads right through the wicket—now floundering through the mud of the *slough*, now journeying beside the hanging hill, now plodding through the low-lying grounds haunted by Apollyon; and what wonder that I should think often and much of one who has passed over the same tract? and who, undeterred by the dark valley or the enchanted grounds, with all their giants and all their wild beasts, has at length set him down amid the gardens of Beulah, in full view of the glorious city. . . . . . . My booksellers in London are Smith and Elder, to whom, should you succeed in procuring a few names for me, the list may be transmitted.'

Allan sent a few sensible and friendly words in reply. 'I am glad that you think of publishing; for there is so much truth and nature and information in your writings, that they cannot fail of doing your name a good turn. A work of the kind set forth in your prospectus will be welcome to all true-hearted Scotsmen, and, though limited in its range, will influence many who live besouth the Tweed. I have laid one of your printed intimations on the table of my bookseller, and desired him to mention it to his visitors. When the work appears I will say a good word for it with all my heart. I mentioned it to some friends here; but you must understand that the Londoners are not accustomed to put down their names for works of a literary nature, whatever the merit may be: but this must not discourage you; almost all authors sacrifice a work or two for the sake of having their merits made widely known. I did this; and now I find purchasers, though I found few at first. I desired our mutual friend Carruthers to place my name among the subscribers long before you wrote to me. Your bookseller must send copies to most of the influential newspapers and reviews; a kind word from them saves an advertisement, and possibly helps the sale of the works. But take a brother-mason's as well as a brother-writer's advice. Don't be too solicitous about being noticed in reviews; let the thing take its course: a worthy work seldom fails.'

One sample will suffice of the letters in which he applied to gentlemen of influence—landed proprietors, clergymen, leading merchants—in his district, to countenance his enterprise.

TO SIR GEORGE MACKENZIE, OF COUL, BART.

'Permit me to submit to your judgment the inclosed prospectus. I am acquainted with only your writings, and the high character which you bear as a gentleman of taste and science; but there is more implied in such an acquaintance than in a much closer intimacy with a common mind; and it is the knowledge I have derived from it which now emboldens me to address you.

'I am one of the class almost peculiar to Scotland, who became conversant in some little degree with books and the pen amid the fatigues and privations of a life of manual labour. For several years past I have amused my leisure hours in striving to acquire the art of the writer, and in collecting and arranging the once widely spread, but now fast sinking, traditions of this part of the country. I have written much, that I might learn to write well, and have made choice, as the scene of my exertions, of a field so solitary and little known that I might not have to contend with labourers more practised than myself; and I have found in this field much that I have felt to be interesting, and much that I deem original—incidents of a structure wholly unborrowed, striking illustrations of character and manners, inventions not unworthy of poetry, and strongly-defined traces of thought and feeling, which might afford employment to the philosopher.

'My amusements at length produced a volume which, though not quite such a work as I had conceived might have been written on the subject, I deemed not entirely devoid of merit. I submitted my MS., through Sir Thomas D. Lauder (a gentleman who has honoured me by his notice and shown me much kindness), to some of the literati of Edinburgh: their judgment regarding it has been favourable beyond my most sanguine anticipations. Still, however, the work is local in its character, and more exclusively so in its title, and in times like the present I can have nothing to expect from the booksellers. But the circumstances which militate most against the general interest of the work must have some little tendency to impart to it a particular interest in the district of country whose traditions it relates, and whose scenery and general character it purports to describe. If, like a convex lens, the focus bears on only a narrow space, in that narrow space the rays must be concentrated. For a work of this kind the mode of publishing by subscription is the only available one; and, after hesitating long—for the scheme has often been resorted to in this part of the country by men of an inferior cast, inferior both in sentiment and intellect, and I was unwilling it should be thought that I had anything in common with them—I now betake myself to it. I would ill like to risk my respectability as a man for the uncertain chance of being a little known as a writer; but there is surely nothing mean in a mode of publication which such men as Pope and Cowper and Burns have had recourse to. The meanness must consist not abstractedly in the scheme itself, but, when the work chances to be a worthless one, in the inveigling the public into what must be regarded as an unfair bargain. I have no eye to pecuniary advantage. My hopes and fears are those of the literary aspirant only; and, little known either as a man or a writer, my eye naturally turns to one whose favourable opinion, holding as he does so high a place in society and letters, would obtain for me the suffrages of the class best able to forward my little plan.'

The disappointment experienced by Miller in procuring the publication of his book was confined to his attempts to induce a bookseller to undertake the risk of issuing it. No sooner did he adopt the plan of subscription, than he met with encouragement and aid on all hands. Without any conscious effort he had succeeded in inspiring every one who knew him with confidence, and those who knew him well were not only confident of his future, and proud of his abilities, but bound to him by strong personal attachment. He had shown himself friendly, and he had found friends who took delight in serving him. At last Mr Black agreed to publish on terms which, under the circumstances, must be pronounced generous. Miller received 400 copies for his subscribers at cost price, and, in the event of profit being realized on the sale of the remainder of an edition of 1250 copies, was to share it with Mr Black. The selling price was fixed at seven shillings and sixpence. On these terms Miller would clear about sixty pounds, even if the unsubscribed copies should not sell. These terms were not arranged until after Mr Black had met Miller in Edinburgh, and it is evident that he also had learned to believe that the Cromarty mason, just developing into a Bank clerk, was a man with a future.

In the spring of 1835, then, the book was in the

hands of the subscribers, and congratulations poured in upon Miller. He was in a mood of quiet satisfaction, wholly unimpassioned; nay, he was not without anxiety as to the loss which might be incurred by Mr Black: his friends were joyful, cordial, exultant. Here is a heart-warming letter from Mr Carruthers.

'Inverness, April 17, 1835.

'Many thanks for your bonny little book. It was delivered to me yesterday evening about six o'clock, and I went through fully three-fourths of it before going to bed. Depend upon it, my dear fellow, you have made a *hit* this time. I don't say that the Legends will lift you into high popularity with all your robes and singing garlands just at once. Your fame will not come rushing on you like a *spate*. But the book will have a steady general sale, and will lay the foundation of a permanent literary reputation, destined, I trust, to go on increasing, and be crowned with many honours.

'You are right in your remark about there being rather too much dissertation, especially in the first two or three chapters. This surplus, like that of the Irish Church, would have, perhaps, been better appropriated to other purposes; yet one soon becomes reconciled to it or ceases to consider it unnatural. I think you lack dramatic power; at least, your sketches of character struck me as inferior to the descriptive and moralizing passages. I should, however, except honest Donald Miller, who is equal to Washington Irving's happiest creations. The great charm of the book is that it is full of original matter,—not concocted from other works, though you have much curious reading too, but fresh and flowing, full of truth and nature. Taste, you know, is a plant of very slow growth, yet you have already outstripped our friend Allan Cunningham in this respect. Allan had better opportunities than you in his early days. His father had an excellent library, was an intelligent man, and mixed with intelligent people. Nay, the poet himself was turned of thirty, had been a reporter for the London press, and was almost necessarily well versed in critical lore, before he tried his hand at prose. Yet even his last work, his Life of Burns, is full of sins against right taste and delicacy of feeling. But, after all, your solitude and seclusion were your best teachers. We may wonder how you got your style—so pure and vigorous, but it was your lonely communings with nature that fixed the matter in your mind, and gave it room to grow. You studied deeply and minutely all you heard, read, and saw, and thus came to your task fraught with thoughts, feeling, and knowledge, pondered over daily for years, and moulded into perfect shapes. Your imagination had merely to supply a *coping* for this depository. But I am getting too *dissertative* myself. If the *Edinburgh Review* is at your command, turn to one of the early volumes for a review of "Cromek's Reliques," and you will find some excellent observations of Jeffrey, on the peculiar position of Burns in his youth. Situation is as necessary for the proper growth of genius as of forest trees; and I cannot help thinking, my dear friend, that though your early lot has been hard, it has been favourable for the development of your mental power.

'It will be your own fault if you do not sail with full and prosperous gale. Your next appearance will be looked forward to with interest, and will secure you good terms with your bookseller. Publishers are a fraternity wise in their generation, and I really think they will be casting out nets for you hereafter. I hope you will go on writing, and accumulating materials. You speak of White's Selborne as a sort of model. Your work resembles Crabbe's Borough, and his general style, much more closely. The same faithful and minute painting of humble objects,—the same love of the sea and all pertaining to it,—fishes, men, and marine scenery. Of course the characters are different, being modified by national and local circumstances. We of Scotland have the advantage in point of morality and staid demeanour. But Crabbe's poachers and navigators, with their strong unbridled passions, are perhaps better fitted for poetry. What do you say to a series of sketches in verse of your Cromarty worthies, their characters, passions, and adventures? Of this, and fifty other subjects, I shall hear you speak, I hope, soon. When the suns get warmer, and spring is leading (as Wordsworth finely says) her earliest green along the leaves, I shall steal away some Friday or Saturday, and ruralize with you on the hill-side over the bay. I hope sincerely that Wilson will shine on you with one of his long, laudatory, imaginative articles in *Blackwood*. Adam Black will take every means of giving you publicity. But I see no fear of your success, so that the pushing of the trade will be the less necessary. I send you a capital review from the *Spectator*, which you may not have seen. Tell me from time to time how you get on, and how the work goes off.'

Miller sent at this time a copy of his book of poems to Mr Robert Chambers, accompanying it with the following letter.

'The Moray Frith has been so blocked up this spring, by the westerly winds, that it is only now an opportunity occurs of sending you the Jacobite Psalm which I mentioned to you when in Edinburgh. It is by no means a very polished composition, but the writer was evidently in earnest; and in the closing stanzas there is an energy and power, united to much simplicity, which he must have owed rather to his excited feelings as a Scotchman and a Jacobite, than to his art as a poet. It has struck me as a curious fact, and one which I do not remember to have seen noticed, that almost all our modern Jacobites are staunch Whigs. Burns was a representative of the class, and I think I see from the verses of the poor Jacobite Psalmist, that had he flourished ninety years later he would have been a Whig too.

'Oblige me by accepting the accompanying volume. It contains, as you will find, a good many heavy pieces, and abounds in all the faults incident to juvenile productions, and to those of the imperfectly taught; but you may here and there meet in it with something to amuse you. I have heard of an immensely rich trader, who used to say he had more trouble in making his first thousand pounds than in making all the rest. I have experienced something similar to this in my attempts to acquire the art of the writer. But I have not yet succeeded in making my first thousand. My forthcoming volume, which I trust I shall be able to send you in a few weeks, will, I hope, better deserve your perusal. And yet I am aware it has its heavy pieces too,—dangerous looking sloughs of dissertation, in which I well nigh lost myself, and in which I shall run no small risk of losing my readers. One who sits down to write for the public at a distance of two hundred miles from the capital, has to labour under sad disadvantages in his attempts to catch the tone which chances to be the popular one at the time,—more especially if, instead of having formed his literary tastes in that tract of study which all the educated have to pass through, he has had to pick them up for himself in nooks and by corners where scarce any one ever picked them up before. Among educated men the starting note, if I may so express myself, is nearly the same all the world over, and what wonder if the after-tones should harmonize,—but alas for his share in the concert who has to strike upon a key of his own!

'All my young friends here, and I have a great many, are highly delighted with your volume of Ballads, and some of the elderly, who have hardly taken up a piece of light reading for the last thirty years, have eagerly renewed by means of it their acquaintance with the favourites of their youth. I have an aunt turned of seventy, who, with the assistance of spectacles, has perused it from beginning to end. It is by far the best collection I have yet seen, and the notes add infinitely to its value.'

To this there came, in due course, the following reply.

'Anne Street, Edinburgh, March 31, 1835.

'I have just received your letter of the 19th inst., with the accompanying volume, of which I have already read a considerable portion. It is fortunate it arrived to-day, as I was about to write for another purpose than the acknowledgment of your letter; and it is better to kill two birds with one stone than a single one only. My object was to mention that I have read your history of Cromarty all to the last two chapters, being, perhaps, the fourth or fifth work of which I have read so much these half dozen years. For a copy which has been sent to me, apparently by your order, I beg to thank you, but I had previously bought one, and was by that time far on in the perusal of it. Further, I have put an extract from it into our printer's hands, with a preliminary notice, in which I express my opinion of it; three weeks, however, must elapse before this can appear. I think you will not be displeased with the terms in which I have spoken of the volume and its author; at least, I am very sure that the notice is meant for the benefit of both. I dwell chiefly on the value which I conceive the book to have, as an example of the operations of a mind of deep reflection and sensibility, reared amidst humble scenes and circumstances, imperfectly educated, and in want of all appropriate material to act upon. Yet, while acknowledging that the reflection and the sensibility are often misspent, I take care to convey the impression that the book is a good one of the kind it professes to belong to, and calculated to afford much amusement to the reader, for I believe it would not be bought as a "psychological curiosity" only. Between ourselves, I think it would have been better to retrench a good deal of the moralizing in the early chapters. I assure you, though not unaccustomed to philosophical reading, I find your thinking pretty hard and solid; it requires a little more time and pains to follow you than the most of us care to expend on a book of what we suppose light reading. The history of Cromarty! you would have made a history of John o' Groat's house philosophical, I believe. Yours seems to be the true sort of mind to make minnows talk like whales. Such powers are not appropriate to topographical narration, or the chronicling of old stories. A playful fancy and a power of whimsical allusion answer these walks of literature much better. You are like a man assorting needles with a gauntlet. I must also mention to yourself that I have found a few little matters in your volume which cannot be traditionary; such, for instance, as a George's Square in Edinburgh some fifteen years before the actual erection of the place bearing that name; the numbering of houses, too, when there were no numbers; and the coming by the head of Leith walk from Queensferry to Edinburgh. You speak of dates of the thirteenth and fourteenth centuries on the front of Urquhart Castle, in connection with architectural styles, I am sure not much earlier than the seventeenth, and when I am equally sure that no dates were carved on houses in Scotland,—at least, far as I have ridden, and much as I have seen in my native country, I never saw a date upon a building earlier than the sixteenth century. This shows that you fill up and round off; and why not? but only such matters must be managed discreetly.

'I am much obliged by the Jacobite Psalm, which is certainly much above the tame poetry of the period. Your remark about the Jacobites has often struck myself. I account for their becoming Liberals in after-times, by the fact of Jacobitism at length becoming identified with a patriotic indignation at the corrupt government of the early Brunswick sovereigns, in which last character it must have very readily associated with modern Liberalism.

'I had thought of it as a duty to endeavour to give you some hints as to your future conduct in literature, such as a metropolitan may be sometimes able to give to a provincial. But now that I see your volume I deem it needless to try. A mind such as you have the fortune to possess, can hardly ever or anywhere be at a loss. I could hope, however, that you may keep in view the advantage, for your own happiness, of advancing into some more conspicuous situation in life, where the powers and tendencies of your mind may find more fitting scope and exercise than at present. For the attainment of such an end, great worldly prudence and, what people are now universally calling tact, are as essentially necessary as the bare possession of talent; and here I hope you will never be found wanting. With the best wishes for your happiness under whatever circumstances, I remain, &c.'

The Jacobite Psalm, referred to by Mr Chambers, is what Miller describes as 'a curious version of the 137th Psalm, the production of some unfortunate Jacobite.' He supposes it 'to have been written at Paris shortly after the failure of the enterprise' of 1745, 'when the prince and his party were in no favour at court; for the author, a man, apparently, of keen feelings, with all the sorrowful energy of a wounded spirit, applies the curses, denounced against Edom and Babylon, to England and France.' Readers will perhaps like to see the verses.

'By the sad Seine we sat and wept,
When Scotland we thought on;
Reft of her brave and true, and all
Her ancient spirit gone.

'"Revenge," the sons of Gallia said,
"Revenge your native land;
Already your insulting foes
Crowd the Batavian strand."

'How shall the sons of freedom e'er
For foreign conquest fight!
How wield anew the luckless sword
That failed in Scotland's right!

'If thee, O Scotland, I forget,
Till fails my latest breath,
May foul dishonour stain my name,
Be mine a coward's death.

'May sad remorse for fancied guilt
My future days employ,
If all thy sacred rights are not
Above my chiefest joy.

'Remember England's children, Lord,
Who on Drumossie day,
Deaf to the voice of kindred love,
"Raze, raze it quite," did say.

'And thou, proud Gallia, faithless friend,
Whose ruin is not far,
Just Heaven on thy devoted head
Pour all the woes of war!

'When thou thy slaughter'd little ones
And ravish'd dames shalt see,
Such help, such pity, may'st thou have
As Scotland had from thee.'

'My legendary volume,' says Miller in the *Schools and Schoolmasters*, 'was, with a few exceptions, very favourably received by the critics. Leigh Hunt gave it a kind and genial notice in his "Journal;" it was characterized by Robert Chambers not less favourably in *his;* and Dr Hetherington, the future historian of the Church of Scotland and of the Westminster Assembly of Divines, at that time a licentiate of the Church, made it the subject of an elaborate and very friendly critique in the *Presbyterian Review.*' We have already referred to the remark on its style made by Baron Hume, and the eager delight of Miller at being recognized as a worthy successor of the Addisons and Goldsmiths, at whose feet he had loved to sit. The book 'attained no great popularity;' but it crept gradually into circulation, and moved off 'considerably better in its later editions than it did on its first appearance.'

These words are likely to prove true for an indefinite period. This is one of those books which has to find its readers, but which, when it has found, retains them by a charm like that of old friendship and of old wine. There is in it an aroma of racy thought and natural home-bred feeling. We may call it a bit of genuine historical literature, for it reproduces with vivid faithfulness the aspect of human life in one particular corner of the planet. The actual fields and waters, crags and woods, green dells and bleak moors, grey castles and thatched cottages, wimpling burns, and broomy braes, and brown sea-shores, beside which Miller has played since childhood, form the scenery of the drama; and amid these life goes masquerading in its coat of many colours—life, with its fitful changes and abrupt contrasts, its heart-wrung tears and grotesque grins, its broken-winged sublimities and grandeurs tempered by absurdity; its queer jumble of tragedy and comedy, and merry scorn of all the unities. The writing is, perhaps, too careful for full display of strength. Miller lingered for many years over his stories, copying and recopying—here polishing down a roughness, there throwing in a touch of colour; now rounding a sentence with more subtle curve, now drawing out a similitude with more elaborate precision, grudging no labour and no time. In this kind of work his arm could not show its sweep and power. In much of his subsequent writing there is a rapid force, a rhythmic energy, which we do not find in the Addisonian periods of his first prose book. But in quiet, delicately-wrought perfection—in beauty fine as the tints of a shell, as the veining of a gem, as the light and shade of a cameo—Miller never surpassed, if he ever equalled, some parts of this volume. The hint of Mr Carruthers as to its defect in dramatic power, is not without pertinency and justice. Hugh had trained himself to narrative, and was comparatively unskilled in dialogue; but the essential element in dramatic power, the ability to realize human character and feeling in different situations, is certainly displayed in the *Scenes and Legends*. The characters live. We see them; occasionally we hear them, and what they say is characteristic: it is mainly the dramatic form, not the dramatic substance, that is wanting.

To illustrate the careful finish of this book, it would be easy to find a number of passages—exquisite descriptions of landscape, specially felicitous similitudes, apt and eloquent reflections; but to select from these one or two brief enough for quotation, and decisively the best to be had, would be exceedingly difficult. It seems preferable, therefore, to take by way of sample a passage which is not remarkable for style, but is fitted to convey a fair general idea of the character, interest, and power of the book. I allude to the account of Donald Miller, referred to with enthusiasm by Mr Carruthers. The story is introduced with a statement, not quoted here, respecting severe storms which, for successive winters, had visited Cromarty.

DONALD MILLER.

'Donald was a true Scotchman. He was bred a shoemaker; and painfully did he toil late and early for about twenty-five years with one solitary object in view, which, during all that time, he had never lost sight of—no, not for a single moment. And what was that one? Independence—a competency sufficient to set him above the necessity of further toil; and this he at length achieved without doing aught for which the severest censor could accuse him of meanness. The amount of his savings did not exceed four hundred pounds; but rightly deeming himself wealthy—for he had not learned to love money for its own sake—he shut up his shop. His father dying soon after, he succeeded to one of the snuggest, though most perilously situated, little properties, within the three corners of Cromarty; the sea bounding it on the one side, and a stream—small and scanty during the droughts of summer, but sometimes more than sufficiently formidable in winter—sweeping past it on the other. The series of storms came on, and Donald found that he had gained nothing by shutting up his shop.

'He had built a bulwark in the old, cumbrous Cromarty style of the last century, and confined the wanderings of the stream by two straight walls. Across the walls he had just thrown a wooden bridge, and crowned the bulwark with a parapet, when on came the first of the storms—a night of sleet and hurricane—and lo! in the morning the bulwark lay utterly overthrown; and the bridge, as if it had marched to its assistance, lay beside it, half buried in sea-wrack. "Ah!" exclaimed the neighbours, "it would be as well for us to be as sure of our summer's employment as Donald Miller, honest man!" The summer came; the bridge strided over the stream as before; the bulwark was built anew, and with such neatness and apparent strength, that no bulwark on the beach could compare with it! Again came winter; and the second bulwark, with its proud parapet and rock-like strength, shared the fate of the first! Donald fairly took to his bed: he rose, however, with renewed vigour, and a third bulwark, more thoroughly finished than even the second, stretched, ere the beginning of the autumn, between his property and the sea. Throughout the whole of that summer, from grey morning to grey evening, there might be seen on the shore of Cromarty a decent-looking, elderly man, armed with lever and mattock, rolling stones, or raising them from their beds in the sand, or fixing them together in a sloping wall, toiling as never labourer toiled, and ever and anon, as a neighbour sauntered the way, straightening his weary back and tendering the ready snuff-box.

That decent-looking elderly man was Donald Miller. But his toil was all in vain. Again came winter and the storms; again had he betaken himself to his bed, for his third bulwark had gone the way of the two others. With a resolution truly indomitable he rose yet again, and erected a fourth bulwark, which has now presented an unbroken front to the storms of twenty years.

'Though Donald had never studied mathematics as taught in books or the schools, he was a profound mathematician notwithstanding. Experience had taught him the superiority of the sloping to the perpendicular wall in resisting the waves; and he set himself to discover that particular angle which, without being inconveniently low, resists them best. Every new bulwark was a new experiment made on principles which he had discovered in the long nights of winter, when, hanging over the fire, he converted the hearth-stone into a tablet, and, with a pencil of charcoal, scribbled it over with diagrams. But he could never get the sea to join issue with him by charging in the line of his angles; for, however deep he sunk his foundations, his insidious enemy contrived to get under them by washing away the beach; and then the whole wall tumbled into the cavity. Now, however, he had discovered a remedy. First he laid a row of large flat stones on their edges in the line of the foundation, and paved the whole of the beach below until it presented the appearance of a sloping street,—taking care that his pavement, by running in a steeper angle than the shore, should at its lower edge, base itself in the sand. Then, from the flat stones which formed the upper boundary of the pavement, he built a ponderous wall which, ascending in the proper angle, rose to the level of the garden, and a neat firm parapet surmounted the whole. Winter came, and the storms came; but though the waves broke against the bulwark with as little remorse as against the Sutors, not a stone moved out of its place. Donald had at length fairly triumphed over the sea.

'The progress of character is fully as interesting a study as the progress of art; and both are curiously exemplified in the history of Donald Miller. Now that he had conquered his enemy, and might realize his long-cherished dream of unbroken leisure, he found that constant employment had, through the force of habit, become essential to his comfort. His garden was the very paragon of gardens; and a single glance was sufficient to distinguish his furrow of potatoes from every other furrow in the field; but now that his main occupation was gone, much time hung on his hands, notwithstanding his attentions to both. First he set himself to build a wall quite round his property; and a very neat one he did build, but unfortunately, when once erected, there was nothing to knock it down again. Then he white-washed his house, and built a new sty for his pig, the walls of which he also white-washed. Then he enclosed two little patches on the side of the stream, to serve as bleaching greens. Then he covered the upper part of his bulwark with a layer of soil, and sowed it with grass. Then he repaired a well, the common property of the town; then he constructed a path for foot-passengers on the side of a road, which, passing through his garden on the south, leads to Cromarty House. His labours for the good of the public were wretchedly recompensed by at least his more immediate neighbours. They would dip their dirty pails into the well he had repaired, and tell him, when he hinted at the propriety of washing them, that they were no dirtier than they used to be.

Their pigs would break into his bleaching greens, and furrow them up with their snouts; and when he threatened to pound them, he would be told "how unthriving a thing it was to keep the puir brutes aye in the fauld," and how impossible a thing "to watch them ilka time they gaed out." Herd boys would gallop their horses, and drive their cattle, along the path he had formed for foot passengers exclusively, and when he stormed at the little fellows, they would canter past, and shout out, from what they deemed a safe distance, that their "horses and kye had as good a right to the road as himself." Worse than all the rest, when he had finished whitening the walls of his pigsty, and gone in for a few minutes to the house, a mischievous urchin, who had watched his opportunity, sallied across the bridge, and seizing on the brush, whitewashed the roof also. Independent of the insult, nothing could be in worse taste; and yet when the poor man preferred his complaint to the father of the urchin, the boor only deigned to murmur in reply, "that folk would hae nae peace till three Lammas tides, joined intil ane, would come and roll up the Clach Malacha" (it weighs about twenty tons), "frae its place i' the sea till flood water-mark." It seemed natural to infer, that a tide potent enough to roll up the Clach Malacha would demolish the bulwark, and concentrate the energies of Donald for at least another season.

'But Donald found employment, and the neighbours were left undisturbed to live the life of their fathers without the intervention of the three Lammas tides. Some of the gentlemen farmers of the parish who reared fields of potatoes, which they sold out to the inhabitants in square portions of a hundred yards, besought Donald to superintend the measurement and the sale. The office was one of no emolument whatever, but he accepted it with thankfulness; and though, when he had potatoes of his own to dispose of, he never failed to lower the market for the benefit of the poor, every one now, except the farmers, pronounced him rigid and narrow to a fault. On a dissolution of parliament, Cromarty became the scene of an election, and the hon. member apparent, deeming it proper, as the thing had become customary, to whitewash the dingy houses of the town, and cover its dirtier lanes with gravel, Donald was requested to direct the improvements. Proudly did he comply; and never before did the same sum of *election*-money whiten so many houses, and gravel so many lanes. Employment flowed in upon him from every quarter. If any of his acquaintance had a house to build, Donald was appointed inspector. If they had to be enfeoffed in their properties, Donald acted as bailie, and tendered the earth and stone with the gravity of a judge. He surveyed fields, suggested improvements, and grew old without either feeling or regretting it. Towards the close of his last, and almost only illness, he called for one of his friends, a carpenter, and gave orders for his coffin; he named the seamstress who was to be employed in making his shroud; he prescribed the manner in which his lyke-wake should be kept, and both the order of his funeral, and the streets through which it was to pass. He was particular in his injunctions to the sexton, that the bones of his father and mother should be placed directly above his coffin;—and professing himself to be alike happy that he had lived and that he was going to die, he turned him to the wall, and ceased to breathe a few hours after. With all his rage for improvement, he was a good old man of the good old school. Often has he stroked my head, and spoken to me of my father; and when, at an after period, he had learned that I set a value on whatever was antique and curious, he presented me with the fragment of a large black-letter Bible which had once belonged to the Urquharts of Cromarty.'

It is perhaps worthy of mention that the value of the black-letter Bible here alluded to may have been somewhat enhanced to Hugh by the circumstance that, in his researches into the history of his native district and of its remarkable men, he had come upon evidence that he had the blood of Sir Thomas Urquhart of Cromarty in his veins. Far too proudly contemptuous of such a title to distinction to specify the fact in his published writings, he nevertheless referred to it, when the matter turned up in conversation, as incontrovertible. The chapter devoted to Sir Thomas in the *Scenes and Legends* describes him as a man of genius and learning, but fantastic, speculative, and eccentric in the highest degree. He flourished in the times of the Covenant and Commonwealth.

CHAPTER III.

DEATH OF MISS DUNBAR.

DURING those months which Miller passed in Linlithgow, his friend Miss Dunbar lay on her deathbed, slowly sinking under intolerable agonies. She retained her faculties unimpaired, and in the intervals of pain manifested that gracious interest in all that concerned her friends which characterized her in health. Her malady was known to be incurable, but it does not appear that Miller was aware of any reason for apprehending that it would soon have a fatal termination. He continued, therefore, to write to her in the light discursive manner he had previously adopted.

'Linlithgow.

'I must try to *coin* time (the phrase is poor Henry Kirk White's, who killed himself in the process), in which to show you that the hurry of my new occupation is as unable to dissipate the recollection of your kindness as the rougher fatigues of my old one. The more I see of life, the more I am convinced that "it is not in man that walketh to direct his steps." Here am I in Linlithgow, acquiring that degree of skill in business matters that may fit me for a bank accountant. Six weeks ago I had as much thought—nay, more—of emigrating to the wilds of America.

'I would much rather have spent in Edinburgh the few weeks I have to pass in this part of the country than here; but it was necessary, in order to acquire the skill of the branch accountant, that I should remove to a branch bank; and as I take care never to quarrel with necessity, I get on pretty well. On the evening of my arrival, Mr Gillon, the Radical M.P., addressed his constituents of Linlithgow in the Town-house. I attended, for I was desirous to ascertain what I had long doubted, whether Radicalism and a powerful intellect be compatible, and regarded Gillon as one of the heads of the party in Scotland. But the doubt is a doubt still. The speech I heard was such an one as might, perhaps, pass without remark in an inferior debating society; but there was a sad lack of taste about the parts in which the speaker attempted to be fine, and a deplorable deficiency of grasp when he strove to be sensible. His audience, too, seemed miserably low, and, with all the good-will in the world, had hardly sense enough to applaud. Wherever I looked I saw only low, narrow foreheads, and half-open mouths. What can such people know of the most difficult of all sciences—politics? You know my opinions on the subject. I am a Whig, and not the less Whiggish now that the party are out (I found I could not quite agree with them in all their measures when they were in). But I see every day that men are not born equal; that "those who think must govern those who toil;" and that those who toil frequently mistake the pressure of the original curse for an effect of misgovernment. Besides, I detest all quackery; and the Radical, if he be not altogether a blockhead, is of necessity a political quack.'

The following letter from Miss Dunbar was written, as I conclude, after she had received the preceding. The expression, 'that clamorous fool, a Radical,' and the words, 'I am a Tory, though I believe your Whiggism and my Toryism are not very dissimilar,' though they do not occur in the way of formal reply to Miller, have the look of being suggested by his political remarks.

'Forres, January 1, 1835.

'"I sat between the meeting years,
The coming and the past,
And I asked of the future one,
Wilt thou be like the last?

'The same in many a sleepless night,
And many a painful day?
Thank Heaven, I have no prophet's eye,
To look upon thy way."

'These lines are Miss Landon's—no very great favourite of mine; but as I lay sleepless and in pain last night they came into my mind, and I found in them my own thoughts and feelings sublimed into poetry. I suffer much. I have many privations: it is not one of the least of these that my right hand should have forgotten its cunning; but I trust I can value the comforts still left to me. It is now two o'clock, and I am but just up and dressed and this is my first occupation. I heard of your appointment from the newspaper, and of your having gone to Edinburgh from an acquaintance. Pardon me, my dear friend: some sad thoughts—I may even call them bitter ones—I will own I had amid all the pleasure which both circumstances afforded me. I have always borne much good-will to my acquaintance and friends in general; there are few whom I absolutely dislike, and not a few whom I really like much; but there was always one who more particularly occupied my heart, and whom

I loved more than all the rest put together, and for the last five years you have been that one; and now that I have so short a time to be here, I can have no hope of ever again seeing you. But I assure you the sad, bitter thoughts were but passing ones; and your letter gave me entire pleasure.

'There is scarcely any one I am sorry to part with but you. God knows how fervently I wish you life and happiness, and your advancement in public favour. I consider your late appointment as very respectable, and as procured for you in the most agreeable and delicate manner, but cannot help regarding your literary pursuits as the main business of your life. When, probably, will your book be out, or is it actually gone to press? Many a time I have wondered to myself if ever I shall see it, and have sometimes hopes that I may; but were I to be guided by my present feelings of pain and discomfort, I should say the thing is not very likely. . . . . You are not aware of Lord Medwyn's high appreciation of your genius. It was to his brother, Mr John Forbes, that Major Cumming Bruce lately transmitted one of your letters and extracts from your *Traditional History*, given him by the Messrs Andersons, in four franked covers. The Major had sent to me, as a thing of course, for your address, that Mr Forbes might have waited on you, but I could give him no clue.

'Sir Thomas [Dick Lauder] seems changed in many things since he left this part of the country. He will never become that clamorous fool, a Radical; but he is certainly far too violent in his politics. But I am a Tory, you know; though I believe your Whiggism and my Toryism are not very dissimilar. . . . . . . Will you not come and see me on your way home? You cannot, surely, return by sea at this season, or by the Highland road; and if you come by the coast, Forres lies quite in your way. And then, or never! My only objection to your new employment is, that it ties you down to Cromarty, so that you cannot visit any of your friends a day's journey away. . . . . . The day is coming when the first people in the land will be desirous of seeing and being acquainted with you.'

Before writing the next letter, Miller heard that Miss Dunbar had just lost by death two near relatives to whom she was much attached. The thought of her bereavement recalled to him his own sorrow for departed friends, and in pensive mood, tenderly sympathetic, deeply affectionate, he took pen in hand, and described to her his experience of grief. If we would understand the enthusiasm of love with which all who knew Hugh Miller well regarded him, we ought to consider carefully the heart-delineation of this letter.

'Cromarty, March, 1835.

'Intelligence of your sad bereavement reached me through the medium of the newspapers;—I cannot express what I felt. I knew that your cup was full before,—full to the brim;—but I saw that, regarding it as mingled by your Father, you were resigned to drink. Now, however, a new ingredient has been added,—an ingredient bitterer perhaps to a generous mind than any of the others. To days of languor and nights of suffering, torn affections and blighted hopes have been added; and those relatives to whom the overcharged heart naturally turns for solace and sympathy are equally involved in misfortune. When we sit during the day in a darkened chamber, we have but to throw open a casement and the light comes pouring in; but it is not so during the night, when there is no light to enter. You were before sitting in the shade,—not, however, in so deep a recess but that at times a ray reached you from without; but I now feel that your sad bereavement must have converted your day into night;—that you are sitting in darkness, and that an atmosphere of darkness surrounds you.

'I am not unacquainted with grief. There are friends separated from me by the wide, dark, impassable gulf whom I cannot think of even yet without feeling my heart swell. Shall I not describe to you that process of suffering of which my own mind has been the subject? There may be some comfort to you in the reflection that what you experience is, to use the language of Scripture, "according to the nature of man." The similarity in the structure of our bodies, which shows us to belong to the same race, obtains also in our minds; and as dangerous wounds in the one are followed in most cases by fevers and inflammations, which bear the same names in every subject, and to which we apply the same remedies, so wounds of the other are commonly followed by similar symptoms of derangement in the feelings, and to mitigate the smart and the fever, philosophy applies the same salves, and religion, when called upon, pours in the same balm.

'There is an analogy between grief in its first stage and that state of imperfect consciousness which is induced by a severe blow. We are stupefied rather than pained, and our only feeling seems to be one of wonder and regret that we should feel so little. We ask our hearts why they are so callous and indifferent, and wonder that what we so prized as the lost should be so little regretted. But we know not that, were we affected less, we should feel more. The chords have been so rudely struck, that, instead of yielding their shrillest notes, they have fallen slackened from the stops, and time must recover their tone ere they vibrate in unison with the event. In this first stage whole hours pass away of which the memory retains no firmer hold than if they had been spent in sleep. Seven years ago, when residing in Inverness, word was brought me that an uncle, to whom I was much attached, and who, though indisposed for some time previous, was not deemed seriously ill, was dead. I set out for Cromarty, and must have been about four hours on the road; but all that I next day recollected of the journey was that the road was very dark (I travelled by night), and that, as I drew near to the town, I saw the moon in her last quarter, rising red and lightless out of the sea.

'Sorrow in its second stage is more reflective. The feelings have in some degree recovered their tone, and we no longer deem them weak or blunted. At times, indeed, we may sink into the apathy of exhaustion, but when some sudden recollection plants its dagger in the heart, we start up to a fearful consciousness of our bereavement, and for the moment all is agony. The mind during this stage seems to exist alternately in two distinct states. In the one it pursues its ordinary thoughts or its common imaginings, but when thus engaged the image of the departed starts up before it without the ordinary aid of association to call it in,—it starts up sudden as an apparition, and the heart swells, and the tears burst out. And this forms the second state. I have remarked as not a little strange the want of connection between the two. Occasionally, indeed, some recollection awakened in the first may lead to the second, but much oftener I have found the commoner principles of association set aside altogether, and the image of the deceased starting up as uncalled for by the previous train of idea as if it were truly a spectre. And oh, the aspect of that image! How graceful its attitude! How kind its expression! How beautiful does the soul look at us through the features! Best, and kindest, and most affectionate, and when we felt with most certainty that we were truly dear to him! And hence the depth of our regret,—the bitterness of our sorrow. Grief, my dear madam, is an idolater. It first deifies, and then worships. It has a strange power, too, of laying hold of the moral sense, so that it becomes a matter of consequence with us to deny ourselves all pleasure, and to reject all comfort, in what we deem justice to the deceased. There is something wonderful in the feeling I have not yet seen explained. It seems to have its seat deep in the mysterious parts of our nature, and constitutes a tie to connect, as it were, the living with the dead. No man who truly deserves the name can desire to die wholly unlamented; and the regret which the heart claims for itself, it willingly—oh how willingly!—renders to another. We weep not for ourselves, but in justice to the lost, and even after exhausted nature cannot yield another tear, there is a conscience in us that chides us for having sorrowed so little. I need not ask you if you have experienced this feeling;—no heart was ever truly sorrowful without the experience of it. It is a sentiment of our nature that lies contiguous, if I may so express myself, to that noble sentiment which leads us, independent of our reasonings, to *feel* that there is a hereafter. For do we not think of the dead to whom we owe so many tears, as a being who exists; and could we owe anything to either a heap of dust or a mere recollection? It may be well, however, to remind you that there is a time when the claims of this moral sense should be resisted. It continues to urge that tribute be given to the dead long after the tribute is fully paid, and spurs on exhausted nature to fresh sorrows, when the voice of duty and the prostration of the energies call it to repose.

'Of grief in its third and last stage I need say little. It forms the twilight of a return to our ordinary frame, and is often more pleasing than the indifference of that everyday mood in which there is nothing either to gratify or to annoy. There is luxury in the tear,—regret has become a generous feeling that opens the heart,—and we can love and praise all that we valued in the departed without feeling so continually that what we so valued we have lost. There is truth in the doctrine of purgatory, when premised not of the departed, but of the surviving friend. There is the brief hurried period, in which we can take no note of what we feel; the middle state, with its unspeakable profundity of suffering; and the after state, in which there is a cessation from pain, and when even our sorrows become pleasant.

'I shall not urge with you the commoner topics of consolation; I know the heart will not listen even when the judgment approves. Grief is a strange thing; it is both deaf and blind. Where could it be more perfectly pure from every mixture of evil and folly than in the breast of our Saviour? and yet even in Him we see it finding vent in a flood of tears, when He must have known that he whom He mourned as dead was to step out before Him a living man. Can I, then, hope to dissipate your sorrow? Can I urge with you any argument of consolation equally powerful with the belief which He entertained? or, were I possessed of some such impossible argument, could I hope that it would have more influence with you than that belief had with

Him? He believed, and yet He wept. May I not remind you, however, that He who sorrowed then can sympathize in our sorrows now; that He loved little children, and declared that of such is the kingdom of Heaven; and that He has enjoined us, through His servant, not to sorrow as those who have no hope?'

At about the same time, perhaps in the very hour, when this letter was put into the hand of Miss Dunbar, the *Scenes and Legends*, for which she had looked with a solicitude more tenderly intense than that of Miller himself, reached Forres. The heart of the sweet and gentle lady thrilled once more amid her anguish with a joy like that of a mother when she knows that a beloved son, whose efforts she has long watched, with whom she has long hoped and feared, whose claim to a place of honour among men she has never questioned, has at last done something which will compel the world to own that he is all *she* knows him to be. Miss Dunbar wrote Miller the following touching letter, probably the last she ever penned:—

'I know you wish to hear from me, and in gratifying you I would gratify myself, for I have much to say to you, but, alas! the power of writing is past. My intervals of ease from most excruciating pain are truly like angel visits; and when they do occur I am in such a state of lowness and exhaustion, as to be incapable of any exertion. I am now raised up, and supported in bed by pillows, while I make this, I fear, last effort to write to you. . . . . . . What can I do, but throw myself on His mercy who is the sent of God? He is my rock, my strength, my hope in life and in death. Often do I wish to see you, and to hear you speak of the things which pertain to eternity. I recollect the light and comfort I derived from your conversation last summer

. . . . . . But to the Book; contrary to all my anticipations, I have lived to have it in my hand! What shall I say of it? It would seem, from the very little of it I have yet read, as if I were quite satisfied with seeing and handling it. I look into every chapter, I glance over the whole, but, somewhat childlike, I feel too happy to read.'

Hugh, with no suspicion that the end was near, had begun his reply to this letter, and finished two or three pages, when he received the following notice: 'Forres, June 30, 1835, Miss Dunbar, of Boath, died here last night at half-past ten o'clock.' Here is his unfinished letter.

'Cromarty, June, 1835.

'I have sitten down to write you at the side of a little cliff, grey with moss and lichens, and half hid in fern, that rises on the northern sweep of the hill of Cromarty. The Moray Frith is at my feet. Towards the north I see it spreading out from the edge of the precipice below, league beyond league, till I lose it in the long blue line of the horizon; while the shores of Moray, with their pale undulating strip of sand, rise over it towards the east. The sun is bright overhead; but the sky is dappled with clouds, and the whole landscape is checkered with an ever-changing carpeting of sunshine and shadow. There is a sail on the far horizon so very bright and so very minute, that I can liken it only to a spark of fire; the tower over Forres is also lighted up, with the old castle beyond; and still further to the west I can see the coast line so thickly inlaid with spark-like mansion-houses and villages, that I can only resemble it to a belt of purple speckled with pearls. A true lover of nature will not love it the less should circumstances render his interviews with it brief and occasional; absence only serves to enhance his passion, and he learns to concentrate in his rare and hurried visits the same amount of enjoyment which in other times was spread over the many and prolonged. Never was I more strongly impressed with the truth of this than at present, or more desirous that I could convey to the solitude of your chamber a transcript of the scene I contemplate, and of the feelings it has awakened. Quiet pleasures are ever the most lasting. How many sources of enjoyment are shut up by time! but that which draws its supply from the wonderful sympathy that exists between the frame of nature and the spirit of man is assuredly not of the number. The child draws from it all unwittingly when, rejoicing in the clear air and the sunshine, it flings itself down for the first time on a bed of flowers; and many long years after, when the seasons of youth and riper manhood have passed away, and a thousand pleasures of after growth have palled on the sense, and then ceased to exist, the heart of the invalid in his sick chamber swells with all the quiet fervour of its earliest attachment, when, from under the open curtain, he sees the foliage of midsummer waving to the cool breeze, and its sun sparkling to the sea. The true religion seems to be the only one that addresses itself to this feeling. The Psalms abound with delightful descriptions; and there are lovely images, that have all the green freshness of nature about them, in the books of the prophets. But there is, perhaps, only one religion that *could* avail itself of the feeling. It is well for the Mahommedan and the Polytheist, who wish to remain such, that they confine themselves, the one to his mosque, and the other to his temple; but he who believes in the God of revelation may look abroad on the glories of nature, and find no discrepancy between the aspects of His character which

His word presents to us and those exhibited in His works.

'I should have written you long ere now, but for the last three months my mind has been in a sort of transition state, and passing from old, firmly fixed habits, to the acquisition of new ones, and my powers of application were so dissipated in the process that I could literally do nothing. I am coming round again, however, and, with a pile of unanswered letters in my desk, dedicate to you the first-fruits of my diligence. I find my new profession will leave me well-nigh as much leisure as my old one; but exercise will claim its part; and as my occupations must be less mechanical than formerly, I shall have less time for thought. If, however, my mind be naturally a buoyant one, and I trust it is, those circumstances which will weigh me down must be more untoward than any I have yet experienced.

'My stay in Edinburgh last spring after my return from Linlithgow was extremely brief, and I had to quit it (a circumstance I shall ever regret) without seeing Mrs Grant of Laggan. With all my haste, however, I might have found time enough for the purpose could I have but found courage, but the fear of being deemed obtrusive held me back. I am the silliest fellow, in this respect, I ever knew. No degree of faith in the assurance of others can give me confidence in myself, and I am certain I must often seem a cold and ungrateful fellow when I am in reality shrinking from the possibility of being deemed an impertinent one. But I cannot overcome the feeling. I have to regret, too, that though I had a direct invitation to spend an evening with Mr Thomson, the friend and correspondent of Burns, I could not avail myself of it. In this case, however, it was a prior invitation, not the deprecated feeling, that interfered;

but in both cases I have lost what, from the great age of the parties, no after opportunity can afford me. I dined one day with Mr Black, and met at his table with his brother-in-law, Mr Tait, the Radical bookseller. He seems to be an outspoken somewhat reckless man, with a good deal of rough power about him, and by no means devoid of sense, but he is more disposed to pick up his arguments from the surface of a subject than to take the trouble of going deeper. I had little conversation with him, for my spirits were rather low at the time, and there were a great many topics on which I knew there was small chance of our being at one. I had seen before leaving Cromarty a volume published by Mr Tait on the Game Laws, by my old antagonist, and as it seemed a miserable production both in point of style and argument, I was curious to know how such a work must fare in the hands of an active bookseller. "Ah, poor ——," said he, in reply to my query regarding it, "he succeeds just like every other man who writes in spite of sense and nature; and yet though invariably unlucky he still persists. In men of a literary cast," he continued, "the will and the power of production are often sadly disjoined. I sometimes meet with persons who, I am certain, could write admirably, but who cannot be prevailed on to take up the pen, while a numerous tribe of others, destitute, like ——, not only of ideas, but even of words, cannot be persuaded to resign it." Some of the few hours I passed in Edinburgh were spent very agreeably in the back parlour of Mr Black,—a most agreeable lounge, where in the course of a single forenoon one may meet with half the literary men of the place. I saw in it in the space of two short hours Mr James Wilson, Professor Pillans, Professor Napier, Dr Jamieson, the author of the Scotch Dictionary, and Dr Irving, the biographer of Buchanan. Through the friendship of Mr John Gordon, of the College, I was introduced to Professor Wilson, and heard him lecture. He received me with much politeness, but I felt a little out, and found almost every time he spoke to me that I had nothing to say in reply. I was collected enough, however, to remark that his head is one of the most strangely formed I ever saw;—it is of great size, immensely developed both in the ideal region and in that of what are called the knowing organs, but singularly deficient for a head of such general power in the reflective part. And the lecture I heard was such an one as the phrenologist would have anticipated from such a composition. I was, besides, introduced to Robert Chambers, and passed a long morning with him,—sitting down to breakfast at the usual hour, and rising from it about twelve o'clock. He is possessed of a fund of anecdote altogether inexhaustible, and is one of the most amusing and agreeable companions I have ever met with.'

In a letter written, a few days subsequently, to Sir Thomas D. Lauder, Miller refers to Miss Dunbar as follows:

'My kind friend, Miss Dunbar, of Boath, is dead; she died on the evening of Monday, the 29th ultimo For the last four years her life has been one of much suffering; but she had a youthfulness of spirit about her that availed itself of every brief cessation from pain. She had learned, too, to draw consolation and support from the best of all sources; and so her latter days, darkened as they were by a deadly and cruel disease, have not been without their glimpses of enjoyment. Her heart was one of the warmest and least selfish I ever knew. It was not in the power of suffering or of the near approach of death to render her indifferent to even the slightest interests or comforts of her friends. I was employed in writing to her, and with all the freedom which her goodness permitted me, when the letter reached me which intimated her death. My thoughts were so cast into the conversational mould that I could almost realize her presence; and had she suddenly expired before me, I could not have been more affected.'

We must not quit this episode in Hugh Miller's life without the remark that it reveals much of what he was. The sister of Sir Alexander Dunbar, Bart., of Boath, moving in the most refined and cultivated society of Scotland, Miss Dunbar was in every sense a lady. Her penetration and sound literary judgment might have convinced her that Miller was a man of genius, and led her to desire his acquaintance; but that that acquaintance should have ripened into friendship—nay, that she should have signalized the journeyman mason as the truest and dearest of all her friends—can be accounted for only on the supposition that there was in him a sterling worth, a delicate nobleness, a beaming purity of soul and dewy tenderness of feeling, which would have marked him out in any class of society as one of nature's gentlemen.

CHAPTER IV.

LETTERS TO MISS FRASER, FINLAY, AND DR WALDIE.

ON returning from Linlithgow to Cromarty, Miller addressed himself with assiduity to his duties as a bank accountant. In the course of the bank's operations a sum of money, amounting to some hundreds of pounds, was transmitted weekly from Cromarty to Tain, and he thought it necessary to act as messenger. He walked the whole way from the northern shore of Cromarty ferry to Tain and back; and as part of the road lay through a deep wood, he provided himself with a brace of pistols, and travelled with them loaded. This was the first occasion of his carrying fire-arms; and he seems to have never subsequently, except, perhaps, for brief periods, abandoned the practice. The resolute intensity of application with which he mastered the details of banking, and the conscientious caution with which he took the road in order to obviate mishaps in the transmission of the money, may be noted as characteristic of our man.

The change in his circumstances, when he thus passed out of what is termed the working-class, was naturally pleasing to his friends. Mr Stewart declared with hearty satisfaction that he was 'at length fairly caught.' For his own part, he took the matter with conspicuous quietness, betraying no consciousness of having risen in life, not altering his demeanour by one jot or tittle, and except in his thoughts of the future of domestic felicity which was now virtually secure to him, not finding himself a happier man. After a day spent in the uncongenial drudgery of running up columns of figures, he did not experience in literary composition that delicious freshness which it had formerly yielded him. 'For the first six months of my new employment,' he says, 'I found myself unable to make my old use of the leisure hours which, I found, I could still command. There was nothing very intellectual, in the higher sense of the term, in recording the bank's transactions, or in summing up columns of figures, or in doing business over the counter; and yet the fatigue induced was a fatigue, not of sinew and muscle, but of nerve and brain, which, if it did not quite disqualify me for my former intellectual amusements, at least greatly disinclined me towards them, and rendered me a considerably more indolent sort of person than either before or since. It is asserted by artists of discriminating eye that the human hand bears an expression stamped upon it by the general character as surely as the human face; and I certainly used to be struck, during this transition period, by the relaxed and idle expression that had on the sudden been assumed by mine. And the slackened hands represented, I too surely felt, a slackened mind. The unintellectual toils of the labouring man have been occasionally represented as less favourable to mental cultivation than the semi-intellectual employments of that class immediately above him, to which our clerks, shopmen, and humbler accountants belong; but it will be found that exactly the reverse is the case, and that, though a certain conventional gentility of manner and appearance on the side of the somewhat higher class may serve to conceal the fact, it is on the part of the labouring man that the real advantage lies. The mercantile accountant or law-clerk, bent over his desk, his faculties concentrated on his columns of figures, or on the pages which he has been carefully engrossing, and unable to proceed one step in his work without devoting to it all his attention, is in greatly less favourable circumstances than the ploughman or operative mechanic, whose mind is free though his body labours, and who thus finds in the very rudeness of his employment a compensation for its humble and laborious character. And it will be found that the humbler of the two classes is much more largely represented in our literature than the class by one degree less humble. Ranged against the poor clerk of Nottingham, Henry Kirke White, and the still more hapless Edinburgh engrossing clerk, Robert Ferguson, with a very few others, we find in our literature a numerous and vigorous phalanx composed of men such as the Ayrshire Ploughman, the Ettrick Shepherd, the Fifeshire Foresters, the sailors Dampier and Falconer, Bunyan, Bloomfield, Ramsay, Tannahill, Alexander Wilson, John Clare, Allan Cunningham, and Ebenezer Elliot. And I was taught at this time to recognize the simple principle on which the greater advantages lie on the side of the humbler class.' The unfavourable influence of his new occupation on his literary activity proved to be of temporary nature. 'Gradually,' he proceeds, 'I became more inured to a sedentary life, my mind recovered its spring, and my old ability returned of employing my leisure hours, as before, in intellectual exertion.'

Once more, therefore, we may pronounce him happy. A time which, to him, seemed doubtless long, was still to elapse before his union with Miss Fraser, but the engagement was now fully countenanced by her mother, and the intercourse of the lovers was constant and unconstrained. William Ross was in his grave; John Swanson was about to leave the district; his friendship with Miss Dunbar had become a tender and exalting reminiscence. He clung all the more closely to her who was yet left to him, in whom he found the affection of Ross, the mental stimulus of Swanson, the sympathy of Miss Dunbar, and who was dearer to him than them all. As Miss Fraser resided almost uninterruptedly in Cromarty, there is not much in the way of correspondence between her and Hugh to throw light upon their intercourse at this period; but we have one or two letters, through which, as through 'luminous windows,' we can see into the 'happy palace' of love and friendship in which these two abode. Here is a note from the lady.

'My own Hugh, I am tired, tired of being away from you. Alas! you have no idea of the frivolous bondage to which sex and fashion subject us. I do nothing all day, and hear nothing, yet I am obliged to take the time from sleep which I devote to you. I have found the young captain whom I threatened you with much handsomer than I described him to you, but a thousand times more insipid. Why, when I look at him, do I always think of you? or why do his black, bright eyes, that would be fine had they meaning, always remind me of those gentle blue ones which I have so often seen melt with benevolence and a chastened tenderness? Why are mankind such slaves of appearances as to admire the casket and neglect the gem? It is degradation to the dignity of thought and sentiment to compare it with a mere beauty of form or colour. Good-bye.

'It is morning, but I am not beside you on the

leafy hill, with the blue water shimmering at our feet. When shall we be there again?'

When this letter reached Cromarty Hugh was in Tain, but he evidently lost no time in replying to it on his return.

'Cromarty, July, 1835.

'I need not tell you at this time of day how much it is in your power to make me happy, and how thoroughly my very existence seems to be bound up in yours. I have but one solace in your absence, my Lydia—that one thought of your return.

'There crossed with me in the ferry-boat a little ragged gipsy boy, the most strongly marked by the peculiar traits of his tribe I almost ever saw. Have you ever observed the form of the true gipsy head? I am much mistaken if it be not the very type of that of the Hindoo. In the line of the nose the forehead is perfectly perpendicular, indicating, I should think, a large development of comparison, but causality is less marked, and the whole contour is one of little power. It is not, however, the sort of head one would expect to find on the shoulders of a savage, more especially of the savage who can continue such in the midst of civilization. On reaching the school-house I learned that John (Swanson) had resigned the school in consequence of an appointment to the mission at Fort William. I find that in a pecuniary point of view he is to gain almost nothing from the change. The salary does not exceed sixty pounds a year, and he is to be furnished with neither house nor garden. But it is to open to him a wider field of usefulness, and to John that is motive enough. He is, in the extreme meaning of the term, what Bonaparte used to designate with so much contempt, an *ideologist*, i. e. a foolish fellow who does good just because it is good, and for the pure love of doing it. I feel, however, very anxious on his account regarding the mission. The part of the country to which he is going is said to be wretchedly unwholesome—full of lakes and marshes, and infested with miasma; and sometimes, when I consider the exhaustive fervency of his spirit and the weakness of his frame, I cannot avoid fearing that I may have yet to think of him in connection with a solitary Highland churchyard and a nameless grave. Poor William Ross! he is now seven years dead, and were I to lose John also, where might I look for friends of the same class,—men who, attached to me for my own sake alone, could regard me in every change of circumstance with but one feeling? And John, too, is more than *my* friend. He is, my own Lydia—and I love him ten times the more for it—he is *ours*.

'I pursued my journey from the school-house in the morning, and in passing through the deep, dreary wood of Culrossie, found myself, as I supposed, quite on the eve of an adventure. I carried with me a considerable sum of money—several hundred pounds,—and that I might be the better able to protect it, had furnished myself with a brace of pistols, when, lo! in the thickest and most solitary part of the wood up there started two of the most blackguard-looking fellows I ever saw. They seemed to be Irish horse-jockeys. One wore a black patch over his eye, and a ragged straw hat; the other a white frieze jacket, sorely out at the elbows; and both were armed with bludgeons loaded with lead. I had time enough ere they came up to cock both my pistols. One I thrust under the breast flap of my coat, the other I carried behind my back, and sheering to the extreme edge of the road with a trigger under each fore-finger, I passed them unmolested. One of them regarded me with a sardonic grin. My posture, I suspect, must have seemed sufficiently stiff and constrained for that of a traveller.'

He next touches upon some book-purchases in which he and his correspondent have a common interest. 'There is a neat pocket-copy of Johnson's Lives that will do well for the beech tree; I have besides got a copy of Paley similar to the one you had from Mrs I—; a copy of Smollett's Humphrey Clinker (my heart warmed to this book,—for, though many years have passed since I last perused it, it was one of my earliest favourites), and a minute copy of Childe Harold. I saw in Douglas's Leigh Hunt's Journal. The notice of our little book is a highly gratifying one;—is it not well that it is the highest names who praise it most? Hunt characterizes it as "a highly amusing book, written by a remarkable man, who will infallibly be well known." I am placed side by side with Allan Cunningham; there is a *but,* however, in the parallel, which I suspect Allan will not particularly like. "But," says Hunt, "Mr M—, besides a poetical imagination, has *great depth of reflection;* and his style is so choice, pregnant, and exceedingly like an educated one, that if itself betrays it in any respect to be otherwise, it is by that very excess; as Theophrastes was known not to have been born in Attica by his too Attic nicety."

'My poor friend, Miss Dunbar of Boath, is dead; she died on the evening of Monday, the 30th June. The severe and ever-recurring attacks of her cruel disease had undermined a constitution originally good, and it at length suddenly gave way under the pressure of what seemed to be comparatively a slight indisposition. She is gone, and I have lost a kind and attached friend. But it would be selfish to regret that suffering so excruciating as hers should have terminated; for months past I could think of her only as a person stretched on the rack, with now and then, perhaps, a transient glimpse of enjoyment, for such is the economy of human feeling that every cessation from suffering is positive pleasure to the sufferer; but what, alas! had she to anticipate in this world save pang after pang in prolonged and direful succession,—nights of pain and days of weariness, and at length the opening of a door of escape, but only that door through which she has just passed. I trust, my own Lydia, that it is well with her. Her heart was in the right place,—it was ever an affectionate one,—perhaps too exquisitely so, but it seems finally to have fixed on the worthiest of all objects. She had learned to look for salvation through Him only in whom it is alone to be found. There are many whom suffering has the effect of so wrapping up in themselves that they can feel for no one else. But it was not thus with Miss Dunbar: she could think, even when at the worst, of the little comforts and interests of her friends; half her last letter to me is occupied with a detail of what she had thought and heard regarding my Traditions. I was engaged in writing her when the note was brought me which intimated her death.

'I have got a rather severe cold, which hangs about me. Never was cold better treated than mine;—it eats and drinks like a gentleman. A shop-keeping acquaintance gives it liquorice, Mr Ross gives it bramble-berry jam, Mrs Denham has given it honey, and now Mrs Fraser has sent it a pot of tamarinds. 'Twill be a wonder if, in such circumstances, it goes away at all. I have begun, but barely begun, my statistical account of the parish; it must, I am afraid, be both dull and commonplace, for I am alike unwilling either to repeat myself, or to anticipate any of my better materials for a second volume of "Scenes and Legends," and the residue is mere gossip. Even were it otherwise, my abundance, like the wealth of a miser, would have the effect of rendering me poor. Had I but a single story to tell I would tell it, but who would ever think of telling one of a hundred!

'I have no words to express to you, my own Lydia, how much I long for your return, or how cold a looking place Cromarty has become since you left it. Ordinary pleasures and lukewarm friendships do well enough for men who have not yet had experience of the intense and the exquisite, but to those who have, they do not seem pleasures or friendships at all. I am amusing myself, however, just as I best can; sometimes picking up a geological specimen for my collection,—sometimes making an excursion to the hill or the burn of Eathie. I accompanied to the latter place on Saturday last Mr Ross and his children, with two of their cousins, the Joyners. We were all thoroughly wetted and thoroughly amused; we told stories, gathered immense bunches of flowers, incarcerated a light company of green grasshoppers, who were disorderly, and ruined two unfortunate born beauties of the butterfly tribe. We, besides, ran down a green lizard. I have picked up of late, in the little bay below the willows, a fossil fish, in a high state of preservation;—the scales, head, tail, fins are all beautifully distinct, and yet so very ancient is the formation in which it was found, that the era of the lias, with all its ammonites and belemnites, is comparatively recent.

'You are fretted, my own dear girl, by the bondage to frivolity, which sex and fashion impose upon you. No wonder you should, when one thinks of the sort of laws by which you are bound. The blockheads are a preponderating majority in both sexes: but somehow in ours the clever fellows contrive to take the lead and make the laws, whereas I suspect that in yours the more numerous party are tenacious of their privileges as such, and legislate both for themselves and the minority.'

In the letter from Miss Fraser, to which our next from Miller is a reply, there occurred several descriptive sketches of the scenery amid which she was at the time, and an allusion to the Rev. Mr Fraser of Kirkhill, who had just lost his life by a fall from his gig. The passage in which Hugh refers to the early writings of David Urquhart and the threatening ambition of Russia is curious when viewed in connection with the issue of Russian scheming in the Crimean war.

'I am thinking long for you, dearest, and for the last week have been counting the days,—counting them in the style of the fool whom Jacques met in the forest. "To-day is the 19th, the 20th comes to-morrow, and the 22nd will be here the day after;" they will creep away one by one, and Lydia will be with me ere they bring the month to an end. My heart is full of you; full of you every hour, and every minute, and all day long. I walked last Saturday on the hill and saw our beech tree, but lacked heart to go down to it; I thought it looked dreary and deserted, and I felt that, were I to lose you, it would be, of all places in the world, the place I could least bear to see. Your grave—but how can I speak of it!—would be a place devoted to sorrow, but to a sorrow not sublimed into agony. I could clasp the green turf to my bosom, and make my bed upon it, but our beautiful beech tree with its foliage impervious to the sun, and its deep cool recess in which we have so often sat under the cover of one plaid,—I could not visit it, Lydia, unless I felt myself dying, and were assured I would die under its shadow. Many, many thanks, dearest, for your kind sweet letter. It is just what a letter should be, with heart and imagination and pretty easy words in it, and yet it is an unsatisfactory thing after all. Instead of consoling me for your absence, it only makes me long the more for you. It is but a pouring of oil on a flame that burnt fiercely enough before.

'I have seen one of the scenes you describe so sweetly,—the bridge of Ardross; but it is a good many years since, and it was after I had just returned from the western Highlands of Ross-shire, where I had visited many scenes of a similar character but on a much larger scale. And so I was not so much impressed by it. I still remember, however, the dark rocks and the foaming torrent, and the steep slopes waving with birch and hazel, that ascend towards the uplands, and the abrupt heathy summit of Foyers overlooking the whole. I trust your guides did not forget to point out to you the two majestic oak trees of this wild dell, that are famed as by far the finest in Ross-shire. One of them has been valued at £100; and not many years since, when the late Duke of Sutherland purchased the estate of Ardross on which it grows, there was a road cut to it that he might go and see. You are now in the parish of Urquhart with good Mr Macdonald. There is not much to be seen in your immediate neighbourhood. Do not omit visiting, as it is quite beside you, the ancient burying-ground of Urquhart. See whether there be not yet an old dial-stone on the eastern wall beside a little garden. I saw it there fourteen years ago, and the thoughts which it suggested have since travelled far in the stanzas beginning "Grey Dial-stone."

'I am happy, dearest Lydia, that you are not going

to Cadboll. Typhus is still raging, I hear, in that part of the country. My own dearest lassie, why am I so much more anxious on your account than on my own? But it is always thus when the heart takes a firm grasp of its object. Man in his colder moods, when the affections lie asleep, is a vile selfish animal; his very virtues are virtues so exclusively on his own behalf that they are well-nigh as hateful as his vices. But love, my dearest, is the fulfilling of the law: it draws us out of our crust of self, and we are made to know through it what it is to love our neighbour not merely as well but better than ourselves. We err grievously in those analogies by which we attempt to eke out our knowledge of the laws of God through an acquaintance with the laws of man; and quite as grievously and in the same way, when we strive to become wise by extinguishing our passions. The requirements of the statute book are addressed to the merely rational part of our nature; and could one abstract the reason of man from the complex whole of which he consists, that single part of him would be quite sufficient for the fulfilment of them. But it is not so with the law of Deity: it is a law which must be *written on the heart*, and it addresses itself to our whole nature; or, to state the thing more clearly, it is not more a law promulgated for man's obedience than a revelation of his primitive constitution; and through grace this constitution must be in some degree restored ere the law, which is as it were a transcript of it, can be at all efficient in forming his conduct. We are mutually pledged, my Lydia, and the law of God commands us to love one another better than we love all human kind besides,—than parents, relatives, or friends; and how does the affection we bear to each other conform to this? Why, it is the very injunction embodied. And would it not be so to us with the requirements of the whole law, were our natures as much restored in all as in part?

'On the evening of Monday week I had a long walk among the woods of the hill with a party to whom I was showing "*the lions*." Lieutenant C— and his young wife were there, Lieutenant W—, too, and his wife, and Miss —— and the Misses ——, with two of the young ——s. The ladies were talking altogether of bonnets and scenery, and London and new patterns, and the two gentlemen were discussing the world afloat, but one of the young people, a fine-spirited boy, was curious about stones and berries and old stories, and so I attached myself to him. You remember the remark of Lamb's brother on the boys of Eton school, "What a pity 'tis that these nice young fellows should in a few years become frivolous members of parliament!" There is as much truth as wit in it. Most of our fine young boys and girls are spoiled in the transition stage when shooting up into men and women, and they do not recover all their lives after. Mere men and women are but poor things, my Lydia; but it is well to love them as well as we can,—they will be better ere the world ends. Your pupil, Miss Harriet, has, for one so young, a great deal of heart about her. Children, it strikes me, are little in their whole minds, in their affections as certainly as in their judgments. There obtains, however, a contrary opinion, from the fact, doubtlessly, that in the present state of the world, children have indisputably some affections, and grown-up people, in most instances, little or none. But this, I am certain, is not according to nature. A warm heart, well cultivated, cannot fail of being warmer in one's twentieth than in one's twelfth year. When a boy I could not love with half the warmth either as a lover or a friend that I can now.

'Are you aware that wild deer sometimes swim across wide estuaries such as the Frith of Cromarty? It was only last week that some of our boatmen found a fine roe swimming across to the Black Isle side, nearly opposite the church of Rosskeen, and fully three quarters of a mile from the nearest shore. It is still alive, and in the keeping of Mr Watson. My uncle tells me that in calm weather deer not unfrequently cross the opening of the bay from Sutor to Sutor; and that when he was a boy there was a fine large animal of this species captured by some fishermen when swimming from the Black Park, near Invergordon, to the Cromarty quarries, where the Frith is fully five miles in breadth.

'I have seen of late some highly interesting articles on the political designs of Russia, by David Urquhart of Braelanguel, a talented young fellow, better acquainted with the details of the question than perhaps any other Briton of the present day. It is wonderful with what art this mighty empire has been extending and consolidating its power for the last century. Should it go on unchecked for half a century more civilized Europe must fall before it, and the world witness, a second time, the arts and refinements of polished life overwhelmed and lost in a deluge of northern barbarism. The democratic principle, says Hume, is generally strongest among a civilized people—the thirst of conquest among a semi-barbarous one. Urquhart shows me that the Russians of the present times are strongly possessed by the latter, and never, certainly, was the democratic principle stronger in civilized Europe than now. Witness the struggles of the antagonist parties in France, Austria, and Italy, and to what extremes Whigs and Tories carry matters among ourselves. And the democratic principle has this disadvantage when contemporary with the other, that it leads men to seek their opponents at home and draws their attention from abroad. And hence they may remain unwarned and disunited until warning and union be of no avail. But forgive me, my Lydia, I am boring you with politics; remember, however, that I do not often transgress in this way.

'Poor Kirkhill! I could hardly bring myself to believe the story of his death. Even in the present age, when every college student arrogates to himself the praise of superior ability, men of real talent are very few. The proportion, too, of the people who are cut off by accident in a quiet well-regulated country like ours, is exceedingly small; and hence, in at least the later annals of talent and genius, we find hardly an instance of other than natural death. The lightning, says the proverb, spares the laurel; and Hume remarks that mankind often lose more in the single philosopher, whom the jealousy of a tyrant cuts off, than in the thousands who perish in an earthquake or conflagration—taking it for granted, doubtless, that thousands may perish so without there being a man of mark among them. But alas for the poor minister of Kirkhill! He has left few clearer heads behind him in the Church. I have often wished, when listening to him, that he mingled more of the philosopher with the theologian; that, instead of always strengthening one Scripture doctrine by showing its correspondence with another not better established than itself, he might have occasionally descended to the broad level of self-evident truth. But no one could admire him more than I did in his own peculiar field. Do you really think our Tory Churchmen will be unable to forgive him his Whiggism even in death?'

Readers will recollect Finlay, the gentle rhyming boy, who had been of the Marcus Cave band, and to whom Miller had been ardently attached. Seventeen years had passed away since he left Cromarty, and it does not appear that any tidings of him had reached Hugh in the interval. One day, however, he was surprised by the arrival of a letter dated Spanish Town, Jamaica, signed by the hand of Finlay. He had often, he said, thought of writing, but he had fancied that Miller had left Scotland, being convinced that, had he remained in his native country, he must have distinguished himself. 'Often,' he proceeds, 'have I looked into the advertising columns of *Blackwood*, *Fraser*, and *Tait*, to see the announcement of a volume of poetry, tales, or something to show that genius was not confined to the south, and at length I was yesterday gratified by seeing your name in a stray number of *Chambers' Journal* for last year as the author of the *Traditionary History of Cromarty*. You have no idea, my dear fellow, how my heart glowed when I read your praises; and with the whole Scotsman running riot in my veins, have I revelled in the story of Sandy Wright (there is some of it like my own, *entre nous*), so like the benevolent heart of my ain Hugh Miller.'

This was a great occasion for Miller. The image of his boy-friend lay in his heart like a coin of pure gold committed to a delicate casket, and when he looked upon it after seventeen years the likeness was bright as on the morning when he bade Finlay adieu. He seized his pen and wrote as follows. As we read this letter can we help loving Hugh Miller?

'Cromarty, Oct. 15, 1836.

'MY OWN DEAR FINLAY,

'Yes, the wise old king was quite in the right. "As cold waters to the thirsty soul, so is good news from a far country." My very hopes regarding the boy-friend, whom I loved so much and regretted so long, have been dead for the last twelve years. I could think of you as a *present existence* only in relation to the other world; in your relation to this one merely as a recollection of the past. And yet here is a kind, affectionate letter, so full of heart that it has opened all the sluices of mine, that assures me your pulses are still beating, and shows me they desire to beat for ever. I cannot tell you how much and often I have thought of you, and how sincerely the *man* has longed after and regretted the friend of the *boy;* you were lost to me ere I knew how much I valued and loved you. I dare say you don't remember that shortly before you left Cromarty you scrawled your name with a piece of burnt stick on the eastern side of Marcus Cave, a little within the opening. I have renewed these characters twenty and twenty times; and it was not until a few years ago, when a party of gipsies took possession of the cave, and smoked it all as black as a chimney, that they finally disappeared. Two verses of the little pastoral you wrote on leaving us are fresh in my memory still—fresh as if I had learned them only yesterday. But I dare say at this distance of time you will scarce recognize them.

> "Ye shepherds, who merrily sing
> And laugh out the long summer day,
> Expert at the ball and the ring,
> Whose lives are one circle of play,
>
> "To you my dear flock I resign,
> My colley, my crook, and my horn;
> To leave you, indeed, I repine,
> But I must away with the morn."

'There they are, just as you left them in the winter of 1819. What, dear Finlay, have the seventeen intervening years been doing with your face and figure? The heart, I know, is unchanged, but what like are you? Are you still a handsome, slender, high-featured boy, dressed in green? John Swanson is a little black *manny*, with a wig; and I have been growing older, but you won't believe it, for the last eighteen years. Great reason to be thankful, I am still ugly as ever. Five feet eleven when I straighten myself, with hair which my friends call brown, and my not-friends red; features irregular, but not at all ill-natured in the expression; an immense head, and a forehead three quarters of a yard across. Isn't the last a good thing in these days of phrenology? And isn't it a still better thing that a bonny sweet lassie, with a great deal of fine sense and a highly-cultivated mind, doesn't think me too ugly to be liked very much, and promises to marry me some time in spring? Do give me a portrait of yourself first time you write, and, dearest Finlay, don't let other seventeen years pass ere then. Is it not a wonder we are both alive? John Layfield, John Mann, David Ross, Andrew Forbes, Adam McGlashan, Walter Williamson, are all dead—yes, Finlay, all dead. Of all our cave companions, only John Swanson survives. John is a capital fine fellow. He was quite as wild a boy, you know, as either of ourselves, and perhaps a little worse tempered; but, growing *good* about twelve years ago, he put himself to college with an eye to the Church, and is now a missionary at Fort William. Dearest Finlay, have you grown *good* too? I was in danger of becoming a wild infidel. Argued with Uncle Sandy about cause and effect and the categories; read Hume and Voltaire and Volney, and all the other witty fellows who had too much sense to go to heaven; and was getting nearly as much sense in that way as themselves. But John cured me; and you may now say of me what Gray says of himself, "No very great wit, he believes in a

God." The Bible is a much more cheerful book than I once used to think it, and has a world of sound philosophy in it besides.

'Do you remember how I stole you from John? You were acquainted with him ere you knew me, and used to spend almost all your play-hours with him on the *Links* or in his little garden. But I fell in love with you, and carried you off at the first pounce. And John was left lamenting! I brought you to the woods, and the wild sea-shore, and the deep dark caves of the Sutors, and taught you how to steal turnips and peas; and succeeded (though I could never get you improved into a robber of orchards—though you had no serious objection to the fruit when once stolen) in making you nearly as accomplished a vagabond as myself. Are not you grateful? "The boy," Wordsworth says, "is father to the man." If so, your boy-father was a warm-hearted *bonny laddie*, worthy of all due honour from you in your present filial relation; but as for mine, I can't respect the rascal, let the commandment run as it please. Don't you remember how he used to lead you into every kind of mischief, and make you play truant three days out of four? A perfect Caliban, too.

> "I'll show thee the best springs, I'll pluck thee berries,
> And I, with my long nails, will dig thee pig-nuts."

But he's gone, poor fellow! and his son, a much graver person, who writes a highly sensible letter, has a thorough respect for all his father's old friends, and steals neither peas nor turnips. Fine thing, dearest Finlay, to be able now and then to play the fool. I wouldn't give my nonsense—to be sure the amount is immensely greater—for all my sense twice told.

'You give me the outlines of your history, and I

must give you those of mine in turn. But they are sadly unlike. You have been going on through life like a horseman on a journey, and are now far in advance of the starting-point. I, on the contrary, have been mounted, whip and spur, on a hobby, and after seventeen years' hard driving, here I am in exactly the same spot I set out from. But I have had rare sport in the fine ups and downs, and have kept saddle the whole time. You remember I was on the eve of becoming a mason apprentice when you left me. The four following years were passed in wandering in the northern and western Highlands, and hills, and lakes, and rivers, one of the happiest and most contented, though apparently most forlorn of stone-masons. I lived in these days in kilns and barns, and on something less than half-a-crown per week, and have been located for months together in wild savage districts where I could scarce find in a week's time a person with English enough to speak to me; but I was dreaming behind my apron of poets and poetry, and of making myself a name; and so the toils and hardships of the present were lost in the uncertain good of the future. Would we not be poor, unhappy creatures, dear Finlay, were there more of sober sense in our composition and less of foolish hope? In 1824 I went to Edinburgh, where I wrought for part of two years. I was sanguine in my expectations of meeting with you. I have looked a thousand times after the college students and smart lawyer-clerks whom I have seen thronging the pavement, in the hope of identifying some one of them with my early friend. On one occasion I even supposed I had found him, and then blessed God I had not. I was sauntering on the Calton on a summer Sabbath morning of autumn, when I met with a poor maniac who seemed to recognize me, and whose features bore certainly a marked resemblance to yours. I cannot give expression to what I felt; and yet the sickening, unhappy feeling of that moment is still as fresh in my recollection as if I had experienced it but yesterday. Strange as it may seem, I gave up from this time all hope of ever seeing you, and felt that even were you dead—and I had some such presentiment—there are much worse ways of losing a friend than by death.

'After returning from Edinburgh I plied the mallet for a season or two in the neighbourhood, working mostly in churchyards—a second edition of Old Mortality—and then did a very foolish thing. I published a volume of poems. They were mostly juvenile; and I was beguiled into the belief that they had some little merit by the pleasing images and recollections of early life and lost friends which they awakened in my own mind through the influence of the associative faculty. But this sort of merit lay all outside of them, if I may so speak, and existed in relation to the writer alone—just as some little trinket may awaken in our mind the memory of a dear friend, and be a mere toy of no value to everybody else. My poems, like the Vicar of Wakefield's tracts on the great *monogamical* question, are in the hands of only the happy few; they made me some friends, however, among the class of men whose friendship one is disposed to boast of; and at least one of them, *Stanzas on a Sun Dial*, promises to live. Chambers alludes to it in the notice to which I owe the restoration of a long-lost friend. The volume which, maugre its indifferent prose broken into still more indifferent rhyme, and all its other imperfections, I yet venture to send you, is dedicated to our common friend, Swanson, but being as tender of his name as my own, the whole is anonymous. In the latter part of the year in which it appeared, I sent a few letters on a rather unpromising subject, the Herring Fishery, to one of the Inverness newspapers. They were more fortunate, however, than the poems, and attracted so much notice that the proprietors of the paper published them in a pamphlet, which has had an extensive circulation. I send it you with the volume. Every mind, large or small, is, you know, fitted for its predestined work—some to make epic poems, and others to write letters on the Herring Fishery.

'I continued to divide my time between the mallet and the pen till about two years ago, when I was nominated accountant to a branch of the Commercial Bank, recently established in Cromarty. I owe the appointment to the kindness of the banker, Mr Robert Ross, whom I dare say you will remember as an old neighbour, and who, when you left Cromarty, was extensively engaged as a provision merchant and shipowner. I published my last, and I believe best, work, *Scenes and Legends of the North of Scotland*, shortly after. Some minds, like winter pears, ripen late; and some minds, like exotics in a northern climate, don't ripen at all, and mine seems to belong to an intermediate class. Sure I am it is still woefully green, somewhat like our present late crops; but it is now twenty per cent. more mature than when I published my former volume, and I flatter myself with the hope that, if winter doesn't come on too rapidly, it may get better still. Read, dear Finlay, my *Scenes and Legends* first; you may afterwards, if you feel inclined, peep into the other two as curiosities, and for the sake of lang syne; but I wish to be introduced to you as I am at present, not as I was ten years ago. The critics have been all exceedingly good-natured, and I would fain send you some of the reviews with which they have favoured me (these taken together would form as bulky a volume as the one on which they are written), but I have only beside me at present the opinion expressed by Leigh Hunt (the friend and coadjutor of Byron, you know), and the notice of a literary paper, *The Spectator*. These I make up in the parcel.

'Where, think you, am I now? On the grassy summit of McFarquhar's Bed. It is evening, and the precipices throw their cold dark shadows athwart the beach. But the red light of the sun is still resting on the higher foliage of the hill above; and the opposite land, so blue and dim, stretches along the horizon, with all its speck-like dwellings shimmering to the light like pearls. Not a feature of the scene has changed since we last gazed on it together. What seem the same waves are still fretting against the same pebbles; and yonder spring, at which we have so often filled our pitcher, comes gushing from the bank with the same volume, and tosses up and down the same little jet of sand that it did eighteen years ago. But where are all our old companions, Finlay? Lying widely scattered in solitary graves! David Ross lies in the sea. John Man died in a foreign hospital, Layfield in Berlin, McGlashan in England, Walter Williamson in North America. And here am I, though still in the vigour of early manhood, the oldest of all the group. Who could have told these poor fellows, when they last met in the cave yonder, that "Eternity should have so soon inquired of them what Time had been doing?"

'Regard, my dear Finlay, my *Scenes and Legends* as a long letter from Cromarty. Do write me a little newspaper. Tell me something of your mode of life. Give me some idea of a Jamaica landscape. What are your politics? What your creed? What have you been doing, and thinking, and saying since I last saw you? Tell me all. Your letter is the greatest luxury I have enjoyed for I know not how long.'

The Dr Waldie of whom we heard as a friend of Miller's in Linlithgow, wrote to him after his return to Cromarty, and received a reply. The science, or sham science, of phrenology, was then making much noise in the world, and Hugh was for some time disposed to trust in it. The examination of his own head by a professional phrenologist did not tend to confirm him in his prepossessions; and as he had argued with Dr Waldie in favour of phrenology, he hastened to lay before the doctor those grounds on which he was now constrained to question its pretensions.

'I think with much complacency of our little discussions on phrenology and geology and all the other *ologies*, and of the relief which I used to derive from them when well-nigh worn out with the unwonted employments of the desk. You will, I dare say, be disposed to smile when I tell you that I am not half so staunch a phrenologist now as I was six months ago, and be ready to infer that my head did not prove quite so good a one as I had flattered myself it should. Nay now, that is not the case, the head did prove quite as good a one, scarcely inferior in general size to that of Burns, and well developed both in front and atop. I am more disposed to quarrel with the science for what it confers upon me than for what it withholds. For instance, few men are so entirely devoid as I am of a musical ear; it was long ere I learned to distinguish the commonest tunes, and though somewhat partial to a Scotch song, I derive my pleasure chiefly from the words. Besides, and the symptom is, I suspect, no very dubious one, of all musical instruments I relish only the bag-pipe. Judge, then, of my surprise to learn from Mr Coxe (and I warned him to be wary) that Nature intended me for a musician. You are aware that 20 is the highest number in the phrenological scale,—the proportional development of music in my head is as 16. But if more than justice be done me as a musician, in other respects I have cause to complain. The organ of language is more poorly developed than any other in the head. One, of course, can't claim the faculty one is said to want with as much boldness as one may disclaim the faculty one is said to possess; but you will forgive me if I produce something like testimony on the point. The effects of an imperfect development of language, say the phrenologist, are a difficulty of communicating one's ideas to another from a want of expression, which frequently causes stammering and a repetition of the same words, and a meagreness of style in writing. But what say the critics in remarking on my little book? "What we chiefly found to admire," says one, "is the singular felicity of the expression." "The wonder of the book," says another, "lies in the execution; there is nothing of clumsiness, and the style is characterized by a purity and elegance, an ease and mastery of expression, which remind one of Irving, or of Irving's master, Goldsmith." The *Presbyterian Review* and Leigh Hunt testify to a similar effect. "But has not vanity something to do in calling in such testimony?" Nothing more likely; still, however, the evidence is quite to the point; and as I have perhaps in some little degree influenced *your* opinions regarding phrenology, I deem it proper thus to state to you the facts which have since modified my own. The rest of the forehead, regarded as an index of mind, has its discrepancies. Causality is largely developed; wit, will you believe it? still more largely; whereas comparison and individuality are only moderate. Now, I know very little of my own faculties if this order should not be reversed. Individuality, or the ability of remembering facts, if I be not much mistaken, takes the lead, comparison comes next, causality follows, and wit at a considerable distance brings up the rear. I find little to remark regarding the rest of the head,—constructiveness, benevolence, conscientiousness, firmness, ideality, and caution are all large, self-esteem is moderate, love of approbation is amply but not inordinately developed, and the lower propensities are barely full. You see it is not altogether my interest to become a sceptic to the reality of the science; but my opinions regarding it have, notwithstanding, undergone a considerable change. Phrenology, however, whatever conclusion the world may ultimately arrive at regarding it, will be found to have had an important use. It has brought the metaphysician from the closet into the world, and turned his attention, hitherto too exclusively directed to the commoner operations of our nature, as these may be observed in the species, to the wonderful varieties of individual character.

'I am glad you have read Edwards. He stands high as a philosopher, even with those who differ with him. Sir James Mackintosh, in his masterly dissertation on Ethics, describes his reasoning powers as "perhaps unequalled, certainly unsurpassed, among men." Nothing so common among thinkers of a low order as what are termed common-sense objections to the Scripture doctrine of predestination; from minds capacious enough to receive the arguments of Edwards, we have none of these. But the man who studies him would need be honest. No sincere lover of the truth was ever the worse for his admirable reasonings, and religious men have been often the better for them; but they may be converted by the vicious into apologies for their indulgence of every passion, and the perpetration of every crime.

'I trust you will recommend me to Mr Turpie. Ask him if he remembers how he used to mar my calculations by getting astride of my shoulders, and my many threats of beating him, which he learned to treat with so thorough a disregard.'

In the course of the summer of 1835 Miller was applied to for contributions to the *Tales of the Borders*, a periodical series began by Mr J. Mackay Wilson, and highly and deservedly popular. Wilson had died, and the publication of the Tales was continued for the benefit of his widow. Hugh consented, and a sufficient number of sketches and tales from his pen to fill a considerable volume appeared in the series. His entire remuneration, as he informs us in the *Schools and Schoolmasters*, was five pounds.

CHAPTER V.

MARRIAGE—CORRESPONDENCE—DEATH OF DAUGHTER—VIEWS ON BANKING.

ON the 7th of January, 1837, Hugh Miller was married to the lady whom he had so long and so ardently loved. Mr Ross, his superior in the Bank and attached personal friend, gave away the bride, and Mr Stewart performed the ceremony. It was a day to make Hugh's heart, calm as he was in all things, profoundly calm as to his own achievements and successes, glow with honourable pride and well-earned joy. He had dared, while in his mason's apron, to aspire to the hand of one who was by birth and breeding a lady. The attractions of personal beauty, enhancing those of a cultivated mind and graceful and animated manners, had led him captive, and for the first and last time, in all the intensity of meaning that can be thrown into the word, he loved. This affection had been for him an inspiration, turning the current of his existence into a new channel and rippling its smooth surface with the genial agitations of hope. He had waited five years, and at times he had been anxious and despondent, for he never wavered in his determination either to marry Miss Fraser into the position of a lady or not to marry her at all. He had now established himself in all points essential to right success in life, and might contemplate the future in a mood of quiet assurance. He had made his mark in the literature of his country. He had passed into the ranks of the brain-workers of the community, depending no longer for livelihood on the toil of his hands. Any bride might now be proud of him. What was very pleasant for him at the time, and is pleasant for us to contemplate from this distance, his ascent had been viewed with unaffected satisfaction by his fellow-townsmen and by all who knew him. He had approved himself a thoroughly friendly man, and he had been rewarded by the good-will and kind wishes of many friends. On the occasion of his marriage his happiness was heartily shared in by the people of Cromarty, and the married pair drove off on their wedding trip in the carriage of Mrs Major Mackenzie, which she had offered, some time before, to her friend Miss Fraser.

'Setting out,' says Miller, 'immediately after the ceremony, for the southern side of the Moray Frith, we spent two happy days together in Elgin; and, under the guidance of one of the most respected citizens of the place—my kind friend, Mr Isaac Forsyth—visited the more interesting objects connected with the town or its neighbourhood. He introduced us to the Elgin Cathedral;—to the veritable John Shanks, the eccentric keeper of the building, who could never hear of the Wolf of Badenoch, who had burnt it four hundred years before, without flying into a rage, and becoming what the dead man would have deemed libellous;—to the font, too, under a dripping vault of ribbed stone, in which an insane mother used to sing to sleep the poor infant, who, afterwards becoming Lieutenant-general Anderson, built for poor paupers like his mother, and poor children such as he himself had once been, the princely institution which bears his name. And then, after passing from the stone font to the institution itself, with its happy children, and its very unhappy old men and women, Mr Forsyth conveyed us to the pastoral, semi-Highland valley of Pluscardine, with its beautiful wood-embosomed priory,—one of perhaps the finest and most symmetrical specimens of the unornamented Gothic of the times of Alexander II. to be seen anywhere in Scotland.'

His wedding gift to his wife was a Bible, on which he inscribed a few stanzas expressive of the pious joy, deep but not exultant, and with its pensive vein, which he felt on putting it into her hand.

'O much-beloved, our coming day
To us is all unknown;
But sure we stand a broader mark
Than they who stand alone.
One knows it all: not His an eye
Like ours, obscured and dim;
And knowing us, He gives this Book,
That we may know of Him.

His words, my love, are gracious words,
And gracious thoughts express:
He cares e'en for each little bird
That wings the blue abyss.
Of coming wants and woes He thought,
Ere want or woe began;
And took to Him a human heart,
That He might feel for man.

'Then oh, my first, my only love,
The kindliest, dearest, best!
On Him may all our hopes repose,—
On Him our wishes rest!
His be the future's doubtful day,
Let joy or grief befall;
In life or death, in weal or woe,
Our God, our guide, our all.'

Under such auspices, Hugh Miller set up his household in Cromarty. His salary was but sixty pounds a year, and the addition which he made to it by literary contributions was as yet small. Mrs Miller continued to take a few pupils. A parlour, bed-room, and kitchen had been furnished, and one servant did the menial work. An attic room was occupied with shelves, on which his few books and fossils, the nucleus of a good library and a valuable museum, were arranged. A table and chair were placed in this room, and it became Hugh's study. It was here that he wrote a number of tales and sketches, published in the continuation of Wilson's *Tales of the Borders*. They are, perhaps, of all his compositions, the least marked by fascination and originality. He had no enthusiasm, he tells us, either for the memory of Wilson or for the publication he had set on foot. The dreary, semi-intellectual routine of a bank clerkship, besides, damped his literary ardour; and when, at a late hour in the evening, he took up for literary composition the pen which an hour earlier he had used to chronicle his monotonous summation of figures, he seemed to have lost the power to conjure with it. Add that the remuneration was wretchedly small,—five pounds for matter to fill a goodly volume,—and enough will have been said to account for the defectiveness of the pieces. His mind gradually regained its elasticity, and he soon found more lucrative employment as a writer for *Chambers's Edinburgh Journal.*

The month in which he was married had not yet come to an end when he learned that John Swanson had lost his mother. To comfort his friends in their distress was always a sacred duty with Hugh, and he at once wrote the following letter.

'Cromarty, January 31, 1837.

'You have lost a very dear relative, and I condole with you. What a world of tender retrospection and fond regret is associated with that one word, mother! How many tender recollections that bear date from the first dawn of memory to the moment of the last farewell! How untiring her solicitude! how innumerable her acts of kindness! How certain our assurance that these rose out of a love which, perhaps the least selfish of all human affections, sought only the welfare of its object! And then to think that she who loved us so much and did so much for us has left us for ever! You will now find yourself much more alone in the world than you ever did before. You have passed, as it were, into a new state of life. For several years you have been the head of your family; but yet, so long as your mother was with you, you must have felt that you were merely residing in *her* house. It was the same person who sat at the head of the table beside you, that had done so when you were a little boy. You had your mother to come between you and the world, as it were; there were some remains of the feeling of confidence in her protection with which, when you were a child, you have laid hold of her gown. But now she is gone, and you will deem yourself stripped of your shelter. There is no longer a breakwater between you and the casualties of life. Even the grave itself seems immensely nearer;—

the generation you spring from has passed away, and you stand in the front rank.

'There is little need, my dear John, that I should address you in the language of consolation. You know that she whom you loved "is not dead, but sleepeth." Nor is yours one of those doubtful cases when he who attempts to console is fain to rest in vague generalities ;—fain to take it for granted without question that the deceased has been living well, and therefore has died safe. I have known Mrs Swanson for the last twelve years. I have seen her character with all its peculiarities in a hundred different aspects. And at what conclusion have I arrived regarding her? Simply this,—That I never knew a woman more thoroughly conscientious. We might deem her at times over wise and prudent, but who ever saw her prudence chill her kindness, or her wisdom set up in opposition to that of God, as displayed in the gospel? Nay, but it is by much too little to say so. Was she not a humble, devoted Christian?

'I am happy,—happier than I was ever in my life before, and my companion is happy too. Love marriages are after all the most prudent of any. Had I taken Lydia's advice, I would have written you long ere now; but believe me, and do not accuse me of selfishness when I say so, I was by much too happy to write. Great enjoyment naturally induces an inaptitude for thought and exertion. I am convinced that, had not Adam fallen and become miserable, the alphabet would have been yet to invent.'

So that Miller held the doctrine which, according to Goethe, Shakespeare proclaims 'with a thousand tongues' in *Antony and Cleopatra*, 'that enjoyment and action (*Genuss und That*) are irreconcilable.'

On the 12th of September of the same year we find him writing to Mr Robert Chambers, offering one or two pieces for the Journal. 'I am leading,' he says, 'a quiet and very happy life in this remote corner, with perhaps a little less time than I know what to do with, but by no means overtoiled. A good wife is a mighty addition to a man's happiness; and mine, whom I have been courting for about six years, and am still as much in love with as ever, is one of the best. My mornings I devote to composition; my days and the early part of the evening I spend in the bank; at night I have again an hour or two to myself; my Saturday afternoons are given to pleasure,—some sea excursion, for I have got a little boat of my own, or some jaunt of observation among the rocks and woods; and Sunday as a day of rest closes the round.

'Your collection of ballads I have found to be quite a treasure,—excellent in itself as a most amusing volume, and highly interesting, regarded as the people's literature of the ages that have gone by. Barbour's Bruce and Blind Harry's Wallace belong, also, to the same library, and must in their day have exerted no unimportant influence on the character of our great-grandfathers. You yourself now occupy the place in relation to the people which the metrical historians and the authors of the ballads did a few centuries ago.'

To this there came a reply warmed by that true-hearted kindness with which Miller's correspondent has cheered so many of the youthful soldiers of literature. 'The account you give me of your domestic condition is necessarily gratifying to one who feels as your friend and is anxious to be regarded in the same light by yourself. Your present circumstances are most creditable to you, and show that your intellect has its true and proper crown,—moral worth. May you ever be thus happy, as you deserve to be! I have sometimes thought of more prominent and brilliant situations for you; but after all, if you can be content with the love of a virtuous woman in a place where you have the chief requisites and a little of the luxuries of life, and where, exempt from the excitements and sordid bustle of a town, you can employ those contemplative powers of mind which I believe to be your highest gifts, you are probably better as you are. Wordsworth has been much laughed at for keeping so constantly in the country, but I believe he is right. There is everything certainly in town that can make the mind active, but it is not the place for doing anything great, and it is not the place for a pure and morally satisfactory life.'

Hugh was gladly welcomed as a contributor to the Journal.

As usual, he replies promptly to Mr Chambers's 'kind and truly friendly letter.' Having stated that Mr Ross had been absent in London, he proceeds :—'Since he left me—now rather more than a month ago—I have been a busier man than I was ever in all my life before; for though the business of the office here is perhaps not very extensive for that of a branch bank, it is sufficiently so for the employment of a single person, and I have been left in charge of the whole without clerk or assistant. How strangely flexible the human mind! Three years ago I was hewing tombstones in a country churchyard, with perhaps a little architectural knowledge, and rather more than the average skill of my brother mechanics, but totally unacquainted with business. And now here I am among day-books and ledgers, deciding upon who are and who are not safe to be trusted with the bank's money, and doing business sometimes to the amount of five hundred pounds per day.

'I have some legendary stories lying by me, which I wrote about a twelvemonth ago, with the intention of giving them to the public in a volume, but the Journal will spread them much more widely. There is a dash of the supernatural in some of them, but I trust that will break no squares with us unless they should lack in interest;—besides, I hope there is philosophy enough in them to save the writer's credit with even the most sceptical of your readers. Superstition, however, is not at all the same sort of thing in these northern districts of the kingdom that it is in those of the south. It is no mere carcase with just enough of muscle and sinew about it for an eccentric wit to experiment upon now and then by a sort of galvanism of the imagination, but an animated body, instinct with the true life. I am old enough to have seen people who conversed with the fairies, and who have murmured that the law against witchcraft should have been suffered to fall into desuetude; and as for ghosts, why, I am not very sure but what I have seen ghosts myself. Superstition here is still living superstition, and, as a direct consequence, there is more of living interest in our stories of the supernatural and more of human nature. When man has a place in them, it is not generic, but specific, man—man with an individual character. The man who figures in an English or South of Scotland legend is quite as abstract a person as the man in a fable of Æsop;—with us he has as defined a personality as the *Rip Van Winkle* of Washington Irving himself. By the way, much of the interest of this admirable story is derived from the well-defined individuality of poor *Rip*.'

One or two other letters or notes passed between him and Mr Chambers, but they are of little importance. Hugh once has this remark on the nature of contentment:—'The content which is merely an indolent acquiescence in one's lot is so questionable a virtue, that it seems better suited to the irrational animals than to man.

That content, on the other hand, which is an active enjoyment of one's lot, cannot be recommended too strongly. And it is this latter virtue, if virtue it can be called, that my papers attempt to inculcate. True, it leads to no Whittington and his cat sort of result, but it does better,—it leads to happiness, a result decidedly more final than a coach and six.' Mr Chambers once suggests that he might advantageously 'shift the scene' from his 'dearly beloved Cromarty,' and that the less he introduces of superstition the better. 'I am at the same time very sure,' he adds, 'that whatever be the nature of your subject, you cannot fail to give it that certain yet indescribable interest which so peculiarly characterizes all that comes from your pen.'

In a letter to Sir T. Dick Lauder, containing much eulogistic criticism of a legendary work, just published by Sir Thomas, Hugh refers incidentally to his maniac friend of Conon-side, mentioning a particular or two which he does not elsewhere touch upon. 'She was a McKenzie, some of whose ancestors, as they had resided for centuries within less than a mile of the chapel of Killiechrist, might probably have fallen victims to the fiery revenge of the MacDonald. I wish you could have heard some of her stories. She was, like Christy Ross (one of the personages of Sir Thomas's book), a wild maniac, and used to spend whole nights among the ruins of the chapel, conversing, as she used to say, with her father's spirit; but her madness was of the kind which, instead of obscuring, seems rather to strengthen the purely intellectual powers (her malady seemed but a wilder kind of genius), and so, mad as she was, I used to deem her conversation equal to that of most women in their senses. I scraped an acquaintance with her when working as a mason apprentice, about sixteen years ago, in the neighbourhood of Killie-christ. She had taken it into her head that I was some great person in disguise, and used to come to me every evening after work was over, to consult me on such questions as the origin of evil, and the eternal decrees, and others of equal simplicity and clearness. I gave her, of course, the benefit of all my metaphysics, whether doubtfully bad, or bad beyond doubt, and got in return sets of the finest old stories I have met with anywhere. How her eyes used to brighten when she spoke of the bloody and barbarous MacDonald, and the fearful raid of Killiechrist! She was a sister to that Mr Lachlan McKenzie of Loch Carron—the modern prophet of the Highlands, of whom you cannot fail to have heard; and in some respects she must have closely resembled him. But she is gone, and many a fine old story has died with her.'

He gives Sir Thomas a bright little sketch of his marriage trip, and subsequent happiness:—'I had the pleasure, last winter, when on a two days' visit to Elgin, of seeing the Priory of Pluscardine, which you have so beautifully described in your tale of the rival lairds; and a painting could not have recalled it more vividly to my recollection. I saw much during my brief visit, and enjoyed much, for the occasion was a joyous one—my marriage—and I had Mr Isaac Forsyth of Elgin for my guide. I saw Lady-hill, with its rock-like ruin, and its extensive view; the hospital, with all its wards; the museum, with its spars and its birds; the splendid institution, so redolent of the showy benevolence of the present age; and the still more splendid cathedral, so redolent of the showy piety of a former time; but above all, it was the hermit-like priory, in its sweet half-Highland, half-Lowland glen, with its trees, and its ivy, and all its exquisite innumerable combinations of the simple and the elegant, that impressed me most strongly. I found, too, that my companion, whose taste had been much more highly cultivated than mine, was quite as much delighted with it. You, who are yourself so happy in your domestic relations, will not be displeased to learn that, after having enjoyed for full five years all that a lover enjoys in courtship, I now possess all that renders a husband happy in a wife. I have now been rather more than eight months married, and am as much in love as ever.'

Pleasant as were Miller's relations with the Messrs Chambers, he experienced the want of a literary vehicle, in which the deepest feelings of his nature, his religious feelings, might have free expression. He accordingly offered his services as occasional contributor to the *Scottish Christian Herald,* a periodical conducted chiefly by clergymen of the Church of Scotland. He forwarded for publication a sketch and a poem. Both were declined. 'The narrative,' said the editor, 'though well written, is scarcely of that description which accords with the design of the *Herald,* our great anxiety having always been that the interest of the periodical should rest, not on love affairs or the operation of mere worldly motives, but on the varied aspects in which Christianity presents itself among men. From beginning to end, in short, we wish not to be merely moral and entertaining, but decidedly *religious.*' He added that, if Miller pleased, he would hand the piece to the Messrs Chambers for publication in the Journal.

Hugh was considerably pained. 'My story,' he wrote in reply, 'turns on a case of conversion. Whoever heard of such a thing in one of our lighter periodicals? *Not* in Blackwood, I am sure, with all his zeal for the Church, nor in Tait, with all his love of the Dissenters. Chambers, too, though beyond comparison less exceptionable than either of these, seems to have a pretty shrewd guess that stories of this kind are not at all the best suited to secure to him his hundred thousand readers. I would fain see a few good periodicals set agoing of a wider scope than either those of the world or of the Church—works that would bear on a broad substratum of religion the objects of what I may venture to term a week-day interest. I can cite no book that better illustrates my *beau ideal* of such a work than the Bible itself. No book more abounds in passages essentially popular in their interest. And yet I question whether some of these passages, were they now to appear for the first time, would find admittance into many of our religious periodicals. There are chapters in the Book of Proverbs that would be deemed merely moral, and the stories of Ruth and of Esther would be thought to rest too exclusively on the operation of mere worldly motives. But this is not at all what I meant to say. I merely intended to remark that I would fain see periodicals established, which would not only be suited to convince the man of the world that it is more philosophic to believe than to doubt the truth of Christianity, but which would also be so written that the men of the world would read them.'

The cup of Miller's happiness was full when a little daughter began to smile upon him from the arms of her mother. All gentle helpless things he loved with a passion of tenderness, and his affection for his own little prattler was inexpressible. He observed her movements with ever fresh interest and charm. 'My little girl,' he wrote once, 'has already learned to make more noise than all the other inmates of the house put together, and is at present deeply engaged in the study of light and colour. She is still in doubt, however, whether the flame of the candle may not taste as well as it looks.' 'She was,' says Mrs Miller, 'a delight and wonder to Hugh above all wonders. Her little smiles and caresses sent him always away to his daily toil with a lighter heart. When he took small-pox, I, of course, slept on a couch in his room, and was with him night and day. But the great privation was, that he could not see her. We ventured, when he was mending, to open the room door, and let him look at her across the entrance lobby, and allow her to stretch out her little arms to him. Her own illness began soon after. It was a very tedious one, connected with teething, and lasting nine or ten months. All our mutual recreations, and many of my employments, including a school for fisher-lads, which I had taught for some years, at eight every evening, had to be given up. In the spring of '39 I had a close nursing of several weeks. Then there was a marked amendment. One lovely evening in April I went out, for the first time that spring, to breathe the air of the hill. When I returned, I found the child in her nurse's arms, at the attic window, from which she used to greet her papa when he came up street. She had been planting a little garden in the window sill of polyanthus, primrose, and other spring flowers. When she saw me, she pushed them away, with the plaintive "awa, awa" she used to utter, and laid her head on my breast. An internal fit came on. The next time she looked up it was to push my head backwards with her little hand, while a startled, inquiring, almost terrible look, came into her lovely eyes. All the time she lay dying, which was three days and three nights, her father was prostrate in the dust before God in an agony of tears. Whether he performed his daily bank duties, or any part of them, I do not remember. But such a personification of David the King, at a like mournful time, it is impossible to imagine. All the strong man was bowed down. He wept, he mourned, he fasted, he prayed. He entreated God for her life. Yet when she was taken away, a calm and implicit submission to the Divine will succeeded, although still his eyes were fountains of tears. Never again in the course of his life was he thus affected. He was an affectionate father, and some of his children were at times near death, but he never again lost thus the calmness and dignity, the natural equipoise, as it were, of his manhood.' This was the first and the last poignant domestic sorrow Miller experienced. He cut the little headstone for his darling, and never again put chisel to stone.

It contributed not a little to his happiness at this period that his relations with Mr Ross, his superior in the bank, were amicable and harmonious in the highest degree. Several years after Hugh had terminated his connection with the bank, he took occasion to bear witness to the 'unvarying kindness' he had experienced from Mr Ross. During the five years when they worked together he had not once heard from Mr R. 'the slightest word of censure or of difference,' and this though his place had been one of 'great trust and occasional difficulty.' Nay, his acquaintance with the character of Mr Ross heightened his appreciation of that of mercantile men as a class. 'During the not inconsiderable period,' he says in the dedication of a pamphlet to Mr Ross, 'in which I enjoyed your confidence, I was conversant with the inner details of your conduct in the various branches of trade which you have prosecuted so long and so successfully, and the effect has been to heighten my estimate of the important class—our men of merchandise and traffic—to which you belong. Of the many thousand transactions in which I have seen you engaged, there was not one of a kind in any degree suited to lessen my well-founded respect for you, as a sagacious man of business, of genial feelings, of nice integrity, and a discriminating intellect.'

Irksome as was the routine of bank business to one who, in his hours of hardest toil, had always previously had the breath of heaven, the cool breeze of the hill-side, to fan his brow, Miller soon found that as a bank accountant he occupied a coigne of vantage from which to conduct his studies of human nature and of human life. If anything was still wanted to secure him against those showy plausibilities and political, social, sentimental extravagances, which are the besetting sins of self-educated men, it was the opportunity afforded him in a bank office of drawing the line between sound and substance, between gold and glitter, between reality and pretence. There is a passage in the *Schools and Schoolmasters* illustrative of this remark, which has always seemed to me to contain a singular amount of practical sense, and of accurate valuable information.

'However humbly honesty and good sense,' thus it proceeds, 'may be rated in the great world generally, they always, when united, bear premium in a judiciously managed bank office. It was interesting enough, too, to see quiet silent men, like "honest farmer Flamburgh," getting wealthy, mainly because, though void of display, they were not wanting in integrity and judgment; and clever unscrupulous fellows like "Ephraim Jenkinson," who "spoke to good purpose," becoming poor, very much because, with all their smartness, they lacked sense and principle. It was worthy of being noted, too, that in looking around from my peculiar point of view on the agricultural classes, I found the farmers on really good farms usually thriving, if not themselves in fault, however high their rents; and that, on the other hand, farmers on sterile farms were *not* thriving, however moderate the demands of the landlord. It was more melancholy, but not less instructive, to learn from authorities whose evidence could not be questioned, bills paid by small instalments, or lying under protest, that the small-farm system, so excellent in a past age, was getting rather unsuited for the energetic competition of the present one; and that the *small* farmers—a comparatively comfortable class some sixty or eighty years before, who used to give dowries to their daughters, and leave well-stocked farms to their sons—were falling into straitened circumstances, and becoming, however respectable elsewhere, not very good men in the bank. It was interesting, too, to mark the character and capabilities of the various branches of trade carried on in the place,—how the business of its shopkeepers fell always into a very few hands, leaving to the greater number, possessed, apparently, of the same advantages as their thriving compeers, only a mere show of custom; how precarious in its nature the fishing trade always is, especially the herring fishery, not more from the uncertainty of the fishings themselves, than from the fluctuations of the markets, and how in the pork trade of the place a judicious use of the bank's money enabled the curers to trade virtually on a doubled capital, and to realize, with the deduction of the bank discounts, doubled profits. In a few months my acquaintance with the character and circumstances of the business men of the district became tolerably extensive, and essentially correct; and on two several occasions, when my superior left me for a time, to conduct the entire business of the agency, I was fortunate enough not to discount for him a single bad bill. The implicit confidence reposed in me by so good and sagacious a man was certainly quite enough of itself to set me on my metal. There was, however, at least one item in my calculations in which I almost always found myself incorrect. I found I could predict every bankruptcy in the district; but I usually fell short from ten to eighteen months of the period in which the event actually took place. I could pretty nearly determine the time when the difficulties and entanglements which I saw *ought* to have produced their proper effects, and landed in failure; but I missed taking into account the desperate efforts which men of energetic temperament make in such circumstances, and which, to the signal injury of their friends and the loss of their creditors, succeed usually in staving off the catastrophe for a season.'

When the financial policy of Sir Robert Peel threatened the one pound note circulation of Scotland with extinction, Miller took part strongly with those who opposed the measure. His views found expression in a series of newspaper articles on Sir Robert Peel's Scotch currency scheme, which were republished in the form of a pamphlet entitled *Words of Warning to the People of Scotland.* Having carefully examined this little work, I am bound to admit that the writer appears to me to have failed to appreciate with perfect accuracy and lucidity the distinction between wealth and the machinery by which wealth is exchanged. That he erred with some of the shrewdest men the world has seen—with Benjamin Franklin, for example, and Sir Walter Scott—is unquestionable, but the error does not change its nature on that account. A passage which Miller quotes from Franklin, with the italics as Miller inserts them, will show wherein, as I conceive, all orthodox economists of the present day would declare both men in the wrong. 'The truth is,' wrote Franklin, 'that the balance of our trade with Britain being against us, the gold and silver are drawn out to pay that balance, and the necessity of some medium of trade has induced the making of paper money, which cannot thus be carried away. Gold and silver are not the produce of North America, which has no mines, and that which is brought thither cannot be kept there in sufficient quantity for a currency. Britain, an independent great state, when its inhabitants grow too fond of the expensive luxuries of foreign countries that draw away its money, can, and frequently does, make laws to discourage or prohibit such importations, and by that means can retain its cash. But the Colonies *are* dependent governments; and their people, having naturally great respect for the sovereign country, and being thence immoderately fond of its modes, manufactures, and superfluities, cannot be restrained from purchasing them by any province law, because such law, if made, would immediately be repealed, as prejudicial to the trade and interests of Britain. *It seems hard, therefore, to draw all their real money from them, and then refuse them the poor privilege of using paper money instead of it.*' So far Franklin, as approvingly quoted by Miller. That a country which has no real money can make money out of paper, is an idea so palpably erroneous that one has difficulty in comprehending how it could have imposed upon Benjamin Franklin.

It was not, however, necessary to penetrate far into the mysterious problems of the currency in order to find sound reasons for leaving the circulating system of Scotland alone. That system had worked well. Sir Walter Scott and Hugh Miller knew the fruits of the tree, that they were good, and they justly concluded that it was a good tree. There was no cause for anxiety on the part of Sir Robert Peel that the Scotch might put too great confidence in their one pound notes; and the opinion of eminent Scottish bankers, that if they were deprived of their note circulation, they would be compelled to close their branch banks, and thus curtail the accommodation they had afforded to industry in the provinces, was conclusive against tampering with a system which might be injured but could hardly be improved. 'We have the authority,' wrote Miller, 'of the first men of the country for holding that our cash credits could not outlive the projected change of Sir Robert for a twelvemonth.' In a provincial bank he had ample opportunities of observing the operation of the Cash Credit system of Scottish banking; and as it would be difficult to find a more lucid and appreciative account of that far-famed system than Miller's, and the passage, apart from biographical considerations, is interesting, I shall here insert it.

SCOTCH CASH ACCOUNTS.

'In at least one department of banking the Scotch have not yet been imitated in any other country. Hume published the second part of his *Essays, Moral, Political, and Literary,* in 1752, just ninety-two years ago; and in this work, in his essay on the balance of trade, we find the first notice of a Cash Credit Account. He describes it as "one of the most ingenious ideas that has been executed in commerce," and as first devised in Edinburgh only a few years previous. And we see that Sir Walter, in describing it a second time, seventy-four years after, had still to characterize it as "peculiar to Scotland." The description of Hume is singularly happy and philosophic. "A man goes to the bank," he says, "and finds surety to the amount, we shall suppose, of a thousand pounds. This money, or any part of it, he has the liberty of drawing out whenever he pleases, and he pays only the ordinary interest for it while it is in his hands. He may, when he pleases, repay any sum so small as twenty pounds, and the interest is discontinued from the very day of repayment. The advantages resulting from this contrivance are manifold. As a man may find surety nearly to the amount of his substance and his bank credit is equal to ready money, a merchant does hereby in a manner coin his house, his household furniture, the goods in his warehouse, the foreign debts due to him, his ships at sea, and can on occasion employ them all in payments, as if they were the current money of the country. If a man borrow a thousand pounds from a private hand (besides that it is not always to be found when required), he pays interest for it whether he be using it or not, whereas his bank credit costs him nothing except during the very moment in which it is of service to him. And this circumstance is of equal advantage as if he had borrowed money at much lower interest." This we deem a very admirable appreciation, so far as it goes, of the peculiar advantages of the Cash Credit Account, a mode of transaction, be it remembered, in which our Scottish banks have invested millions of money, but we do not deem it a full appreciation. It is doubtless a great matter to our traders of Scotland that they should be able to coin, through its means, their houses, their furniture, their ships, and the debts owing to them. But it enables them to do much more: it enables them to coin their characters should they be good ones, even should houses, ships, and furniture be wanting. True, the banks must be satisfied that there is property in every individual case to coin, but the holder of the account may have none, it is enough that what *he* wants his sureties have. He may have character only, and this character his sureties enable him to transmute into gold. The Cash Credit becomes thus a premium on integrity. Few men have risen in trade from the lower walks whose character for honesty and persevering diligence some kind friend has not enabled them thus to coin into money; but in Scotland alone does there exist an ingenious system, tested by the experience of a century, which provides for accommodations of this nature, as a part of the regular business of the country. Benjamin Franklin devoted by will two thousand pounds to be laid out in loan to young industrious workmen of the cities of Boston and Philadelphia, whose moral character was such that there could be found at least two respectable citizens willing to become sureties in a bond with the applicants for the repayment of the money so lent, with interest. "I myself," he says in the same document, in rendering a reason for the bequest, "was assisted to set up my business in Philadelphia by kind loans of money from two friends there, which was the foundation of my fortune, and of all the utility in life that may be ascribed to me; and I wish to be useful even after my death, if possible, in forming and advancing other young men that may be serviceable to their country." Now, such was the device of a wise and benevolent man. But our Scottish scheme, though a matter of mere business, possesses a mighty advantage, both in wisdom and benevolence, over the device of Franklin, and this advantage we find admirably indicated by Hume. A hundred pounds borrowed from the Franklin Fund would bear interest on its full amount against the borrower from the moment in which it was consigned over to him till it was repaid. And if, during some pause in the commercial world, he suffered it to lie unemployed beside him, it would be a positive disadvantage to him at such a time to possess it. It would be at least, during the pause, not a means of bettering his circumstances, but a means of reducing them; and this very conviction would be a spur that, in times of doubtful speculation, might provoke to the unwise employment of it. But lent on the Cash Credit scheme, no portion of it would bear interest save the portion in present use; and if not only the whole was relodged with the bank during some languid interval in which the wheels of trade moved heavily, but also an additional sum, that additional sum would bear interest to the holder's advantage. A few figures, however, arranged in the form of a Cash Credit Account, may give our readers a much better idea of the nature of the device so admired by the philosophic Hume for its ingenuity, than any mere verbal description :—

Cash Credit for £100, A. B., Dr. to the —— Bank of Scotland.
C. D. Surety.
E. F. Surety.

1844.			Dr.	Cr.	Balance Dr.	
			£	£	£	Days
Nov.	1	To	10	—	10	2
„	3	To	20	—	30	2
„	5	To	10	—	40	3
„	8	To	12	—	52	2
„	10	By	—	40	Dr. 12	2
„	12	By	—	20	Balance Cr. 8	2
„	14	By	—	20	Cr. 28	

The nature of the business represented by these simple figures,—part of the page of a bank ledger,—is as follows :—A. B. holds in one of our Scotch banks a Cash Credit Account to the amount of a hundred pounds, for which C. D. and E. F. are sureties, and on which he began to operate on the first day of November last, by drawing on it to the amount of ten pounds. These ten pounds were straightway entered in both the "Dr." column and the "Balance Dr." column, and bore interest against him in the bank's behalf for two days, when he again operated on his account by drawing twenty pounds more, which raised the sum of his debit in the "Balance Dr." column to thirty pounds; and thus he continued operating for nine days, when the balance at his debit amounted to fifty-two pounds. And of these nine days he had to pay interest during two of the number for ten pounds,—during two more for thirty pounds,—during three for forty pounds—and during two, yet again, for fifty-two pounds. He then lodged in the bank forty pounds, which were entered to his credit in the "Cr." column, and the balance to his debit in the "Balance Dr." column sank to twelve pounds, which a second lodgement, in two days after, of twenty pounds, extinguished altogether, leaving a balance of eight pounds in his favour, and these eight immediately began to bear interest in his behalf. Yet another lodgement raised the eight to twenty-eight pounds; and thus the account runs on, on terms immensely more favourable to A. B., the holder, than if he had been a borrower on the Franklin Fund. But what the holder of the Cash Credit gains in this way, the bank, it may be said, loses: the scheme, though the best possible for the borrower, must be an indifferent one for the lender. Far better for the bank it may be said, had A. B. taken out the hundred pounds at once, and continued to pay full interest upon them. Nay, not so fast. What is best for A. B. in this case is not by any means worst for the bank. The account here, as shown by the dates, is actively operated upon. Though the balance at the holder's debit must at no time exceed a hundred pounds, many hundreds may be drawn upon it in the course of the twelvemonth, and these all in the bank's notes; and from the notes kept in circulation through its means, the bank derives a profit of three per cent. It draws interest for its money from two distinct sources,—in the first place, from A. B. for the cash which it advances to him, and, in the next, from the notes in which its advances to A. B. are made. In no other part of the world are the interests of borrower and lender so admirably adjusted as in Scotland under this scheme.'

The essential advantage of this Cash Credit system is that, in the department of trade, commerce, and agricultural enterprise, it puts tools into the hands of those best fitted to use them. Men of character and brains—men who can discern openings for capital, and whom other men can trust in the attempt to avail themselves of those openings—attain that position in society which nature has prepared them to occupy. The country is benefited in many ways,—vitally, incalculably benefited. The right men do the work, therefore it is effectually done; the right men earn the reward of competence and respect, therefore there is comparatively little grumbling. It is manifest that such a system promotes that genial circulation of talent from the low places to the high places of society which does so much to prevent social stagnation, to give geniality and healthfulness to the relation of class to class, to knit firmly yet freely together the entire social framework. Hugh Miller knew Scotland, as few men have ever known her, and it was his deep conviction that, in certain important particulars, she had the advantage of England. 'Scotland,' he said, 'is still truly a nation,—not a mere province: her institutions are diverse from those of England,—her interests distinct,—her character different.' It is fair to add that the system of Cash Credits, with its distinctive feature of enabling men to coin their character, experience, and talents, is obviously better adapted to a small country, where each man may know his neighbour, and where the possibilities and limitations of enterprise admit of easy calculation, than to a large one.

It may be an exaggeration to say that it is from a Bank parlour you can best observe modern society, but it is the exaggeration of a truth. No one can understand modern society who does not apprehend to what an extent every form of romance or enthusiasm peculiar to other ages is drying up in the intense and scorching thirst for gold. Our notion of modern society will always be more or less fanciful, more or less of a morning dream rather than an afternoon reality, if we do not appreciate in its fulness of meaning the banker's definition of 'a good man.' The banker's definition; and, with but a half-hearted, ineffectual, semi-honest protest, the definition of literary, artistic, and clerical society. We have to remark, also, that though as a Bank Accountant Hugh Miller had the best opportunities of witnessing the infinite importance attached by society to money, he was never conscious even of a temptation to devote himself to the task of becoming rich, and he never admitted into his breast that feeling of bitterness and exasperation, on account of the unequal distribution of the gifts of fortune, which, according to Mr Carlyle, instilled its subtle, maddening poison into a heart so genial, healthy, and generous as that of Robert Burns.

CHAPTER VI.

SCIENCE IN THE ASCENDANT.

DURING these last quiet years of his residence in Cromarty, when Miller was putting the last touches to the *curiosa felicitas* of his style, and choosing irreversibly the form in which his higher intellectual activity was to be exerted, the question came often directly or indirectly before him whether his supreme devotion should be to literature or to science. Poetry had been as good as abandoned. He did, indeed, as his wife and one or two of his most confidential friends were aware, cherish the resolution to return to verse, and had visions of bringing even his science ultimately to minister to the Muse. But for the present his critical faculty in the poetical department had outstripped his productive faculty, and he wrote almost exclusively in prose. We find, however, from his correspondence, that the legendary tales and biographical sketches to which he had so long devoted attention, had ceased to interest him as formerly, and that he contemplated a transference of his allegiance from literature to science. Literature has been called the science of man; science may be called the literature of nature. If the hackneyed quotation from Pope as to man being mankind's noblest study has become hackneyed on account of its truth, and if Sir W. Hamilton's favourite lines about man and mind being the only great things on earth are not rhodomontade, and if Shakespeare is higher than Newton among the moderns and Homer higher than Aristotle or Plato among the ancients, it would seem to follow that literary art, as displayed in history, poetry, the drama, and prose fiction, takes legitimate precedence of that inquiry into the sequences of the physical world, which bears specifically the name of science. But it was under the influence of no abstract considerations that Miller determined in favour of the latter. Literature, as it presented itself to his mind, did not afford scope to his abilities. The traditions, the legends, the history of his native place,—the characters of the men he had known since boyhood,—did not appear to furnish materials out of which important literary works could be constructed. The vein was worked out.

It is perhaps surprising that he did not, so far as can be discovered, think of Scottish history as a field in which to employ himself. He might, I think, have written a history of Scotland which the world would have placed among the acknowledged masterpieces in this species of composition; a history in which the distinctive character of the Scottish race, and the distinctive contribution of Scotland to modern civilization, would have been justly defined, and in which the patriotic enthusiasm of the writer would have swelled at times into the grandeurs of epic poetry. The view taken of Scottish history by Miller was essentially the same as that taken by Burns, by Scott, by Wilson, by Carlyle, and he was more profoundly in sympathy with the religious genius of his nation than any one of these. His strength as a stylist lay in description, and all his books, particularly the *Schools and Schoolmasters*, and *First Impressions of England and its People*, afford proof of his skill in continuous narrative.

In the history of literature, again, he was fitted to excel. He could have given us, for example, a critico-biographical work on Pope, Addison, and their contemporaries, which might have been unique and superlative of its kind. He knew those men as if he had lived and talked with them. They were the models whose chastened beauties appeared to him more worthy of emulation than the passionate and metaphorical writing in vogue in his own day.

No such employment of his powers, however, seems to have occurred to him, and it is certain that neither of the works indicated could have been produced at so great a distance from the original sources of literary information as Cromarty. Science invited him to an unbeaten path—to an assured originality—and the scene in which her wonders were to be sought had been his playground since infancy. That devotion to science was attaining in his mind the power of a ruling passion is attested in that chapter of the *Scenes and Legends* in which he presents himself to his readers as the 'Antiquary of the world.'

There is a curious interest in observing how much this chapter contains of what, a few years subsequently, made Miller famous as a geologist throughout the world. As yet he is groping his way, feeling darkly round a cave which he surmises to be coated with jewels and precious metals, but able only to guess vaguely what may be their nature and their value. He has already noticed those 'flattened nodules of an elliptical or circular form,' of which all the world was one day to hear. He has laid some of them open and found them to contain, one 'the remains of a fish, scaled like the coal-fish or haddock,' another, 'the broken exuviæ of a fish of a different species, which, instead of being scaled like the other, seems roughened in the same way as the dog-fish, or the shark,' a third, 'a confused, bituminous-looking mass that has much the appearance of a toad or frog,' a fourth, 'a number of shining plates like those of the tortoise,' a fifth, 'a large oval plate, like an ancient buckler, which seems to have formed part of the skull.' Such were Hugh Miller's first impressions of the pterichthys and the coccosteus.

The tenacity with which ideas and images, once conceived, clung to his mind, is evinced by the fact that some of the illustrations which we have in this chapter recur on all subsequent occasions when he touches upon the subjects here treated. He never tired of explaining the advantage to the geologist of having strata uptilted by volcanic agencies by reference to 'the ease which we find in running the eye over books arranged on the shelves of a library, contrasted with the trouble which they give us in taking them up one after one when they are packed in a deep chest.' The materials of that description of the structure and aspect of the great Caledonian valley, which he subsequently wrought out with a breadth and a vividness belonging to the highest order of literary art, may all be discovered in this chapter, though they as yet lie about in comparative disorder, the dissevered fragments which the master-hand is one day to arrange into a symmetrical structure. The imaginative faculty by which he threw a mantle of beauty over the scenes that ocular observation, or the restoring processes of scientific reflection, afforded him, is displayed in this essay with a delicate splendour of effect which he hardly surpassed in after-days. 'I remember one day early in winter,' he writes, 'about six years ago, that a dense bank of mist came rolling in between the Sutors like a huge wave, and enveloped the entire frith, and the lower lands by which it is skirted, in one extended mantle of moisture and gloom. I ascended the hill; on its sides the trees were dank and dripping; and the cloud which, as I advanced, seemed to open only a few yards before me, came closing a few yards behind, but on the summit all was clear, and so different was the state of the atmosphere in this upper region from that which obtained in the fog below, that a keen frost was binding up the pools, and glazing the sward. The billows of mist seemed breaking against the sides of the eminence; the sun, as it hung cold and distant in the south, was throwing its beams athwart the surface of this upper ocean,—lighting the slow roll of its waves as they rose and fell to the breeze, and casting its tinge of red on the snowy summits of the lower hills, which presented, now as of old, the appearance of a group of islands. Ben Wyvis and its satellites rose abruptly to the west; a broad strait separated it from two lesser islands, which, like leviathans, raising, and but barely raising, themselves above the surface, heaved their flat backs a little over the line; the peaks of Ben Vaichard, diminished by distance, were occupying the south; and far to the north I could descry some of the loftier hills of Sutherland, and the Ord-hill of Caithness. I have since often thought of this singular scene in connection with some of the bolder theories of the geologist. I have conceived of it as an apparition, in these latter days, of the scenery of a darkly remote period,—a period when these cold and barren summits were covered with the luxuriant vegetation of a tropical climate, with flowering shrubs, and palms, and huge ferns; and when every little bay on their shores was enlivened by fleets of *nautili* spreading their tiny sails to the wind, and presenting their colours of pearl and azure to the sun.'

Nor is it only of his capacity as an observer and skill as an imaginative describer that this remarkable chapter gives us an earnest. It reveals with impressive distinctness the spirit in which he would cultivate science. Not as a mere collector of facts, or word-painter of geological landscape, would he work, but in full view and constant recollection of every momentous question relating to the nature and destiny of man on which science might touch. Religion had become the central and regulating force in his character; religion he believed to have been the source of all that was best and most enduring in the character of his country; religion, turning the regards of man to a Divine Father, and opening the human eye on a pathway of eternal advance, appeared to him inseparably involved with the majesty and well-being of man as a species. Inspired with a passionate devotion both to science and to religion, he could not but seek to put the hand of the one into the hand of the other, and to mediate between the two. The following passage might serve as a preface to that whole series of works in which he subsequently made it his effort to interpret the language of the rocks in a manner which would not contradict the statements of Holy Writ. 'Let us quit,' he says, as he turns from those scenes of primeval creation on which he has been expatiating, 'this wonderful city of the dead, with all its reclining obelisks, and all its sculptured tumuli, the memorials of a race that exist only in their tombs. And yet, ere we go, it were well, perhaps, to indulge in some of those serious thoughts which we so naturally associate with the solitary burying-ground, and the mutilated remains of the departed. Let us once more look round us and say whether, of all men, the geologist does not stand most in need of the Bible, however much he may contemn it in the pride of speculation. We tread on the remains of organized and sentient creatures which, though more numerous at one period than the whole family of man, have long since ceased to exist; the individuals perished one after one, their remains served only to elevate the floor on which their descendants pursued the various instincts of their nature, and then sunk, like the others, to form a still higher layer of soil; and now that the whole race has passed from the earth, and we see the animals of a different tribe occupying their places, what survives of them but a mass of inert and senseless matter, never again to be animated by the mysterious spirit of vitality,—that spirit which, dissipated in the air, or diffused in the ocean, can, like the sweet sounds and pleasant odours of the past, be neither gathered up nor recalled? And oh, how dark the analogy which would lead us to anticipate a similar fate for ourselves! As individuals we are but of yesterday; to-morrow we shall be in our graves, and the tread of the coming generation shall be over our heads. Nay, have we not seen a terrible disease sweep away in a few years more than eighty millions of the race to which we belong? and can we think of this and say that a time may not come when, like the fossils of these beds, our whole species shall be mingled with the soil; and when, though the sun may look down in his strength on our pleasant dwellings and our green fields, there shall be silence in all our borders, and desolation in all our gates, and we shall have no thought of that past which it is now our delight to recall, and no portion in that future which it is now our very nature to anticipate? Surely it is well to believe that a widely different destiny awaits us; that the God who endowed us with those wonderful powers, which enable us to live in every departed era, every coming period, has given us to possess these powers for ever; that not only does He number the hairs of our heads, but that His cares are extended to even our very remains; that our very bones, instead of being left, like the exuviæ around us, to form the rocks and clays of a future world, shall, like those in the valley of vision, be again clothed with muscle and sinew; and that our bodies, animated by the warmth and vigour of life, shall again connect our souls to the matter existing around us, and be obedient to every impulse of the will. It is surely no time, when we walk amid the dark cemeteries of a departed world, and see the cold blank shadows of the tombs falling drearily athwart the way,—it is surely no time to extinguish the light given us to shine so fully and so cheerfully on our own proper path, merely because its beams do not enlighten the recesses that yawn around us. And oh, what more unworthy of reasonable man than to reject so consoling a revelation on no juster quarrel than that, when it unveils to us much of what could not otherwise be known, and without the knowledge of which we could not be other than unhappy, it leaves to the invigorating exercise of our own powers whatever in the wide circle of creation lies fully within their grasp!'

His literary essays and his legendary tales had drawn upon Miller the attention of men eminent in the world of literature; this chapter of his book constituted his introduction to circles interested in the pursuit of science. 'I may mention,' he says in a letter to Mr Robert Chambers, 'that a geological chapter in my little volume of *Scenes and Legends* has attracted more notice among the learned, than all the other chapters put together. Mrs —— had a hit at me in *Tait* for introducing such a subject; I could now tell her, however, of Fellows of the Geological Society and Professors of Colleges whom my chapter has brought more than a day's journey out of their route to explore the rocks of Cromarty.' The Old Red Sandstone was at this time a comparatively unknown region to geologists, and the palæontological discoveries to which Miller was feeling his way excited the keenest interest. Dr John Malcolmson, who had recently arrived in this country from India, visited Cromarty, discussed geological problems with Hugh, and examined with him the geological sections of the neighbourhood. The Cromarty geologist began to correspond with Sir Roderick, then Mr Murchison, and with M. Agassiz. Fleming was at this time professor of natural science in King's College and University, Aberdeen, and he hastened to Cromarty to look with the only eyes he ever trusted in matters of observation, his own, into the wonders of Eathie burn and Marcus cave. It was doubtless of great service to Miller at this stage in his geological studies to be brought into converse with the author of the *Philosophy of Zoology*, and to have his theories, just beginning to take shape, overhauled by one of the acutest, most searching, most philosophically sceptical intellects of the century. His controversy with Dr Fleming on the old Scotch coast-line, the existence of which the latter denied to the last, probably commenced at this period, and twenty years afterwards, when the eminence and authority of both were acknowledged in the Geological Society of Edinburgh, the debate remained unfinished. But in none of its stages did it do anything else than add zest to the cordiality of their friendship. Miller knew how to value the trenchant logic, sharp analysis, and severe inductive *cross-examination* of Fleming; and to find a worthy antagonist, whom he might bring under the raking fire of his argumentative batteries, was one of the choicest pleasures Fleming could find in life. He was now in the prime of his faculties; and his brilliant, incisive talk, touching, often with caustic humour, on a thousand men and things, is remembered by Mrs Miller as very pleasantly enlivening their quiet life in Cromarty.

The following passage from a letter of this period to Dr Malcolmson, with its critical notes upon the speculations of Agassiz and Lyell, has some interest as showing Miller's geological whereabouts at the time.

'Mrs Miller has read to me in very respectable English Agassiz' paper on the Moraines of Jura and the Alps. I have been much interested in it. The phenomena it describes are as new to me as the theory founded upon them, and both serve to fill the imagination. But where am I to seek a cause for the intense cold which would cover Europe with one immense sheet of ice from the pole to the shores of the Mediterranean? and why infer that the earth in cooling down, instead of gradually sinking in temperature according to the ordinary and natural process, should sink by sudden fits and starts, and reach at each leap a much lower degree than it was subsequently able to maintain? The theory, I question not, accounts well for the appearances which suggested it; but then, like the tortoise, on whose shoulders your old friends the Brahmins support the globe, it seems sadly to lack footing for itself. I have got a copy of Lyell's late work, the elementary one, and find that he, too, in accounting for erratics, introduces the agency of ice. His theory, if I may venture to decide, seems much less open to objection than that of Agassiz,

—it is a good feasible theory, with a host of every-day experiences to support it. I cannot square it, however, with some of the facts. Why have all the *travellers* of this part of the country come from the west, and none from the opposite quarter? We have masses in abundance of blue schistose gneiss and micaceous schist from the hills of Ross-shire, and blocks of basalt even from the Hebrides, but we have no granites from Aberdeen, no granitic gneisses or grauwackes from Banff or Moray, no yellow sandstone from Elgin or Broughhead. The hills of these countries are nearly as ancient as those of Ross-shire or the Hebrides, and we have no winds so violent as those which blow from the east; but these winds have failed to convey to us a single block from those hills. There is another rather puzzling fact. The conglomeration which in this part of the country forms the inferior bed of the Old Red Sandstone, contains in some localities large water-rolled blocks, scarcely inferior in size to our more modern erratics. Now, there was surely no ice in the days of the Old Red Sandstone. But truce with the subject. Lyell's theory seems excellent so far as it goes, though it may doubtless be pushed too far. Ice was most probably but one of several agents,—the most ancient of which, such as floods and currents, must have existed in the earliest eras. As for the appearances described by Agassiz, do you not think the glaciers which produced them might have existed in those comparatively modern times when the greater part of Europe was covered by forests, and the climate, even to the coasts of the Mediterranean, was chill and severe? Hume, in his Essay on the populousness of ancient states, tells us that the Tiber was frequently covered with ice even so late as the days of Juvenal.'

From several letters of a geological character, which,

as containing information subsequently published by Miller in a more mature shape, it is unnecessary to print here, the following is selected as a good illustrative sample:

'TO M. AGASSIZ.

'Cromarty, May 31, 1838.

'HONOURED SIR,

'I have just learned from my friend, Dr Malcolmson, that you have expressed a wish to see one of the fossils of my little collection. I herewith send it you, and a few others, which you may perhaps take some interest in examining.

'I fain wish I could describe well enough to give you correct ideas of the locality in which they occur. Imagine a lofty promontory somewhat resembling a huge spear thrust horizontally into the sea, an immense mass of granitic gneiss forming the head, and a long rectilinear line of Old Red Sandstone the shaft. On the south side are the waters of the Moray Frith, on the north those of the Frith of Cromarty,—the Portus Salutis of the ancients. The clay-stone beds, which contain the fossils, occupy an upper place in the sandstone shaft, covering it saddle-wise from frith to frith. A bed of yellowish stone about sixty feet in thickness lies over them, except where they are laid bare by the sea, or cut into by two deep ravines; a bed of redder stone of unascertainable depth, but which may be measured downwards for considerably more than one hundred yards, lies beneath. The beds themselves average from ten to thirty feet in thickness. They abound everywhere in obscure vegetable impressions and fossil fish; but in some little spots these last are much better preserved than in the general mass. All my more delicately-marked fossils have been furnished by one little piece of beach, hardly more than forty square yards in extent.

'Of all the fossils of these beds, the one with the tuberculated surface seems least akin to anything that exists at present. I have split up many hundred nodules containing remains of this animal, for in the times of the Old Red Sandstone it must have existed by myriads in this part of Scotland. All the larger ones I have found broken and imperfect—the nodules in general are too small to contain more than detached parts of these, and, besides, the hard plates of the animal, separated apparently by open sutures, seem to have offered a less equal degree of resistance to the accumulating weight than those continuous coats of scales, which covered the various fish of the period, and which seem in consequence to have been thrown less out of their original forms. It is only the smaller varieties that I find perfect enough to furnish me with anything like an adequate idea of the proper shape of the animal. In these, however, though the general outline be better preserved, the plates are invariably obscure. In the larger specimens, on the contrary, they are as well defined as the bits of a dissected map, but I lack skill to piece them together. The form of the body seems to have somewhat resembled that of a tortoise, the plates, like those of that animal, received their accessions of growth at the edges which form the fins of the fish, the body of the creature, as shown by the plates, must have preserved considerable thickness at the sides. I have found specimens from which I infer that a transverse section must have formed an oval. At the shoulders of the animal there were two arm-like fins or wings. All my better specimens agree in the form and position of these. Each terminated in a sharp point, and there was a slight curve in each, like that of an elbow a very little bent.

'I can say almost nothing regarding the head of the animal. Almost all the specimens I have yet found show it variously. In some the fossil has a decapitated appearance, in others there are faint markings which merely show that a part of the creature extended above the arm-like fins, while yet a third kind shows a tolerably well-defined head. But even in specimens such as the last all is obscure. Among the detached plates, however, I occasionally find one of a peculiar form which seems to have covered the upper extremity of the animal. In shape it somewhat resembles the snout of a ray. It is arched transversely, and a longitudinal depression runs through the centre, beginning a very little from the point. All my specimens of this plate are rather too weighty to send you, but Dr Malcolmson writes me that one which I picked up for him you have brought with you to Neufchatel. . . . . In specimens 5, 6, and 7 you will find the remains of a small fish a contemporary of the tuberculated one. Unlike the fish with a single large spine in each fin which I find in the same beds, it was furnished with teeth. The fins, too, were not scaled like those of the other, and instead of being fastened to single spines, somewhat like the sails of a boat to the yards, the membrane of each fin seems to have been supported in the commoner way by a number of rays. I find a striking difference, too, in the appearance of the head;—the bones are strongly marked, and the external surface much roughened; whereas the head of the fish with the spines has always a smooth, flat appearance, except at the neck, where what seems to have been the gills, or perhaps rather a sort of fringe which overlapped these, may be traced. In both fishes the scales were ridged longitudinally, but in the toothed one these ridges are so much more strongly marked that, when examined through a glass, each scale resembles a little bunch of thorns. Pardon me, honoured sir, that I am thus minute in describing these differences to you, who observe better than any one else, and can make a better use of what you observe. I have not succeeded in convincing some of our northern geologists that we have two varieties of small scaled fish in our beds, and I am now appealing to you as our common judge, and thus showing the grounds of my appeal. Besides, as I cannot send you my specimens by hundreds, I deem it best, though it may seem presumptuous in one so unskilled, to send you in this way the result of my examinations of the whole. One single specimen sometimes furnishes a characteristic trait regarding which perhaps fifty illustrative of the same fossil may be silent. Among all my specimens of the fish with the spines, only one shows me that the animal was marked by a lateral line.

'The consistency of style, if I may so express myself, which obtains among the fossil fish of these beds, is highly interesting. In no single fish do I find two styles of ornament,—there is unity of character in every scale, plate, and fin. In the large glossy-scaled variety, for instance (Dipterus), the whole external covering of the animal was marked by minute puncturings, so that every scale and plate—every ray of the tail and fins,—presents, when viewed through the glass, a sieve-like appearance. Another variety of large-scaled fish, apparently of an entirely different family, has its distinct and consistent style of ornament also. Each scale furnishes a broad ground (faintly lined by concentric markings crossed by radiating ones) on which there are raised a series of longitudinal ridges,—waved just enough to give them a sort of pendulous appearance; and, as if to conform with this character, the rays of the fins and tail have a pendulous appearance too. In the fish with the spines, the silvery smoothness of the fins must have harmonized well with the smoothness of the head, and the lightly-marked ridges of the scales with the light fringe of the neck and the elegant lines of scales which covered the tail. And the small-scaled toothed fish had its harmony of ornament with the rest. Without bearing perhaps a single thorn, it must have everywhere presented a thorny appearance. Each scale seems a bunch of minute prickles; the head and neck present the same rough appearance, and the rays of the fins are all jagged at one edge. I find that, unlike at least two varieties of fish furnished by the beds,—the fish with the larger punctured scales and the fish with the large ridged ones,—the caudal fin spread equally on both sides the tail. I am afraid, however, that when thus communicating the results of my petty observations, I am but gaining for myself the reputation of being a tedious fellow.

'Do I ask too much, honoured sir, when I request a very few lines from you to say whether the formation in which these fossils occur be a fresh-water one or otherwise, and whether the small-scaled fish with the teeth be of a kind already known to geologists, or a new one? I am much alone in this remote corner, a kind of Robinson Crusoe in geology, and somewhat in danger of the savages, who cannot be made to understand why, according to Job, a man should be making leagues with the stones of the field! But I am sanguine enough to hope that the good-nature of which my friend Dr Malcolmson speaks so warmly may lead its owner to devote a few spare minutes to render these leagues useful to me. I am, I trust, sufficiently acquainted with geology rightly to value the decisions of its highest authority.'

It is an illustration of the carefulness of Miller's habits of study, that he not only rewrote, with minute verbal emendations, one passage in the preceding letter, but copied the second as well as the first version into the book in which he preserved his letters.

The fossils intended for the inspection of M. Agassiz were forwarded, in the first instance, to Sir Roderick Murchison. Miller wrote at the same time to the latter, inviting him to read the unsealed letter to Agassiz, repeating his question as to whether the Old Red Sandstone was a marine or a fresh-water formation, and stating his intention to draw up an account of the geology of the Cromarty district 'for a widely-circulated periodical.' He received the following reply:—

'London, June 23, 1838.

'Though exceedingly occupied in the effort to bring out my ponderous volumes with their copious illustrations, I lost little time in opening your box and examining its treasures as they passed to M. Agassiz. I invited my friend, Sir Philip de Grey Egerton, who franks this letter, and is stronger in fossil-fishery than most of us, to be present at the examination. The result of our view is, that we have no hesitation in pronouncing your small-scaled fish to belong to the genus *Cheirolepis Agass.*, the most striking character of which is the dorsal fin being placed far backwards, and immediately above or opposite to the posterior portion of the anal fin. "The fins of this genus," says M. Agassiz, "have no spinose ray on their anterior edge. Small slender 'rayons,' very much pressed against each other, and imbricated like the scales alongside of the great anterior 'rayons' of these fins, replace in this case the spinose rays which support the fins of the two genera, *Acanthodus* and *Cheiracanthus*." We believe that your specimens belong to the *C. Traillii Agass.*, found in Pomona, Orkney, by Dr Traill. There may be two species in those you have sent, but it seems doubtful.

'As to the fossils 1, 2, 3, we know nothing of them except that they remind me of the occipital fragments of some of the Caithness fishes. I do not conceive they can be referable to any reptile, for if not fishes, they more clearly approach to *crustaceans* than to any other class. I conceive, however, that Agassiz will pronounce them to be fishes, which, together with the curious genus *Cephalaspis* of the Old Red Sandstone (described at length from the Scotch and English specimens), form the connecting *links* between crustaceans and fishes. Your specimens remind me in several respects of the Cephalaspis.

'Although my work was intended to be exclusively devoted to Silurian (or transition) rocks of England and Wales, I have made a few allusions to other tracts, and among these to the Old Red Sandstone of Scotland, in doing which I have, in the descriptions of the organic remains, briefly alluded to your labours. Now that I know the fidelity and closeness of your research, I shall endeavour to introduce another allusion in the Appendix, which is all that remains unprinted.

'I am delighted with your clear and terse style of description, and beg to assure you that if you could send us in the course of the summer any general and detailed account of both the Sutors and all their contents, I shall have the utmost pleasure in communicating it to the Geological Society, to be read at the November meeting.

'You write and observe too well to waste your strength in newspaper publications, and a good digest of what you have done ought to be preserved in a permanent work of reference. I can give you no positive answer as to whether the Old Red Sandstone of Scotland was formed in a lake or in the sea. I have, however, strong reasons for believing that it is a marine deposit, for in England we find marine shells in it to a considerable height above the uppermost beds of underlying Silurian rocks. Besides, it is so analogous in structure and component parts to the New Red Sandstone, which also contains concretionary limestone, spotted marls, fishes, and marine shells, that I cannot but believe that the one as well as the other of these great Red systems were formed beneath the sea, though doubtless, as in the coal-measure periods, there may have been local exceptions and partial lacustrine deposits.

'The oldest lake deposit as yet known (*i.e.* purely lacustrine) is in the upper zone of coal, and is described by myself. Burdie House is another example, but the deposit seems to be of a mixed character.

'I much long to revisit the shores of Caithness and Cromarty with my increased knowledge, and with the conviction that I should learn so much from you, but I fear it is hopeless.'

One of Miller's correspondents at this time was Mr Patrick Duff, of Elgin, an ardent geologist. In a letter addressed to Mr Duff in December, 1838, we meet with the following passage, descriptive of an attack of illness, under which Hugh had recently suffered:—'During the whole of November I was toiled almost to death at the bank with our yearly balance, and I have been confined by small-pox ever since, with a face doubtless a good deal less handsome than usual, and surrounded by faces uglier than even my own. There were faces on the bed-curtains, and faces on the walls, and faces in abundance on my wife's tartan gown; and when I shut my eyes to exclude them, I just saw them all the more clearly. I strove hard to call up more agreeable pictures. The tree-ferns and saurians of the lias, or the half-tailed fish and coccostei of the Old Red Sandstone, it would be worth while getting into a fever to see; but I called upon them as vainly as Hotspur did upon the spirits of the "vasty deep." I saw faces, faces, faces, and saw nothing else. The phenomena of mind as exhibited in disease have, I suspect, been studied a great deal too little. Can you tell me how a person affected by fever can be both a man and a magic lantern at the same time, and marvel exceedingly in his capacity of spectator at what he exhibits to himself in his character of showman? I am as much a geologist as ever,—a huge breaker of stones; and I expect, when I have broken up a few hundred cart-loads more, to know something of the matter. I am fighting my way, all alone, by main strength,—the very antitype of Thor and his hammer, and find that I have not been fourteen years a mason for nothing. . . . . . I must set myself in right earnest, sometime next summer, to draw up an account of the geology of this part of the country. I have picked up in a desultory way a good many facts, some of them of value enough to be preserved; and I am of opinion, besides, that the geology of Cromarty, well understood, may serve, in part at least, as a kind of key to that of Moray and of the various localities in which there occur fish-beds of the same kind with ours. More splendid sections are to be found nowhere. The burn of Eathie is a study in itself.'

In a critique on the poems of a local poet, named MacColl, published by Miller about this time in the *Inverness Courier*, there occurs a comparative estimate of Celt and Saxon, as represented by the Highlanders and Lowlanders of Scotland. Miller would probably have agreed with Professor Huxley that physiologically there is no important difference between the races; but it is evident that he considered them as widely differing in those psychological characters which, though taken no account of by the physiologist, are of immense importance to the historian. Here is the passage:—'The Celt is essentially a different person from the Saxon in the very constitution of his mind. Both are shrewd, but each in his own way. The Highlander is characterized by the shrewdness of *observation*, the Lowlander by that of *inference;* the Highlander is delighted by the external beauty of things, the Lowlander in diving into their secret causes; the Highlander feels keenly, and gives free vent to his feelings, the Lowlander, on the contrary, cautiously conceals every emotion, unless it be a very potent one indeed; the Highlander is a descriptive poet, the Lowlander a metaphysician. Our Adam Smiths and David Humes are types of the one class,—the Ossian of MacPherson is no unmeet representative of the other. The Gaelic as a language is singularly rich in the descriptive, and comparatively barren in the abstract. Phrenologists remark nearly the same thing of the Celtic head; the reflective organs are always less prominently developed than the knowing ones. It may be adduced, too, as a corroborative fact, that the native literature of the Highlands abounds in poems which contain nothing but pure description from beginning to end. One of MacPherson's acknowledged pieces, entitled the *Cave*, which may be found in a well-known critique on Laing's Ossian, by Sir Walter Scott (see Edin. Review, No. 13), and MacColl's address to Loch Lomond, are of this sort of picture-gallery class. We have nothing quite like them in the poetry of the Lowlands, where natural imagery is made to hold a subordinate place to sentiment and narrative, and where description at its best serves but to mark the scene on which some reflection is grounded or in which some action is performed. Perhaps of all our Saxon poets, Professor Wilson approaches nearest in this respect to a Celtic one.'

CHAPTER VII.

THE CROMARTY FAST-DAY—FRIENDSHIP WITH REV. MR STEWART.

WHILE science was more and more absorbing the attention of Miller, a backward glance being now and then cast upon literature, he did not cease to take the same lively interest in the affairs of Cromarty which we witnessed in connection with the 'Cromarty Chapel Case.' In the summer of 1838 rejoicings took place throughout the country, on account of the recent accession of Queen Victoria to the throne. The day appointed for celebrating the coronation coincided with that on which the Cromarty Kirk Session had decided to hold the fast which, in Presbyterian Scotland, usually precedes the celebration of the Lord's Supper. The Church authorities declined to alter their day, and another was appointed for the loyal demonstration. An outcry was instantly raised throughout the North of Scotland on the subject, and the people of Cromarty were proclaimed a parcel of old-fashioned bigots, the slaves of their ministers. A minute clique in Cromarty took the view of the vociferous remonstrants, and, under the animating leadership of a gallant major, a small proprietor of the neighbourhood, got up a loyal display of their own at the time when the majority of the inhabitants were in church. Miller found himself and his townsmen denounced in prose and verse far and wide. He took pen in hand, and addressed a letter in their and his own defence to the *Inverness Courier*. Adopting a tone of good-humoured banter, admirably fitted to the occasion, he brought out the extreme absurdity of the transports of indignation into which the newspaper poets and tap-room loyalists had lashed themselves. The major he struck off in a phrase felicitously descriptive of a large class of men :—' No one ever spoke well of his judgment or thought ill of his heart.' Space must be found for a few sentences from the letter.

' The day of the fast arrived ; and not on the Sabbath itself do our streets present a more solitary appearance, or are our churches better filled. There was no music outside to disturb the congregations, for the respectable and talented musicians, who had been so vehemently urged to lend their aid in desecrating the day, had gone to attend service in a neighbouring parish, and the old veteran, who had been so earnestly solicited to parade his drum through the streets, had indignantly refused. There were sounds of distant firing heard by the people who sat nearest the windows,—the others heard only the preachers, as they poured forth their fervent and impressive prayers that their Sovereign might reign long and prosperously over willing and affectionate hearts,—that she might enjoy all the good of this world and all the happiness of the next. After Divine Service was over, the ghost of a forlorn and miserable pageant glided through our streets. Every door was shut, as if an enemy had landed ; a glimpse might now and then be caught of some anxious matron drawing her too curious urchin from a window ; a few boys had broken loose, but they were wonderfully few indeed. The major walked in front, supported by two placard bearers. He was neatly dressed in what he had been pleased to term "the garb of his Queen," with, of course, the substitution of inexpressibles for the petticoat, and really, for a gentleman of his standing, looked exceedingly well. The suit he wore was certainly a very pretty suit; all agree, however, that he might have chosen a more appropriate cap, and that the addition of a few bells would have been a mighty improvement. About ten o'clock there was a bonfire lighted, at which some of the town children attended, and the major harangued. He was witty on the clergyman and Kirk Session, and more than a little severe on the treasonable disloyalty of attending church. There was much benevolence, however, shown in his concern for the town; and much good nature in his so kindly informing the decent fisherwomen, whom curiosity had brought within earshot, that the rejoicings were in honour of "a lovely young queen, just like themselves."

'The people of the procession have been represented as Liberals, those opposed to them as Conservatives or Tories. Nothing more untrue. The parish had agreed to postpone their coronation rejoicings till Wednesday last; and on that day, in a procession by far the most numerous and respectable ever witnessed in Cromarty, and headed by our magistrates and ministers, I had the pleasure of walking side by side with one of the most intelligent and liberal Whigs of the place. At the demonstration of the evening I had the satisfaction of drinking to her Majesty's ministers,—a toast given from the chair. And I, too, am a Whig. I was a Whig when Major—— was a zealous Tory, and should we both live for a few years longer, I shall be a Whig when the major may have become a Tory again.'

The victory clearly remained with Miller and the clerical and magisterial respectabilities of Cromarty. His friendship with the Rev. Mr Stewart had now become close and confidential. Stewart was no writing man. Wholly devoid of ambition, shrinking with the fastidiousness of an intellectual recluse from celebrity on the one hand and from sustained effort on the other, he permitted his fine intellect to work in its own way and at its own time. On one subject he was deeply in earnest, and on one only—the subject of religion. When he spoke to his people from the pulpit every energy of his soul was roused, and the intense, penetrating, subtle spell of his genius enthralled the hearer. I heard him preach once in my boyhood, and no second time; but the memory of that one sermon will never fade from my mind. His manner was in marked contrast to that which has commonly prevailed among fervid Highland preachers, to that, for example, of Dr MacDonald of Ferintosh. There was little action, no ejaculation; the fire rose from deep inner fountains, not in a lava torrent, but in keen, concentrated, high-mounting jets of flame. He was an evangelical of the old school in his theology, and it was by a curious combination of qualities that sermons theologically as old as the Shorter Catechism were imbued by him with a true originality. He was biblical, textual, in the highest degree, resting on the infallible inspiration of the *ipsissima verba* of Scripture; but his imagination was so vigorous, his inventive faculty so quick and fresh and fertile, that the words of Scripture suggested to him what would have occurred to no other man. The subject on which I heard him preach was the flood, and I recollect how, by casting every word of the Scriptural record into the alembic of his mind, and melting out of it every possible implication and suggestion, he realized the scene with a vividness and an intensity of effect of which I had not previously dreamed. 'The Lord shut him in.' This was a very rich nugget from the Scriptural mine. 'Noah did not close the door. There are works which God keeps for Himself. The burden of them is too heavy for the back of man. To shut the door on a world about to perish would have been too great a responsibility for a son of Adam,—the stress of it would have borne too heavily on a human heart. Another moment, and another, and another might have been granted by the patriarch, and the door might never have been shut at all. And would he have done the work conclusively, even if he had, in the first instance, closed the door? Who knows but that, when the waters rose, and he heard the wailing around, and friends whom he loved held towards him their little ones and shrieked to be taken in, he might have relented, and opened, and a rush might have been made, and the ship that carried the life of the world might have been swamped? He dared not open a door which God had shut; perhaps he could not open it. We never hear that he opened the door even when the earth was drying. God told him when to go. And so it is in the ark of salvation. It is not the Church, it is not the minister, that shuts or opens the door. These do God's bidding: they preach righteousness, they offer salvation, they gather in; it is God that shuts and that opens the door. And what a sound was that, when, in the listening ominous hush of earth's last evening, God shut the door! There have been sounds as well as sights to make the boldest heart quail and the flintiest heart melt; the cry that has gone up from cities given over to fire and sword, the shuddering throe of earthquake which hurries myriads to death; but except the cry on Calvary, which corresponded to it, no more solemn and melancholy sound has been heard by human ears than that which passed into the evening stillness when the broad green earth was left to be the grave of mankind, and God shut the door of the ark. Once again God will shut the door. Man will not do it. Angel will not do it. But oh, what a sigh and shudder will pass through the listening universe when God will shut the door of the heavenly ark upon the lost!'

Such, or something like such, was Stewart's manner of preaching. He had brief outlines before him in writing, and he brought out his picture, touch by touch, as he went along. His strokes were few, decisive, memorable. I shall not say that all I have given above was his; I never put my impressions of his discourse into black and white until now; but this is my honest attempt to reproduce from the fossil bed of memory so much of the living sermon. And a living word it was. The preacher, speaking in clear, intense, low, yet rich and ringing tones, held a large congregation, for, I should say, at least an hour, in attention as fixed and silent as that of a child hearing a wondrous tale from its mother.

The two men of whom Miller used to speak as having done most to form his character and powers were Chalmers and Stewart. It was the general feeling in the Church of Scotland at this period that these were her most remarkable preachers; and considering the remoteness of Cromarty, and the shy and retiring nature of Stewart, this fact must be held to demonstrate that he was a man of a high and peculiar order. After the disruption of the Scottish Church in 1843, when it was proposed to transfer Dr Candlish from his pulpit to a professorial chair, the common sentiment of the Free Church settled upon Stewart as best fitted to succeed him as pastor of the most influential and cultivated congregation in the Free Church. Stewart's profound conscientiousness, finding in the selection a providential indication of duty, overpowered his reluctance to accept the call; but the thought of Edinburgh pressed on him, to use his own phrase, like his gravestone, and one day, before proceeding to the south, he was found dead in his bed.

BOOK V.

EDITOR.

'And who can say:—I have been always free,
Lived ever in the light of my own soul?
I cannot . . . .
But I have not grown easy in these bonds,
But I have not denied what bonds these were!
Yea, I take myself to witness,
That I have loved no darkness,
Sophisticated no truth,
Nursed no delusion,
Allowed no fear!'

CHAPTER I.

NON-INTRUSION.

IF we wish to have a correct appreciation of Hugh Miller, and not to substitute an image of our fancy for the living man, we must clearly apprehend and perfectly admit two propositions: first, that he was, in the deepest foundations of his character, a religious man; secondly, that he was, distinctively and with his whole heart, a Scotchman.

Few passages in the works of Mr Carlyle have been more frequently quoted or more justly admired than that in which, in criticising the cosmopolitan tendencies of Schiller, he affirms the worth of patriotic feeling. 'Nature herself,' says Mr Carlyle, 'has, wisely, no doubt, partitioned us into "kindreds, and nations, and tongues:" it is among our instincts to grow warm in behalf of our country, simply for its own sake; and the business of Reason seems to be to chasten and direct our instincts, never to destroy them. We require individuality in our attachments: the sympathy which is expanded over all men, will commonly be found so much attenuated by the process, that it cannot be effective on any. And as it is in nature, so it is in art, which ought to be the image of it. Universal philanthropy forms but a precarious and very powerless rule of conduct; and the "progress of the species" will turn out equally unfitted for deeply exciting the imagination. It is not with freedom that we can sympathize, but with free men. There ought, indeed, to be in history a spirit superior to petty distinctions and vulgar partialities; our particular affections ought to be enlightened and purified; but they should not be abandoned, or, such is the condition of humanity, our feelings must evaporate, and fade away in that extreme diffusion. Perhaps, in a certain sense, the surest mode of pleasing and instructing all nations *is* to write for one.' During the most active and stirring years of Hugh Miller's life, he emphatically wrote for his country; his literary activity formed part of the history of Scotland. The sympathies which linked him to humanity at large were rooted in his native soil.

From the point of view of the philosophical historian, two principal causes are discernible as having combined to make Scotland what she has been in modern times. The first is the assertion of Scottish independence, under Wallace and Bruce, in the commencement of the fourteenth century; the second, the predominance of the Reformed Church in Scotland, from the middle of the sixteenth century downwards. On this point also we may profitably listen to Mr Carlyle. In *Past and Present*, he thus refers to the Scottish war of independence:—'A heroic Wallace, quartered on the scaffold, cannot hinder that his Scotland become, one day, part of England; but he does hinder that it become, on tyrannous unfair terms, a part of it; commands still, as with a god's voice, from his old Valhalla and Temple of the Brave, that there be a just real union as of brother and brother, not a false and merely semblant one as of slave and master. If the union with England be in fact one of Scotland's chief blessings, we thank Wallace withal that it was not the chief curse. Scotland is not Ireland: no, because brave men rose there, and said, "Behold, ye must not tread us down like slaves; and ye shall not—and cannot!"'

Among the laws embodied in the moral government of the world, few seem more clearly legible, and none more stern, than this, that the nation which inflicts wrong upon another nation, and the nation which sinfully submits to the wrong inflicted, shall both of them, through long ages of suffering, expiate their crime. The nation which did not rise and tear its fetters from its limbs before they were welded on unalterably, has suffered the pangs of immedicable hatred and of burning thirst for revenge, pangs which distemper the whole body politic. The nation which, in mere lust of conquest, trampled its neighbour into the dust, has found its best subsequent efforts to do justice and show kindness frustrated by the embittered spirit of the vanquished. Infinite, therefore, is the obligation under which both nations concerned lie to those men who saved them from centuries of penal animosity. The gratitude due by Scotchmen to Bruce and Wallace and their heroic compatriots is but one degree more ardent than that due to them by Englishmen.

The other factor in Scottish history is the Reformation of the Church under Knox. In his enthusiastic recognition of what the Reformation did for his country, Mr Carlyle appears to forget for the moment what he had said of Wallace. 'The history of Scotland,' he tells us in his lecture on the Scottish Reformer, 'contains nothing of world-interest at all, but this Reformation of Knox.' His own words, which we have just seen, are an eloquent assertion of the world-interest and world-significance of the Scottish war of independence; and what he says of the influence of the Scottish Reformed Church, he would doubtless permit us to regard as the complement of his previous statement. 'Scotch Literature and Thought,' he exclaims, 'Scotch Industry; James Watt, David Hume, Walter Scott, Robert Burns: I find Knox and the Reformation acting in the heart's core of every one of these persons and phenomena; I find that without the Reformation they would not have been. Or what of Scotland? The Puritanism of Scotland became that of England, of New England. A tumult in the High Church of Edinburgh spread into a universal battle and struggle over all these realms; there came out, after fifty years' struggling, what we all call the "*Glorious* Revolution," a Habeas-Corpus Act, Free Parliaments, and much else!'

No Church has been more thoroughly popular than the Reformed Church of Scotland, and, if we apprehend the character and history of that Church, we shall have no difficulty in understanding how this has been the case. The Scottish people embraced the principles of the Reformation with unanimity and ardour; and from the time when the Reformed Church gained the ascendant in Scotland down to the period of the Revolution Settlement, the Church was always identified with the party of political freedom and national independence. The prosperity of the Church was coincident with the vindication of the national honour and the spread of the national influence; the humiliation of the Church was accompanied by the political debasement of the people. These things go together in the history of those times with a singular precision and constancy; the one is to the other, in the windings of Scotland's fortune, as the bank is to the stream. The aggressive pride and self-confident enthusiasm of Scotland never towered so high as when the covenanting army marched to assist the Puritans against Charles I.; few worse civil governments have ever existed than that which oppressed Scotland and the Church of Scotland together under Charles II.; and in the interval between the reigns of Charles I. and Charles II., when England, Scotland, and Ireland alike lay prostrate at the feet of Cromwell, the Church of Scotland had the felicity to represent a pitch of national hatred against the victor of Dunbar keener than was ever known in Scottish bosoms against the race of Stewart. Thus, in Scotland, 'patriot' and 'Presbyterian' became convertible terms.

It may strike English readers as a mystery that a Church intensely loved by a nation should have been a Church of the loftiest ecclesiastical pretensions. Two reasons make this appear strange to observers south of the Tweed: first, that the Church of England has been more willing to sacrifice the nation to its kings than any Church in Christendom; and, secondly, that the word 'Church' has been immemorially connected in the English mind with the priesthood and not with the people. Under cover of the theory that the sacraments can be rightly administered only by men ordained by episcopal successors of the apostles, the Anglican priesthood have had no difficulty in identifying themselves with the Church, and they have been practically so identified by the English laity, whenever question arose as to the spiritual powers of the Church. In the Reformed communions, the word 'Church' conveyed no such impression. For them the Church had reverted to her apostolic acceptation as the assembly of the faithful, not the members of a sacerdotal caste. The only reformer, in the strict and complete sense of the word, was Calvin. Luther, far more amiable and human-hearted, had no such grasp of principles, no such intrepidly destructive and constructive genius. Contenting himself with rescuing from the Papacy those doctrines which he deemed essential to human salvation, Luther left the Church to constitute herself, or to be constituted by princes, as circumstances might direct. The mediæval Church of England became the Anglican Church of modern times in virtue of one essential fact, that she surrendered, under the mask of an evasion, her spiritual liberties to the sovereign of the realm. This has been conclusively established by Hallam, Macaulay, and others, and has been treated as a self-evident truth by the highest legal authorities in the England of to-day. That Henry VIII. should assume spiritual supremacy need not surprise us if we reflect that he and his generation believed kings to be divinely endowed beings, capable of working miracles by their touch. Neither Luther nor the fathers of Anglicanism, in digging the foundations for the Reformed Church, cleared out the debris of Romanism and of feudalism. They retained enough of Romanism to change the Christian pastorate into a priesthood, enough of feudalism to exalt kings, nobles, and 'upper' classes generally into an Unchristian pre-eminence and to depress the body of the people. Calvin alone went *sans phrase* to the Bible and to God, with a view to re-constituting both the pastorate and the congregation on the apostolic model. From the authoritative documents of the apostolic age, he drew the doctrine, the discipline, the rights, the powers, of the Christian Church. Whether he fell or did not fall into errors of his own is not now the question; it is clear, at all events, that this was reformation in a deeper sense than that in which the word applies either to the Lutheran or the Anglican communions; and the instinct of the sixteenth century,—

that vague sense of the general mind which seldom errs,—both as manifested in the Church of Rome and beyond her pale, recognized the fact by signalizing the Church of Geneva, the Church of the Huguenots, the Church of Holland, the English Churches of the Puritans, and the Presbyterian Church of Scotland, as branches, not of the Protestant, but of the Reformed, Church.

The Reformed Church claimed spiritual independence as a right bestowed on her in the Word of God; but her congregations found in this no cause why they should dread her ecclesiastical assumptions, since the Church meant the body of the faithful, not a mystically endowed sacerdotal order. Recognizing the sacred importance of the congregation as an integral portion of the Church, the Reformed communions naturally respected congregational rights, and laid it down as a principle of practical administration that no pastor could be appointed to minister to a congregation against their will.

All this was mother's milk to Hugh Miller. From his earliest years his eye had flamed and his heart beat quick at the names of Bruce and Wallace, and in the Scottish Church he saw an institution which incarnated more of the ancient spirit and freedom of his native land than any other. Like all sane Scotchmen, he regarded the union with England as a crown of blessing for Scotland; but he felt that those political questions which derived importance from their bearing on the interests of the empire at large, had a comparatively remote and indirect connection with the welfare of Scottish towns and villages, Scottish peasants and artisans; whereas the preaching of the Gospel of Christ with purity and fervour, and the influence of a trusted, loved, and genially earnest pastorate, and the feeling of honest pride and self-respect with which Scotchmen looked upon their clergy as not forced on them by the will of man, but sent by God to bless them, and called by themselves to minister to them in spiritual things, he believed to affect, in the closest and most penetrating way, the happiness and well-being of ten thousand Scottish homes. He denied with scorn the allegation that the Church of Scotland gave up, at the time of the union with England, her cherished independence. The burst of eloquent enthusiasm with which Chalmers asserted, in his celebrated London lectures in defence of Church establishments, the spiritual independence of the Scottish Church expressed his clear conviction. 'It should never be forgotten,' said Chalmers, 'that, in things ecclesiastical, the highest power of our Church is amenable to no higher power on earth for its decisions. It can exclude, it can deprive, it can depose at pleasure. External force might make an obnoxious individual the holder of a benefice; but there is no external force in these realms that could make him a minister of the Church of Scotland. There is nothing which the State can do to our independent and indestructible Church, but strip her of its temporalities. *Nec tamen consumebatur;* she would remain a Church notwithstanding,—as strong as ever in the props of her own moral and inherent greatness. And though shrivelled in all her dimensions by the moral injury inflicted on many thousands of families, she would be at least as strong as ever in the reverence of her country's population. She was as much a Church in her days of suffering, as in her days of outward security and triumph,—when a wandering outcast, with nothing but the mountain breezes to play around her, and nought but the caves of the earth to shelter her,—as now, when admitted to the bowers of an establishment. The magistrate might withdraw his protection, and she cease to be an establishment any longer,—but in all the high matters of sacred and spiritual jurisdiction, she would be the same as before. With or without an establishment, she, in these, is the unfettered mistress of her doings. The king, by himself or by his representative, might be the spectator of our proceedings; but what Lord Chatham said of the poor man's house is true in all its parts of the Church to which I have the honour to belong: "In England every man's house is his castle." Not that it is surrounded with walls and battlements. It may be a straw-built shed. Every wind of heaven may whistle round it,—every element of heaven may enter it, but the king cannot, the king dare not.'

These views of Chalmers, unchallenged at the time, are accordant with the constitutional theory of the Church of Scotland. The decision of Hallam, based on express and unequivocal provisions of the Treaty of Union between England and Scotland, is conclusive on the point. The Presbyterian Church was accepted as an integral and essential part of the constitution of the United Kingdom, and the Confession of Faith was enrolled among the fundamental statutes of the realm. But neither Chalmers nor Hugh Miller perceived, until the cold iron of experience sent the death-like truth to their hearts, that the sardonic smile with which Hallam records that 'the Moderator dissolves the Assembly in the name of the Lord Jesus Christ, the Head of the Church, and by the same authority appoints another to meet on a certain day of the ensuing year,' while the Royal Commissioner subsequently performs the same ceremony in name of the sovereign, had a minatory and potent significance. By the Treaty of Union it was decreed that the spiritual province and the ecclesiastical province should remain independent of each other in Scotland, and that the

Church should, in the spiritual province, be supreme. But there was one thing which the framers of the Treaty of Union omitted to specify. They did not say who would decide in the event of there being a difference of opinion between the representatives of the civil power and the representatives of the spiritual power, as to the limits of their respective provinces. To leave this point undetermined was inevitably to give it on the side of the stronger. A lady contracts marriage under favourable conditions. These are inscribed in the marriage settlements, and the swords of her kinsmen, casting their gleam on the parchment, make it certain that, at the moment, those settlements have a real value. But they put their swords into the scabbard, there to rest and rust, nevermore to be again drawn in the lady's quarrel. They betake them to their farms and their merchandise, become comparatively indifferent to her affairs, and dream no longer of risking life or land for her sake. Then a dispute between husband and wife arises. He declares that he wishes to fulfil with loyalty the provisions of the marriage settlements. She maintains that he essentially misinterprets them, that the question affects her honour in a vital point, that, if he persists, the marriage will, *ipso facto*, be dissolved. He persists. She calls on her kinsmen. They are languid and lukewarm; are inclined, on the whole, to agree with her; but do not recur even in thought to the idea of defending her claims by force. The husband has the power in his hands; he decides in his own sense; and the lady announces to him and to all the world that the connection between them is at an end. *Voila*, the history of the Disruption of 1843. The Church of Scotland, at the time of the Union with England, was the lady of our allegory. The Revolution settlement, the Treaty of Union, secured her spiritual independence as well as pen, ink, and parchment could secure it. To have refused the guarantee would have been to encounter the risk of a war between the nations. As old Earl Crawford said at the time, an army of twenty thousand men would not have preserved the peace in Scotland if an attempt had been made to trample on the Church. But the Union was completed. A hundred and thirty years passed away. Scotchmen learned to feel that the interests of Scotland had been drawn into the mighty current of the interests of the British empire. Scotchmen, side by side with Englishmen, had defended the common cause on many a bloody field from Assaye to Waterloo. Then the dispute between the Church of Scotland and the Legislature of the empire—for to that it soon came—arose. The Church interpreted the compact in one way, the State in another. The representatives of Scotland in the British House of Commons were, by a majority, in favour of the claims of the Church. But the State, with all the power of the empire at its back, insisted on its view being taken; resistance by the Scotch members or their constituents was out of the question; and the Church, as represented by the party which had swayed her councils for eight years, dissolved her connection with the civil Government.

The battle-field on which the conflict between Church and State in Scotland was fought out was exactly that which, since the battle had become inevitable, the Church would have chosen. Had the majority in the General Assembly been checked by the civil tribunals while engaged in the rigorous enforcement of Calvinistic doctrine or of domestic morality,—while deposing some unfortunate Arminian preacher, or suspending some genial divine who in his cups had not been true to his moderate principles,—the cry of priestly oppression could hardly fail to have been raised against them, and it is inconsistent with all we know of human nature to suppose that they should have been encouraged by the plaudits of popular enthusiasm. But in the struggle which issued in the disruption of the Church of Scotland, the cause of the Church was the cause of the people. The spiritual independence of the former was a shield held forth to defend the congregational privileges of the latter. The jurisdiction of the Church would have been a poor rallying word even in Scotland, if it had not been linked with the cause of non-intrusion. The right of the people to have no pastor forced upon them was that which the Church brought out all her artillery of suspension and deposition to protect. It is unquestionable that if the majority in the Church courts could have been induced to betray the people,—to make common cause with the patrons,—to put off congregations with some plausible make-believe,—they might have availed themselves of inviting opportunities to secure their own predominance. But no artifice of statesmen, no subtle influence of aristocratic favour, no comfort of pleasant manses and fixed incomes, no pride of State establishment, dear to the clerical bosom, could tempt the Church to prove false to the ancient league, or to abandon that guardianship of congregational rights which had made her the darling of the Scottish commonalty from the days of Knox to the days of Chalmers.*

* It is interesting to observe that Sir Walter Scott, who had a marvellously exact comprehension of the Presbyterianism of Scotland, represents David Deans as sensitively anxious on the subject of his son-in-law's 'real harmonious call' to the parish of Knocktarlity and as ready to maintain 'the right of the Christian congregation to be consulted in the choice of their own pastor' as 'one of the choicest and most inalienable of their privileges.' David Deans is expressly drawn by Scott as a patriarch of the Kirk, and Dr William Cunningham himself could not have defined with nicer exactness the Non-intrusion claim of the Church,—not that the people should choose their pastors, but that they should be 'consulted in the choice' of their pastors,—than old Davie.

CHAPTER II.

LETTER TO LORD BROUGHAM—CALL TO EDINBURGH—LEAVES CROMARTY.

THE most plausible of the make-believes with which the Evangelical majority were tempted by statesmen to put off the people was the concession to congregations of a right to specify certain objections to presentees, which, if supported by adequate proof, would afford valid grounds for their being set aside by Presbyteries. If the pastor or probationer selected by the patron could be proved to be unsound in doctrine, defective in literary acquirement, or lax in morals, the rights of the patron were, in that instance, to lapse. At first sight, it may appear to many that the liberty of congregations would thus have been abundantly fenced. What more could they ask than that their pastor should be accurate in doctrine, accomplished in letters, irreproachable in life? The reply is simple and conclusive. They could ask that he should be one whom they could personally love, and from whose preaching and ministering their souls could derive nourishment; and the reasons why one man might answer to this description, and another man might not, are obviously of a nature to evade distinct apprehension even in thought, and which it might be quite impossible to set down in black and white. Members of the Anglican Church, or of the Episcopal Church of Scotland, are hardly in a position to appreciate the importance, as conceived by a member of the Reformed Church, of positive acceptance by a congregation of its pastor. The Episcopal Church retains, in the first place, the mediæval doctrine of a peculiar sanctity and power of blessing attached to episcopally ordained clergymen. The Reformed Church in all her branches, certainly not least in her Scottish branch, discards the idea of a mystic power inherent in the priesthood, and rests the value of the clergyman on his personal qualities. The Anglican Church, in the second place, has a Prayer Book, and the congregational devotion does not depend on the clergyman. The pastors of the Reformed Church of Scotland compose the prayers which they offer up in the name of the congregation, and though comprehensive, minute, and admirable directions for congregational prayer are given them in the formularies of the Church, it is inevitable that the character of their prayers will be in large measure dependent on their personal qualifications. But if we would get at the very heart of that invincible objection to intruded pastors which the communicants of Scotland have shown in every age, we must call to mind the intense personal religiousness of the Scotch. The religious Scot is not, like the corresponding type in England, demonstrative and emotional. He shrinks with sensitive dislike from those effusive recitations of spiritual experience which Wesley found congenial to large classes in England. He rejects the notion, practically acted on by a proportion of the English Congregationalists, and a still larger proportion of the English Baptists, that presbytery or parson, elder or deacon, has the right or the capacity to investigate a man's religious state before God and to say whether he is or is not converted. Not the less but the more, however, for his thus deeming conversion an inscrutably sacred process transacted in the secret places of a man's soul between himself and his Maker, does the Scottish religionist of the historic type regard conversion as infinitely important, and desire, as the one essential thing, that his pastor shall be a man of God. His judgment on the point may be narrow, uncharitable, superstitious; but it will depend mainly upon sympathies of the soul's life which have never shaped themselves even to himself in an articulate whisper, and which he feels it to be utterly preposterous to attempt to express under specified heads of objection to a presentee. Never, therefore, for one moment were the congregations of Scotland beguiled into the belief that the right of objecting to pastors selected for them by patrons was equivalent to being certified that no pastor should be 'intruded' upon them.

It was Lord Brougham's adoption and enforcement of the view that the right to specify objections to a presentee was the sole concession made by law to the congregations of the Church of Scotland, which roused Hugh Miller from his domestic repose in the shadow of the hill of Cromarty. The patron of the parish of Auchterarder, Lord Kinnoul, had presented the Rev. Mr Young to the charge. About forty of the parishioners abstained from positively opposing him; nearly three hundred expressly declined to have him as their pastor; and only three individuals, two of them not members of the congregation, signed the call. Under these circumstances the Presbytery rejected Mr Young. His solicitor, not carrying the case to any of the higher courts of the Church, turned at once to the Civil Authority.

The case was elaborately and very ably argued in the Court of Session. Eight of the judges were of opinion that the rejection of Mr Young by the Presbytery was illegal. Five were of opinion that the Church had a right to reject the presentee, and among these were Cockburn, Jeffrey, Moncreiff, and Glenlee. The decision of the Court against the Church was pronounced on the 10th of March, 1838. In May, 1839, it was confirmed by the House of Lords, and on this occasion Lord Brougham delivered his famous speech on the Scottish Church question. He dismissed, as wholly untenable, the proposition that congregations of the Scottish Church had a right to choose their pastors.

'Could I do nothing,' writes Hugh Miller in describing his feelings at the time, 'for my Church in her hour of peril? There was, I believed, no other institution in the country half so valuable, or in which the people had so large a stake. The Church was of right theirs,—a patrimony won for them by the blood of their fathers, during the struggles and sufferings of more than a hundred years; and now that her better ministers were trying, at least partially, to rescue that patrimony for them from the hands of an aristocracy who, as a body at least, had no spiritual interest in the Church,—belonging, as most of its members did, to a different communion,—they were in danger of being put down, unbacked by the popular support which in such a cause they deserved. Could I not do something to bring up the people to their assistance? I tossed wakefully throughout a long night, in which I formed my plan of taking up the purely popular side of the question; and in the morning I sat down to state my views to the people, in form of a letter addressed to Lord Brougham. I devoted to my new employment every moment not imperatively demanded by my duties in the bank office, and, in about a week after, was able to despatch the manuscript of my pamphlet to the respected manager of the Commercial Bank,—Mr Robert Paul,—a gentleman from whom I had received much kindness when in Edinburgh, and who in the great ecclesiastical struggle took decided part with the Church.'

This is the letter, addressed to Mr Paul, which accompanied the pamphlet.

'Cromarty, 13th June, 1839.

'I feel deeply interested in the question which at present agitates the Church. It is a vital one; and unless the people can be roused to take part in it (and they seem strangely uninformed and woefully indifferent as yet), the worst cause must inevitably prevail. They may perhaps listen to one of their own body,—to one who combines the opinions of the old with those of the modern Whig, and who, though he feels very strongly on the question, has no secular interest involved in it.

'In this hope I have written the accompanying letter, which I now submit to your perusal. May I request you, should you deem it fitted in some degree to accomplish what I have intended, to put it into the hands of some publisher interested in the welfare of the Church, and of influence enough to secure to it an extensive circulation. Humble as my name is, I think it will secure to me a good many readers in our northern districts;—in those of the south I am of course less known, but as I occasionally contribute to *Chambers's Journal*, the readers of that extensively circulated periodical will recognize me as an old acquaintance, and may be led by curiosity to listen to what I have to say on the subject. A fair hearing on the part of the people of the Church,—of the common people, at least,—is surely all the cause requires to secure their support. The price, should my letter suit a publisher's views, may be very low;—the writer's interest in it will not add to the expense;—it is an offering for the altar.'

The circumstances which followed upon the arrival of the Letter to Lord Brougham in Edinburgh will be best conceived from an account of them given by Dr Candlish in a letter to Mrs Miller, dated 3rd December, 1860.

'I remember them,' he writes, 'as if they had occurred yesterday. Walking along the streets one forenoon, I met quite accidentally Mr Robert Paul. After conversing on general topics, and as I was leaving him, he said to me, in a sort of casual, off-hand manner, "By-the-by, I have a manuscript in my pocket which I wish you would read at your leisure. It is on the Church question. The writer is a friend of mine in a bank in the north." I suppose he mentioned Mr Miller's name, but I had not heard of him before. I took the manuscript home and laid it aside; intending to give a cursory glance over it in the course of a day or two, and then return it; certainly expecting to find nothing remarkable in it,—nothing beyond what might perhaps be useful in the writer's own neighbourhood and circle. That same evening, being alone,—for I recollect that my family were in the country,—and being somewhat dull and listless, I thought I might as well look at the manuscript Mr Paul had given me. I began to read it in a thoroughly indifferent mood. I never can forget the rapture,—for it was nothing short of that,—into which the first pages threw me. I finished the reading in a state of great excitement; so much so that, though it was late, I could not rest till I had hastened with the manuscript to Mr Dunlop, beseeching him to read it that very night. The following day Mr Dunlop and I met with Mr Paul and a few friends, and either then, or within a day or two thereafter, it was agreed to ask Mr Miller to become editor of the *Witness* newspaper, then about to be started. We had been looking out for an editor. Whenever I had read the manuscript, which was, I need not say, that of the Letter to Lord Brougham, I came at once to the conclusion that we had found the man. So did Mr Dunlop when he had read it. The thing was thus immediately and enthusiastically settled.'

The Letter to Lord Brougham was at once published. Hugh Miller has left behind him no more masculine, idiomatic, close-knit, or melodious piece of writing, nor is any of his productions more strongly featured with the characteristics of the man. He opens with a glowing, yet nowise fulsome, tribute to Lord Brougham. 'I have been no careless or uninterested spectator of your lordship's public career. No, my lord, I have felt my heart swell as I pronounced the name of Henry Brougham.' He then makes an admirable point by referring to Brougham's prowess as a Parliamentary Reformer and the relative capacities of Scottish communicants, on the one hand, to elect members of the British Parliament, and, on the other, to choose their parish ministers. 'Surely the people of Scotland are not so changed but that they know at least as much of the doctrines of the New Testament as of the principles of civil government, and of the requisites of a gospel minister as of the qualifications of a Member of Parliament!' His argument on behalf of the Church and her congregations he bases upon the original constitution and distinctive character of the Church of Scotland. 'I read in the First Book of Discipline (as drawn up by Knox and his brethren), that "no man should enter the ministry without a lawful vocation; and that a lawful vocation standeth in the *election of the people*, examination of the ministry, and admission of them both." I find in the Second Book, as sanctioned by our earlier Assemblies, and sworn to in our National Covenant, that as this liberty of election was observed and respected as long as the primitive Church maintained its purity, it should be also observed and respected by the Reformed Church of Scotland; and that neither by the King himself, nor by any inferior person, should ministers be intruded on congregations, contrary to the will of the people.' His expression of devotion to his Church may be not uninstructive to those who cannot conceive of a Church which has a definite and comprehensive creed, as anything else but a grinding intellectual tyranny. 'To no man do I yield in the love and respect which I bear to the Church of Scotland. I never signed the Confession of her Faith, but I do more,—I believe it; and I deem her scheme of government at once the simplest and most practically beneficial that has been established since the time of the Apostles. But it is the vital spirit, not the dead body, to which I am attached: it is to the free popular Church, established by our Reformers,—not to an unsubstantial form or an empty name,—a mere creature of expediency and the State: and had she so far fallen below my estimate of her dignity and excellence as to have acquiesced in your lordship's decision, the leaf holds not more loosely by the tree when the October wind blows highest, than I would have held by a Church so sunk and degraded.'

A large part of this masterly tractate is necessarily devoted to a consideration of the law of patronage enacted by the British Legislature in the reign of Anne. The incompatibility of that law with the fundamental principles of the Church of Scotland, the baseness of the intrigue, hiding treason to the House of Brunswick under very genuine hatred of the Presbyterian clergy, by which it was carried through Parliament, the opposition offered to it during the earlier part of last century by those members and ministers of the Church of Scotland who incarnated her ancient spirit, its effect in driving into involuntary exile her most earnest pastors and most pious adherents, are demonstrated and enforced with extraordinary vigour. But there is one thing Hugh Miller does not assert in his Letter to Lord Brougham, and the omission is of importance. He does not say that the British Legislature, in restoring to the patrons of benefices in the Church of Scotland those rights of patronage which the Church, in the plenitude of her power, had swept away, did not *mean* to transfer to the patrons what the Church had conferred on the people. In order to do justice to the majority of the judges of the Court of Session, and to the Moderate party in the Church of Scotland, we are bound to admit that the rational interpretation of a law restoring patronage is that it conferred upon the patrons real power, and that, by implication, it disallowed, what the Church was ready to agree to, an assignment of the income of the parish to the patron's minister, and the appointment of another minister, approved of by the people and the Presbytery, to be its pastor. In making this admission, we do not invalidate the main argument of the Evangelical majority; but we set the matter in its true historical light; and we render it intelligible that men of eminent ability, men whom it would be unpardonable, except in the actual heat and dust of the struggle, to describe in any other way than as upright and honourable, pronounced the claims of the Church indefensible. The argument of the Evangelical majority rested inexpugnably on this position,—

that the spiritual independence of the Church and the non-intrusion rights of congregations were essentials of the ancient ecclesiastical constitution of Scotland. If this statement is incorrect, the whole world has mistaken the character of the Kirk; Knox, Melville, Henderson have been totally misunderstood; and such Presbyterians as Dr Thomas McCrie and Dr William Cunningham knew nothing of the subject which was the study of their lives. Whatever version you may give of the proceedings of the Church, in the eighteenth and part of the nineteenth century, in relation to the Patronage Act, you cannot impair the force of an appeal to the fundamental principles of the Scottish Reformation. Is your theory that the Church did her best to neutralize the Patronage Act and to resist and protest against it? Then the Evangelical majority had a right to carry on the contest until it became hopeless. Do you affirm that the Church failed in her duty and practically accepted the law of patronage? Then the sooner she roused herself and harked back on nobler days and native principles the better. Will you admit that the Church, on the whole, shilly-shallied,—that she was content with laying a thin film of plausibility and compromise over a wound in which lurked poison? Once more, the party representing her fundamental principles had a clear right to content themselves with no half-measures, with no prudent evasions, but to breathe a new spirit into the policy of their fathers, and to proclaim that the patronage bane was to be removed at all hazards.

The intention of the Veto Act, passed by the Church as soon as the Evangelicals took the lead in her councils, was to annul the Patronage Act of Queen Anne. In putting the veto into the hands of congregations, the Church did openly and decisively what, for many years after the passing of the Patronage Act, she had attempted indirectly and irresolutely. Dr Robertson, in initiating the policy which he pursued during his leadership of the Church, took this position, that, as the Patronage Act was law, and the Church did not avowedly resist it, her consistent and honourable course would be to render it a frank and complete obedience. The majority of the judges of the Court of Session, bound to give to Acts of Parliament the effect which their terms required, were convinced that it was their duty to enforce the Patronage Act. The Church, by her Veto Law, declared, on the other hand, that, irrespectively of the proceedings of her own courts for a hundred years, irrespectively of any statute passed by the British Legislature, irrespectively of whatever decisions the Court of Session might pronounce, the principles of Knox, Melville, and Henderson were her principles, and by her principles she would abide. It complicated, and did not strengthen, the argument of the Evangelicals to maintain that the law of the State as well as of the Church was on their side. No statesman or legal authority could be expected to attach weight to the assertion that an Act of Parliament, which had been on the Statute Book for a hundred and thirty years, had infringed the constitution of the realm and was therefore null and void.

The champions of the Evangelical majority in the Church of Scotland held themselves bound, during the whole continuance of the conflict, to maintain two theses: first, that the spiritual independence of the Church and the non-intrusion rights of congregations were essential to the Presbyterian constitution; second, that these could be secured although the Church maintained her connection with the State. The theory that the civil authority must and will, sooner or later, usurp the powers of the Church which it establishes and endows, they rejected not without scorn, and the Established Church of Scotland they pointed to as a visible demonstration that a State Church might be free. We have found Hugh Miller declaring in his Letter to Lord Brougham, that, if the Church of Scotland had acquiesced in his lordship's decision in the Auchterarder case, 'the leaf holds not more loosely by the tree when the October wind blows highest,' than he would have held to 'a Church so sunk and degraded.' He ultimately made good his words; but in his 'Whiggism of the Old School,' a pamphlet by which the Letter to Lord Brougham was immediately followed, he vehemently maintains the Establishment principle against 'Voluntaries,' and seems to say that, though the decision of the Courts 'has practically determined the law,' the clergy ought not to sever their connection with the State. 'The duty of our ministers'—these are his words—'is not the less clear. They owe it to themselves and to their people, to their country and to their God, that they neither obey this iniquitous law, nor yet quit the Establishment.' The second pamphlet is an inferior performance to the first. It is more ponderously controversial, and lacks the racy vigour and stern imaginative glow of the famous 'Letter.'

Without question, Miller, now and subsequently, was a decided maintainer of the State-Church theory. It was with inexpressible reluctance that he brought himself ultimately to admit that the Church of Scotland had no alternative but to sever her connection with the civil power. Nay, I have a distinct recollection that, in earnest talk with me after the Disruption, he hinted a wish that the leaders of the majority had been somewhat less imperious in their dealings with clergymen who obeyed the civil law rather than the law of the Church, somewhat less fiery and impatient in urging the matter to an issue. Hugh Miller had seen much of infidel voluntaryism, and little of that power of religion in a state of freedom which has in recent times frightened the Pagan party into swift abjuration of the principle of a free Church in a free State, and cordial support of civil establishments of religion. The conception of a State Church as a bed on which Christianity, ascertained by cultivated men to be moribund, may die soft, had not dawned upon the intellect of Hugh Miller. If convinced that the choice lay between disestablishment on the one hand, and a State Church on the other, which was specially recommended by its usefulness in controlling the might of free religious principle, he would with passionate indignation have declared for the former. He lived and died, however, a believer in the soundness of the State-Church theory.

Nor was this the only circumstance which might have disposed him to look with indulgence upon the Moderate party. His literary sympathies, his pride in the modern literature of Scotland, must have pleaded strongly in their behalf. If he could have granted, which he could not, that the ministers of a Christian Church can claim approval or applause on any ground save that of preaching, in season and out of season, the Gospel of Christ, he would have found it difficult to repress his enthusiasm for that party which placed the Church of Scotland in sympathetic harmony with all that was refined and intellectually progressive in the literature, the science, the art, of Scotland. One knows not where to look in ecclesiastical history for a party, of which the nucleus consisted of clergymen, so loyal to the higher aims of the human spirit, so ardent in its love of knowledge, so free from sectarian bigotry, so genial and tolerant in its habits of thought and feeling, as the Scottish Moderates. Under their influence the rugged face of old Scottish Presbyterianism mantled with a smile of calm and bright intelligence which drew upon Scotland the astonished gaze of Europe. Robertson was in his own day known in Paris almost as well as in Edinburgh; his books had an honoured place in the library of Voltaire; and there is no University in Europe at this hour in which his name is not held in esteem. He was the friend and historical rival of Hume, who, *except* in his speculative philosophy, was a true Moderate. If it is to be held a disgrace to the party that its chiefs were in friendly intercourse with the prince of iconoclasts, we ought to remember that there went out also from the Moderate camp those champions who, both in the field of pure philosophy and in that of apologetic divinity, challenged Hume to the combat, and, in the judgment of a world certainly not prejudiced in favour of parsons and against philosophers, put Hume to his mettle. I allude of course to Principal Campbell and Dr Reid. The philosophy of Reid assaulted Hume along his whole line of battle. Interpreted as Hamilton interprets it, that philosophy is one in fundamentals with every great constructive system of spiritual thought from that of Plato to that of Kant. It consists, in one word, of an intelligent and critical appeal to the common spiritual nature—call it the *communis sensus*, call it reason, call it intuition, call it what you will,—of the human race. This remains and must remain, whatever the dialect in which you express it, the sole philosophical refuge from universal scepticism. It is characteristic of the noble tolerance and candour and high intellectual serenity of those old Moderates, that Reid forwarded to Hume in manuscript his reply to the philosopher. A finer proof of a desire to deal fairly with an opponent, and of an absolute rejection of every weapon of personal or vituperative controversy, cannot be imagined. But the Moderate party of the Scottish Church can claim not only Campbell, Reid, and Robertson, but one who would now be placed by many higher than either of the three, the author of The Wealth of Nations. Smith's Theory of Moral Sentiments is a thoroughly Moderate book, moderate in its eloquent and high-toned moralizing, moderate in its lucidity and logical coherence, and also perhaps in its want of intensity, enthusiasm, and penetrating, exhaustive power. It was under the genial auspices of the Moderate party that the Scottish Universities attained to such renown that young Palmerston and Russell went from England to study in their halls. It was under the auspices of the Moderate party that Edinburgh became the Weimar of Great Britain, that the most important publications of the time in Europe, the *Edinburgh Review* and *Blackwood* and the *Quarterly*, arose. When John Murray conceived the scheme of the *Quarterly*, his first step was to start for the North, to confer with Scott and the literati of Edinburgh. The leaders of Moderatism showed a wise and a nobly patriotic spirit in never trying to exclude Seceders from the Universities. They indulged the shrewd preference of the Scottish commonalty for 'college-bred ministers,' and spared their country the stunted and acrid growths of illiterate Dissent. It may doubtless be argued that Moderatism was itself but part of a wider phenomenon, to wit, the prevalence and predominance, throughout society, not Scottish and English alone, but European, during the eighteenth century, of literary and scientific tastes and ambitions as contrasted with those of a religious nature; but the truth of this statement does not neutralize the fact that, under the reign of Moderatism, a literary and philosophical lustre was thrown over Scotland for which, in the more earnest epoch of dominant Evangelicism, we look in vain.

It may be asked how we can hold, first, that 'Scotch literature and thought' in modern times owed their superiority to 'Knox and the Covenanters,' and secondly, that the ecclesiastical party under whose influence the literature and thought of Scotland have fallen from that ascendency which they attained under Moderate sway, were in deeper sympathy with 'Knox and the Covenanters' than their Moderate antagonists. The answer, if we will reflect for a moment, is not difficult. Sir Walter Scott painted a hero in Claverhouse; Burns was 'the fighting man' of the Moderate clergy of his district, and did not scruple to sneer in his satirical verses at the 'holy folks' who 'believe in John Knox:' yet Carlyle expressly mentions Scott and Burns as having become what they were through the accomplished work of the Scottish Reformers and Covenanters. And he is right. These, under God, made the nation strong of spirit and proud of heart, capable of producing a Burns or a Scott. In like manner the Moderatism of Scotland, with its mellow splendour of intellectual light, was a fruit of that tree which had been planted in the olden time amid the storms of war and the fierce contendings of faction. It would not be pleasant to think that the good there was in Moderatism has been lost to Scotland and the world. One may be permitted to hold *both* that the Evangelical majority which led forth the Free Church in 1843 bore with them the old Scottish ark of the covenant,—the essential principles of the Scottish Reformed Church,—and that the cosmopolitan sympathies and high intellectual tastes of the Moderate party will once again prevail within the Church of Scotland. A few State cobwebs and a few abstract propositions are all that prevent the branches of that Church from again growing visibly on one stem; and were this the case, religious Scotchmen would be able to think, as they ought, with equally diffused gratitude and pride, of the unrivalled systematic theology of Hill, the sturdy dogmatism and unconquerable logic of Cunningham, the historic fame of Robertson, the artistic genius of Thomson of Duddingston, and the moral grandeur of Chalmers.

In religion Hugh Miller was an Evangelical, but in literature he belonged essentially to the Moderate school. His literary ideal was that of grace and elegance, and quiet glow of imaginative fire; from declamatory vehemence and the strut and stare and swagger of the modern 'earnest' school, he shrank with sensitive repugnance. To become what Moderate critics would have pronounced a classic was an object of ambition incomparably nearer his heart, than to dazzle with splendid metaphors, or to produce a temporary sensation. Hence the satisfaction with which we found him mentioning that 'Baron Hume, the nephew and residuary legatee of the historian,—himself very much a critic of the old school,'—had described his *Scenes and Legends* as a work 'written in an English style, which he had begun to regard as one of the lost arts.' The literary skill with which Hugh Miller was to assail Moderatism had been acquired at the feet of its own Gamaliels.

The conclusion that the author of the Letter to Lord Brougham was the man to edit an Edinburgh non-intrusion paper had, as we saw from Dr Candlish's letter, been speedily arrived at by the Evangelical leaders in the capitâl of Scotland, but the proposal was not immediately made to Miller. A note reached him from Mr Robert Paul, manager of the Commercial Bank in Edinburgh, an ardent non-intrusionist, requesting him to visit Edinburgh 'in the course of the summer,' in order that he might be 'brought into communication with some gentlemen,' who thought that his talents could be employed 'in some literario-Christian objects'—as Mr Paul, with cautious indefiniteness, puts it. There is no hint of a newspaper, no allusion to the Church question, but, under the circumstances, Miller may have read all that between the lines.

Mr Ross, his judicious and friendly superior in the Bank at Cromarty, acquiescing in the arrangement, he proceeded, in company with Mrs Miller, to the south. They resided with Mr Paul. It was the third time Hugh Miller had come to Edinburgh. The first time, he was in quest of employment as a journeyman mason; the second, he wanted to qualify for the post of banker's clerk; he now (1839) found himself 'looked upon as a lion—a sort of remarkable phenomenon.' The leaders of the popular party in the Church welcomed him as a valuable ally. 'There was, of course,' writes Mrs Miller, 'a famous dinner party, at which we were introduced to Dr Cunningham, Dr Candlish, Dr Abercromby, and others.' The precise nature of the enterprise in which it was proposed that Miller should be engaged was now explained to him, but Mr Paul found that the business could not be got forward so speedily as might be wished. Hugh still had his doubts, his hesitations, his fears. Though his nature was stirred to its inmost depths by sympathy for the cause of the Church and the people, and though the success of his Letter to Lord Brougham had been splendid, that timidity, that caution, that habit of despondent brooding and of looking at things on their dark side, which blended so curiously in his composition with a high estimate of his own powers and a courage which, when roused, no opposition could daunt or peril dismay, held him back. True prophets have always shrunk, like Moses, from the work assigned them, and only the false prophet, intent not on the message to be delivered but on the rewards and honours to be earned in delivering it, rushes into momentous tasks with frivolous hardihood. Miller, however, ultimately accepted the Editorship of the projected paper, and returned to Cromarty to resign his situation in the Bank and to arrange for permanent settlement in Edinburgh.

His mind was far from cheerful. A pleasant episode, however, occurred about the time of his return to Cromarty. His friend Finlay had returned from the West Indies, and, making a tour through England and Ireland, bent his steps to the North to revisit the scenes of his youthful gipsyings with Miller. From Stratford-on-Avon, whither he had gone to do homage at the shrine of Shakspeare, he wrote to his friend. The letter is so full of kind intelligence that we must make room for the greater part of it.

FROM ALEX. FINLAY.

'Stratford-on-Avon, 4th September, 1839.

'I little thought the last time I wrote you that the next letter should have been dated from this classic spot. My last was from Jamaica about three years since, in answer to your excellent letter which I have in my portmanteau, and which I have treasured most carefully, assuring me, as it did, that I had one heart at least in the world that beat in unison with my own. I have suffered much sickness since I received it, and am travelling to my native land to try the climate in removing some chronic affection of the stomach which causes spasms and cramps that on occasions nearly deprive me of existence. I left Jamaica on the 1st June, and in St Iago de Cuba was seized with the yellow fever which happened to rage there at the time, and was as nearly as possible left there. Two died of it in the house I was in, and I was insensible. An American doctor, however, rescued me from the Spanish Sangrados, and by bleeding and calomel I recovered, but was as yellow as a guinea. After visiting all the principal places in Cuba, I left Havana for Charleston, in South Carolina, and visited Philadelphia, Washington, Baltimore, and other chief towns in the States, and sailed or rather steamed from New York in the British Queen on the 1st August. Our departure from the Hudson was the most enthusiastic scene I ever saw. The city was emptied of its inhabitants. Upwards of 150,000 people lined the banks of the noble river, manned the rigging of the forest of masts, and crowded about thirty steamboats, which accompanied us nearly 20 miles down the river with bands of music, singing, and huzzas; when to this was added the beautiful green islands, the romantic banks, and the unclouded flood of light poured from an American August sun, and the noble steamship herself,—it was really singularly pleasing. I arrived at the Isle of Man on the 15th, and at London on the 16th. The leviathan city confounded me, notwithstanding all I was prepared to expect, especially in the river. . . . . . . I had an interview with his Grace the Duke of Wellington on West India affairs; he was pleased to say he was obliged by my information, mentioned his being intimate with my father, expressed willingness to serve me, and invited me to pass some days with him in winter at Walmer Castle. I also saw Lord Normanby. I hope something may be done for our unfortunate Colonies. I was in the House of Lords when Lord Lyndhurst, the Duke, and your friend Lord Brougham gave the wretched ministry such a thrashing at the close of the session. I have not read your Letter to Lord Brougham, but suppose you take part with the Assembly, in which case I fear you are on the wrong side. . . . . And now here I am within a few yards of the spot where Shakspeare breathed existence and not far from where he expired. I have walked over his grave. I have just returned from the churchyard (10 P. M.), and the wish was ever uppermost in my mind that you had been with me to enjoy the half-superstitious awe with which I eyed the tombstones and the abutments of the handsome old church as they glimmered in the imperfect light; and so I have been constrained to write to you. I do not know why, but I only half enjoy the pleasure of a scene unless I have some one I love along with me to show it to. . . . . This is a beautiful rural village; some parts very like what they were in Shakspeare's time. I passed a few miles off an old house which has the reputation of being the scene of the poet's high jinks. Adieu, my dear Hugh, till I visit Scotland. I do not feel at home although all my family are in England, nor will I till my eye is blessed by the sight of the mountains of dear old Scotland, and my hand is grasped by thine.'

When this letter was put into Miller's hand, he was in the act of repelling one of those pieces of insolent annoyance with which the Radical and semi-Radical section of Cromarty politicians loved to tease him. Belonging as he did to the extreme right of the Whig party, and seeing, in its half-Radical left, 'the plague-spots and running sores of the party,' he was naturally to them an object of hostility. Here is his reply to Finlay.

'I received your kind-hearted letter at a time when I was in no kind-hearted trim. I was engaged amid the turmoil of a Court of Appeal into which I had been summoned at the instance of a political shop-keeper who indulges in the ambition of becoming great in the Town Council of Cromarty, and who, regarding my vote as an obstacle in the path, would fain have swept it away. I was a good deal annoyed by the fellow and somewhat angry to boot, for nothing, as the event showed, could be more frivolous than the objection; but your letter, which I first perused amid the hum of the Court, brought me to myself at once, and made me ashamed that I could be either. I am less a politician than most men. I love peace, I love leisure, I love my friends, and when let alone I never think of my enemies; but I have been born under a vile, bustling, political, knock-down sort of planet, and, like Yorick's starling, can't get out. Well, I shall be quiet enough one day, and till then, since fate is fate, I must just try and keep up the quarrel on the right side.

'And so you are in our own country once more,—in the country of green fields and temperate suns; and after a separation of twenty years I am to have the pleasure of grasping your hand as of yore. We shall wander among the woods together, and visit the caves, and climb the Sutor roadie and MacFarquhar's Bed. The rowans and blae-berries will be all gone, and the crows'-apples are a failure this year, but you will be in time for the brambles and the sloes; it will go hard, too, if we won't be able to steal a pocketful of potatoes out of one of the Square parks, and get them roasted among the rocks, and we shall at least try whether two old fellows, on the wrong side of thirty-five, can't be as foolish and happy for a day or two as when they were twenty years younger. My wife, whom I dare say you will deem worth liking, is nearly as anxious to see you as I am myself. She knows all about our early intimacy. I have shown her the cave in which we have had so many happy days' experience of the ease with which man can lay down the usages of civilized life and take up the savage. I have told her of the little closet in which we used to draw vile, libellous landscapes without knowing they were libellous, and read foolish books; and of the deep pit which we dug in one of the thickets of the hill when we intended becoming robbers, and were furnishing ourselves with bloody-minded daggers, fashioned out of table-knives, to frighten the girls who came to the wood for sticks. You, I presume, are still a single man, and the married have but one advice for such; but I am not a giver of advices, nor a taker of them neither, and so I spare you. There is little danger of men becoming too happy in any state. It was only last week I placed a tombstone over the grave of my infant daughter, a sweet little girl, who in less than two years had found means to lay strong hold on the hearts of both her parents, and who left us when our hopes for her were at the highest. But we must just live on, thinking as much of duty and as little of enjoyment as we can; which, after all, is, I believe, the way in which most is to be enjoyed. Philosophy, however, is of no use on such occasions; it just serves now and then to point a sentence, and that is all.

'It is probable that in little more than two months I shall leave Cromarty for Edinburgh, where I am invited by a most respectable and influential body of men to conduct a Whig newspaper on the side of the Church. I do feel somewhat like a shrinking of heart when I think of the undertaking. There is incessant labour and many a hard battle before me; and all my hopes of happiness have been wrapped up in dreams of some quiet retirement with abundance of leisure to devote to literature and science. But my present choice is not between literary leisure and a course of literary agitation. It is between ceaseless employment of a less, for ceaseless employment of a more, congenial character; and I am just making up my mind for the decision. You fear I am in the wrong in taking part with the General Assembly in its present quarrel. I have luckily no such fears for myself, however; and as I know you too well to suspect that you will condemn me unheard, I have some hopes of yet seeing you on the side of the Assembly too. I have written two pamphlets on the subject;—the Letter to Lord Brougham, to which you allude, and the Whiggism of the Old School. The first, after passing through four editions in as many weeks, has been stereotyped,—the second, though it contains a larger amount of thought, is circulating more slowly.

'It is a sad thing to suffer among strangers,—sickness is sickness everywhere, but in a strange land it is something more. I have experienced just enough of it to know what it is, and to sympathize with those who know more of it. Had you died in St Iago de Cuba, it is probable I would have been left to mere surmise and conjecture regarding your fate;—it would have been one of the many sad secrets of the same kind which await the last day. I trust you will be able to lay in a large stock of health and strength in Scotland;—the air of our hills is keen and shrewish, but the same pair of lungs may breathe it for a hundred years, and there is no sky beneath which there are better braced muscles or spirits of a more vigorous tone.

'But what apology am I to make you for not writing you on return of post as you wished me? Just this, that, as Burns says, "my nose has been held so closely to the grindstone" ever since, that I have not had more than time to write you a hurried note, and I was desirous to write you a letter. I hate Chartists and Radicals in the gross from my very heart, but I have great sympathy with the poor Chartists and Radicals who, having to work sixteen hours per day for a meagre livelihood, avenge their hard fate on all and sundry when they break loose. It does not surprise me that a man who has no other amusement should take a longing to cut throats.'

The visit thus eagerly anticipated proved as pleasant a translation of hope into fact as is ordinarily experienced. The friends, it is true, did not reach that climax of resuscitated boyhood which Miller imagined. No potatoes were stolen or roasted. Time failed, and perhaps heart also, even for the kindling of fires in the caves. The men found, when they engaged in confidential talk, that they had notes to compare which eclipsed in interest the memories of their boyhood. Finlay had been engaged to a young lady who perished at sea, and the tender-hearted man remained single for her sake. His affairs in Jamaica had not prospered,—affairs in Jamaica seldom did in those years. In one word, Hugh also having a weight on his mind, there was probably as much of pensive reflection, prospective and retrospective, in the intercourse of the friends, as of gaiety and mirthfulness. This was the last meeting of Finlay and Miller. The former returned to Jamaica, was elected to the House of

Representatives, advocated a wise and liberal policy, based on the necessity of educating the Negroes in the duties of free men and elevating them to the level of the whites, and died in three years.

Before the arrival of Finlay, Miller wrote to Mr Dunlop, in answer to a request from that gentleman to furnish a prospectus for the paper to be started in Edinburgh. The letter consists mainly of despondent expressions as to the risk run by the writer in undertaking the editorship. His situation in Cromarty is, at least, he represents, a certainty, a permanent certainty, and he requests information as to what he may expect in the event of the paper proving a complete failure.

Mr Dunlop replied to this letter with assurances fitted to dispel the apprehensions of his correspondent, at least on the head of pecuniary venture. 'We shall in the first place require security for your salary in all events for three years, so that, whether the plan succeeds or not, you will have some time to look round for another situation. Then, assuming it to fail, although I could not point out anything so definite as you might wish, yet I can scarcely conceive the possibility of our not being able to procure for you a situation equal at least to what you possess.' For the rest, he encourages Miller to 'summon courage to make an advance,' and concludes in these brave words:—'I do believe that when a man has an opening through which to do essential service to the cause of religion and his country, he may very generally enter with confidence that he will not be a loser.'

Mr Dunlop enclosed certain resolutions, adopted at a meeting of the promoters of the forthcoming newspaper, and embodying the 'leading principles' which it was to represent and advocate. It was to be in the common sense a newspaper, comprising political and general intelligence; to be 'pervaded by a spirit of decided piety;' to espouse the cause of no one political party, but be 'scriptural and constitutional;' to maintain the spiritual independence of the Church, combat Erastianism, and 'place the connection of Church and State upon its true and Scriptural footing.' Of non-intrusion, or the choice of their pastors by congregations, there is no express mention in these resolutions. Taking them as his leading lights, Miller composed a prospectus for the new paper. The name he suggested was the 'Old Whig,' and by this name he designated it in his prospectus. He followed his instructors in merging the special interest of non-intrusion in the more general one of spiritual independence. The prospectus was 'shortened' by the Edinburgh Committee before being printed.

In the note which accompanied the prospectus to Edinburgh, he remarks that the production is 'unique in one respect, perhaps, as it is not merely the only one its author ever wrote, but also almost the only one he ever read. I have set my best leg foremost,' he adds, 'and made my bow,—a stiff enough sort of bow, I dare say, as becomes my breeding, but made with hearty good-will and a great deal of respect.' Though 'as much a coward as ever,' he has now, he says, 'determined on making the leap.' He refers to a Pamphlet on the Intrusion side by the Dean of Faculty, from which he has seen some extracts. 'If he could have printed his wig and his silk gown,' thinks Miller, 'it would have been doubtless a splendid affair; but bad prose and little ill-packed arguments that look for all the world like bundles of pointless needles are not very serious matters. It is well that the Dean is no Jeffrey. Wit would do us a world of mischief; and good wit, unlike good argument, can be made to fight on the worst side.'

Having forwarded his prospectus to Edinburgh, Miller waited with eagerness for tidings from the South. Day by day, week by week, dragged itself away, and no answer came. To increase his anxiety some vague rumour crept northwards that a hitch had occurred in the newspaper project, and that it was likely to fail after all. He saw no public advertisement of the paper, and it was evident that his prospectus had not been printed. At length his patience gave way, and on the 20th of September, exactly two months after the date of Mr Dunlop's last letter, he wrote to Mr Paul, requesting a few lines of explanation. Alluding to the 'vague report of some misunderstanding' which had arisen, 'I cannot help feeling,' he proceeds, 'that the elements of such a misunderstanding exist in sufficient amount among our friends.' He is 'somewhat uneasy.' Not to be too severe on 'our friends,' he couches his rebuke in the terms of a philosophical reflection. 'It is, I am afraid, a too certain fact that, the more honest any party is, the surer it is of being ill-organized and full of conflicting opinions. The Jesuits have but one heart and mind among them;—the Evangelicals of the Church of Scotland, on the contrary, differ as much among themselves on minor points as they do from their opponents on the truly important ones.' He returns to the subject of the Dean's pamphlet, of which his opinion has not improved. Its arguments 'have neither the dignity of vigorous thought nor the charm of elegant expression to recommend them,—they are tedious without the recommendation of being just, and sophistical without the merit of being ingenious.' How exactly Miller had at this time caught the tone of the Queen Anne critics! 'It is a fatal sign,' he adds, 'in a literary man when he cannot cease being professional. Who would ever have guessed from his lighter writings that Currie was a physician, or Rogers a banker, or Sir Walter a clerk of session, or Jeffrey an advocate? But the Dean, a man of a very different stamp, is nothing apart from his *trade;*—he can write law papers, and nothing else;—his pamphlet is a piece of mere special pleading; —he is a lawyer, and a lawyer only.'

This of course brings a prompt reply. Mr Paul, deeply grieved, has been trusting to Mr Dunlop to keep Miller informed; Mr Dunlop, deeply grieved, has been trusting to Mr Paul. On the main point, however, all is right. The capital which the publisher who was to have undertaken the practical part of the enterprise could divert from his general business and devote to the paper had been deemed inadequate, and the affair had hung back. But 'a respectable young man with a few hundred pounds' had consented to join him in part-proprietorship of the paper, and to give his whole attention to its management. The young man thus described by Mr Dunlop was Mr Robert Fairly, bred a printer, experienced in the printing department of a newspaper, whose steady industry, high moral character, and eminent common sense had placed him in a position to make this offer. Mr Fairly was prepared to enter into friendly relations with Miller. He had seen the Lines to a Dial in a Church-yard, and was greatly struck with them. From the time when Miller came as editor to Edinburgh to the day of his death, Mr Fairly continued his warmest admirer and most affectionate friend, and now cherishes his memory with a reverence and tenderness which would satisfy Mr Carlyle's utmost requirements in the way of hero-worship.

In Mr Paul's letter to Miller on this occasion the inevitable pamphlet of the Dean turns up. Mr Cunningham, Mr Candlish, Mr Dunlop, and Dr Chalmers have set about replying to it. 'The whole *skep*,' remarks Chalmers, 'will be overturned upon the Dean,' so that, as Mr Paul hopefully adds, 'he will be absolutely stung to death.' It may be stated for the information of curious readers, that the production of which Miller spoke so contemptuously, but which raised so fierce and general a buzz in the non-intrusion hive, was entitled 'A Letter to the Lord Chancellor, on the claims of the Church of Scotland in regard to its jurisdiction, and on the proposed changes in its polity, by John Hope, Esq., Dean of Faculty.' It has been described as 'a very leviathan among pamphlets, extending to no fewer than 290 pages.' This fact will, I trust, be accepted as my apology for not having made its closer acquaintance, or attempted to estimate the effect of the stings inflicted by the aforesaid bees upon its extensive surface.

It more concerns us to note that Miller is at last ready to take flight for Edinburgh. The name suggested by him for the paper has been discarded, and that of *The Witness* adopted in its stead. On the 23rd of December he once more writes to Mr Dunlop from Cromarty, in answer to a note from that gentleman telling him that all difficulties have been vanquished, and that the sooner he appears in Edinburgh the better. 'I still,' he says, 'feel occasional shrinkings of heart when I think of the untried field on which I am so soon to enter. "Tremble thus the brave?" asks one of Ossian's heroes when on the eve of his first battle. But I think of the past and take courage,—of the past in my country's history, with its clear unequivocal bearing on the cause in which I am to be engaged,—on the past, too, in my own experience of life. I have seen much of the goodness of the Almighty. Twenty years ago I was a loose-jointed boy in rather delicate health, taxed above my strength as a labourer in a quarry. It is surely a much better thing to be employed as an advocate of principles which I have ever regarded as sacred, and of whose importance the more carefully I examine I am convinced the more.'

Hugh Miller did not quit the North of Scotland without some public recognition of his worth and talents. His name was already well known throughout the northern counties, and in Cromarty and its neighbourhood he was not only looked upon with pride as the literary wonder of the place, but regarded with affectionate confidence as a judicious counsellor on local affairs and a ready friend and help in every little business which his townsmen might have in hand. His friends were many. He had not sought them, but they had come, and he had never lost one. Sensible men in the middle and upper classes are apt to be shy of workmen who emerge into local celebrity by writing or speechifying. In the vast majority of instances they turn out to have no real basis of information or talent, and to possess no better claim to distinction than vehement volubility in the expression of political or religious opinions of an extreme type. But Miller's poems were marked by no extravagance; his *Letters on the Herring Fishery* were full of well-selected facts, and were written with animation, picturesqueness, and good taste; his *Scenes and Legends* were remarkable for the quiet elegance of their style and the vigorous simplicity and home-bred force of their thinking. His scrupulous sense of honour in money matters and pride of independence were felt by all who became acquainted with him; and it is a painful but unquestionable fact that when a man of assured position in society finds himself approached by a person who has been transformed from an incompetent artisan into a shabby-genteel hanger-on of literature, he has an instinctive suspicion that the gentleman of the press will beg. No such suspicion in relation to Hugh Miller could hold its place in the mind of any one who was three minutes in his company. His keen interest, and considerable acquirements, in geological science, had brought him into intercourse with the Messrs Anderson of Inverness, and made him known to every enthusiast in geological or antiquarian research in the north of Scotland. Mr Carruthers will stand well for the literary culture of the northern counties, and he was not the sole cultivator of literature in those parts who had formed high expectations of Miller's future career as a man of letters. He had now been for five years connected with the Bank, and this sufficed to place him behind the scenes with regard to all business operations in the district, and to make his name known to its men of capital. Enough: Hugh Miller was already a public man in the north of Scotland; and, ere he departed for Edinburgh, a number of his friends and admirers entertained him at a public dinner in Cromarty, and presented him with a tea-service in plate. Seldom has a demonstration of the kind attested a warmer or more sincere feeling on the one hand or been more honourably earned on the other. Those who had watched Miller most closely and who knew him best stepped forward to declare that they loved him and considered him a credit to them.

The best room in the best Cromarty Hotel was tastefully laid out and lighted for the occasion. The dinner and wines gave complete satisfaction. The chair was taken by George Cameron, Esq., Sheriff Substitute of the eastern district of Ross-shire, supported by General Sir Hugh Fraser, Captain Mackay Sutherland of Udale, the Rev.

Messrs Stewart and Mackenzie of Cromarty, Dr Macdonald, Mr Fowler, younger, of Raddery, Mr Hartley of the 93rd Regiment, Mr Grigor, bank agent, Invergordon, Mr Joyner of Cromarty, Mr Murray, Mr Middleton of Davidston, Mr Taylor of Westfield, Mr D. Ross, merchant, &c. &c. The croupiers were Robert Ross, Esq., banker, and Mr John Taylor, sheriff-clerk. On the right of the chairman sat Mr Miller; near him his friend Mr Carruthers. After the cloth was drawn, the usual loyal toasts drunk, and an air played by the musical band in attendance, the silver service intended for presentation to the guest of the evening was brought into the room and placed on the table.

The Chairman then called for a special bumper, and spoke to the following effect:—'Before calling upon you, gentlemen, to drink your glasses, it becomes my duty, as you have done me the honour of placing me, however unworthy, in this chair, to state that Mr Miller, being about to leave this, his native town, to enter upon a wider field of literary labour in Edinburgh, it was deemed fitting to invite him to this public entertainment, and to present him, at the same time, with some substantial proof of the high estimation in which he is held where he is best known. (Cheers.) The articles on your table have accordingly been procured for that purpose, and procured, I am happy to say, by means of a contribution as free and cordial as the most sensitive mind could desire. It is for this reason, rather than on account of their intrinsic value, that I now, sir (addressing Mr Miller), in name of this most respectable company, and of the absent subscribers, request your acceptance of them, with our best wishes for your prosperity and happiness in that other part of the kingdom where Providence has cast your lot. As, in the language of the poet,—

"'Tis distance lends enchantment to the view,"

we do not anticipate that you will think less frequently or warmly of this district, or admire its physical or moral beauty the less, because you cease to reside in it; but should you, amidst the toils and cares of business, forget it for a time, these things will recall it to your recollection, and tend to strengthen the ties by which you must ever continue bound to a place which you have illustrated by your writings, and the inhabitants of which, while they contemplate your past career with pride, will watch your future progress with mingled feelings of interest and affection. (Loud cheers.) I need only add, in the words of the inscription, which express shortly the motives that prompted the gift, that this plate is "presented to Hugh Miller, Esq., by his friends in the shire of Cromarty and Easter Ross, in testimony of their admiration of his talents as a writer, and of their respect and regard for him as a member of society." Permit me now, gentlemen, before I sit down, to request that you dedicate this bumper to the health of our excellent guest. (Cheers.) For my own part, I feel peculiar pleasure in proposing this toast, because we are assembled not to do honour to a man of rank or title, however highly we may think of those distinctions, but to pay the homage of our respect to a man who has been truly designated by an eminent individual one of Nature's nobles. Fortunately for me, this subject requires little to be said to ensure to it from you a hearty reception. Here, where he is so well known, and where his merits are so justly appreciated, the mere mention of Mr Miller's name suggests to the mind all that is pure, and honourable, and exemplary in conduct, associated with talents and acquirements of no ordinary kind. This has, indeed, been the birthplace and residence of men remarkable for their intelligence, sound sense, and public spirit. None can think of George Ross of Cromarty, and of his liberality and patriotism, by many proofs of which we are now surrounded, without respecting his memory. Nor of William Forsyth, whose enlarged views, commercial enterprise, and active benevolence, extended in their efforts over the whole of this northern part of the kingdom, without a feeling of exultation, that this was the scene of so much excellence. But in Mr Miller we see kindred qualities combined with gifts of a rarer, if not a higher order. Passing over his *Scenes and Legends*, with which you are all acquainted, his genius and industry have enabled him, without the aid of a scholastic education, to present to us the history of the lives of those men in a style as pure, and elegant, and captivating, as that of similar works by authors better known to fame. The same genius has enabled him to read the book of nature, and by means of geological investigations in this locality, to add to the general stock of knowledge, in a department of the greatest interest and importance. His recent writings on a debateable subject have attracted a good deal of attention throughout Scotland. There are now hearing me gentlemen who concur in his views on ecclesiastical questions, and others who dissent from them ; but however widely our opinions may differ on this point, I infer, from our presence here on this occasion, a unanimity of sentiment as to the honest earnestness, the originality of thought, the beauty of composition, and the ability by which those productions are distinguished. (Great cheering.) These things establish a strong claim to our respect, and we do but a simple act of justice in acknowledging it. Gentlemen, you have marked this brief and feeble expression of my feelings in regard to Mr Miller, by your concurrence and approbation, and I shall not detain you longer. It only remains for me to call upon you to drink his health with all the honours, and success to him in the Scottish metropolis.' Mr Miller's health was drunk with tremendous cheering.

Mr Miller then rose to return thanks, and was received with loud cheers. He said, 'Mr Chairman and gentlemen, the vocabulary of feeling is much more limited than that of thought. We have many words to express what we think, but comparatively few to express what we feel, and often, too, the more intense our feelings, those few words become the fewer. Never have I so forcibly experienced the truth of the remark as this day. There is a fine line in one of our modern poets, worth whole pages of ordinary poetry—

"The old familiar faces."

I feel myself among them, and feel that I am not to be among them long. My heart is full when I think of the kindness you have shown me, and the honour you have done me—so unmerited on my part; yet I do feel proud and gratified when I reflect that the kindness and honour I have experienced have been rendered by individuals for whom I have ever cherished the kindest feelings, and entertained the highest respect. There is, I believe, no mind, however humble its powers or its views, that does not cherish some spark of ambition, and I have cherished my own particular wishes and hopes. They were not fixed very high. Burns tells us of an ambition which had animated him from his earliest days—an ambition founded on intense love of his country—the love which led him when he saw "the rough burr thistle spreading wide" to "spare the symbol dear." (Cheers.) His characteristic wish was—

"That he, for puir auld Scotland's sake,
Some useful plan or book might make,
Or sing a sang at least."

My wish was of a more humble kind, and more in accordance with my powers. Even at that early age when our estimates of ourselves are highest, and our hopes most sanguine, it rose no higher than that I might be able to give some name in literature to the interesting district of country in which we live, of which scarce any one had written before. (Cheers.) I found it a new and untrodden field, full of those interesting vestiges of past times which are to be found—not in the broken remains of palaces and temples, but in the traditional recollections of the common people. These traditions formed my earliest literature. (Cheers.) My humble wish of making this district of Scotland known in literature has been but partially accomplished; but my enjoyment has been great, if not my success. I have found literature to be truly its own reward. My advice to every one desirous to increase his happiness would be, Cultivate your mind. (Cheers.) It is natural for man to look beyond his present circumstances. It is no doubt, in its original integrity, an instinct of his constitution, bearing reference to his spiritual nature, and to a future state. Both his imagination and his hope—and we are all, in some degree, creatures of hope and imagination—have their seat, not in the present time or the existing circumstances, but in the past or in the future. In the field of literature, above every other, this principle finds its fullest scope. The shepherd in Ramsay boasts that, when tending his flock upon the hills, he could converse with kings. It is, indeed, no small privilege to be admitted to the converse of the true kings of the human race, the men of the most comprehensive minds and the most exalted sentiments—to the profundities of a Bacon or a Locke, the high imaginings of a Shakspeare and a Milton. (Cheers.) There is another principle to which I would advert. The love of novelty is inherent in man, and it is natural for him to go on in acquirement, adding idea to idea, and one species of knowledge to another. This is one of the grand distinctions between the human intellect and the instincts of the inferior animals. In accordance with this principle, there are few who would not be travellers if they could. The great bulk of mankind, however, are tied down by circumstances to some one particular locality, and it is fortunate that there is no locality in which the love of novelty may not be gratified—in which, however tied down, we may not become travellers, and enter, time after time, on a new field ; and this, not by changing place, but by connecting it with some newly-acquired science, and thus changing, as it were, its nature. I have strikingly experienced this with regard to this district of the north. Having exhausted it with respect to its connection with the history of our country and those remnants of the belief of our forefathers, which let us more thoroughly into the thoughts and feelings of the past, I have set myself to exhibit it as a locality in which the naturalist—a White of Selbourne, for instance—might have delighted : and I straightway found that I had travelled into a new district. Objects, before unnoticed, or but slightly regarded, rose into interest. Even the spiky leaves and light florets of the thirty or forty sorts of the humble family of the grasses, which I met with in my short walks, every insect that enjoyed itself on the breeze or the stream, grew into beauty and importance. Still more was I interested when, passing from the present creation, I found the locality a rich museum of the remains of former creations. Set your imaginations to work, conceive the most wonderful and unheard-of things, and the history of the district in which we now are will go beyond your extremest conceptions. Would that I could raise the curtain, as it rises in a theatre, and show you scene after scene as it arose ! (Cheers.) We are surrounded by the well-known creatures of the present creation ; and the remains of two older creations, each different from the other, and both from the present, are under our feet. Where we now are, in the remote past there existed an immense lake, more extensive, perhaps, than the Lake Superior in North America, filled with the strange uncouth creatures of the second age of the world, creatures whose very type is lost. This state of things passed ; untold ages went by, and where we now are there extended a wide ocean, filled with its peculiar inhabitants. The master reptiles of a later time and the strangely beautiful plants of a tropical climate flourished upon its shores. (Cheers.) But enough of this. It has been well remarked by Burke, that by fixing the mind strongly on any set of ideas, the sense of present evils may cease to annoy us. This is one of the great advantages resulting from the pursuit of literature. It raises the thought above the annoyance of the present time, and makes those troubles which come to all sit upon us more lightly. I trust I may say I have had experience of this ; but I have dwelt upon the subject too long, and shall not further detain you. I have ever owed much to the kindness of friends, and never before perhaps did I see so many of them assembled together.' (Cheers.) Mr Miller again expressed his grateful acknowledgments, and sat down amidst the cheers of the company.

The toast of the Church of Scotland called up the Rev. Mr Stewart, who, having shortly referred to the Church, added his warmest wishes for the welfare of Mr Miller. By his departure he would lose a valued friend and a most enlightened hearer, but he rejoiced that he was to enter on a more extensive scene of usefulness. (Cheers.)

Mr Carruthers returned thanks for the toast in honour of the press, and stated the great gratification he enjoyed in being amongst them that evening. He had early discovered in their respected guest the germ of those fine talents and moral dispositions which had elevated him in the scale of society, and rendered his progress in life and the development of his mind an object of the deepest interest and solicitude. In him the manifestations of original genius had not been accompanied with any of those debasing and revolting alloys which were sometimes mingled with it; he had advanced onwards by a straight path and a spotless life. There was a fine passage in Pope, where the reproaches and detractions which envy sometimes throws around merit were compared to the vapours that rise around the sun; instead of obscuring the great luminary of day, they only added to its brilliancy, and formed a new temple for its effulgence. In like manner, the early difficulties of their friend in the acquirement of knowledge, his solitary hours of study, amidst hardships, toil, obscurity, and neglect, but enhanced the merit of his present position,

> 'For even those clouds at length adorn his way,
> Reflect new lustre and augment the day.'

He trusted his friend was destined to run a long career of usefulness and honour, and he might rely upon it, that his townsmen and his early friends could never become indifferent to his future fortunes.

Speeches alternated with songs, and cheers were liberally allotted to both, until, after an evening spent with utmost hilarity and good humour, without one jarring word or circumstance, the party broke up about ten o'clock. Mr Miller next day left Cromarty for Edinburgh, carrying with him, we need hardly add, the prayers and good wishes of his friends and townsmen.

Such, very considerably abridged, is the account of this festivity, surely one of the best of its kind, and that a good kind, which adorned the columns of the *Inverness Courier*.

CHAPTER III.

AT THE EDITORIAL DESK.

IN the last days, then, of 1839 Hugh Miller proceeded to Edinburgh to edit the *Witness*. He stepped into the arena alone. His wife and infant daughter he left for the present at Cromarty. Taking lodgings in St Patrick's Square, in the old part of the town, he applied himself with ardour and assiduity to his task. 'In weakness and great fear,' diffident of his power to maintain the conflict against 'well-nigh the whole newspaper press of the kingdom,' he was nevertheless 'thoroughly convinced of the goodness of the cause,' and willing to devote to it the whole energies of his mind. 'I found myself,' he was soon able to say, 'in my true place.' The *Witness* started with a circulation of about 600, but the high character of its articles at once attracted attention, and it became evident in an exceedingly brief period that an immense accession had been made to the power with which the majority in the Church acted on the body of their countrymen. And from the first, the personality of Hugh Miller was felt to be too massive and original to be absorbed in the anonymity of journalism. The voice of the *Witness* was known to be his voice, and the name of Hugh Miller was mentioned with affectionate enthusiasm, as that of the people's own champion, who among the laymen of the conflict was what Chalmers was among the clergy. I may be permitted to quote here a few sentences which I have written upon the subject elsewhere. 'Of the influence exerted upon the public mind of Scotland by Hugh Miller's articles in the *Witness* on the Church question, there are thousands still living who can speak. A year or two before the Disruption I passed a winter in a Highland manse. I was too young to form a distinct idea of the merits of the dispute. But there was a sound in the air which I could not help hearing. It seems as if it were in my ears still. Never have I witnessed so steady, intense, enthralling an excitement. And I have no difficulty, even at this distance, in discriminating the name which rang loudest through the agitated land. It was that of Hugh Miller,—the people's friend, champion, hero!' It was appropriate that a self-educated man should speak for the commonalty of Scotland. It suited the stubborn independence and self-helping vigour of the race. The popular imagination, besides, ready always to be moved by adventitious circumstances, found an additional charm and picturesqueness in his having been a stone-mason, one who had actually 'bared a quarry' and hewn in a church-yard. But this rugged plebeian, who stood forth to fight the people's battle, was not one who required the indulgence of refined critics. No pen wielded on either side in the controversy was more classic than that of Hugh Miller.

He shared the excitement which he contributed so largely to produce. Not only was he animated by the clearest sense of duty, and profoundly convinced that the cause was that of conscience, liberty, and Scotland, but he was conscious that the fray was not without its spectators. 'The series of events which terminated in the Disruption'—the words are his own—'formed a great and intensely exciting drama, and the whole empire looked on.' He shared the excitement of his countrymen; but he also, it need scarcely be added, suffered from it. Never did Hugh Miller toil as during these first three months of his editorship of the *Witness*. He wrote not merely the leading articles, but a large proportion of the remarks introductory to the reports of public meetings, paragraphs on the decease of eminent men, and so on. The paper was published twice a week, and Miller would often have more than one regular leader in each number. His brother combatants, his personal friends, the buzz of applause arising throughout Scotland, cheered him on. In nothing was his temperament more characteristically the temperament of a man of genius—of literary genius—than in his susceptibility to the influence of praise. It was once truly said of him that 'he was like a horse which can be urged by the voice of encouragement beyond its power of living exertion.' Soon also the new paper was attacked by one or other of its many rivals of the opposite side, and with all his gentleness Miller was, when roused, a terrible foe. Michelet holds that there is a trace of ferocity in the artist's temperament. 'He is kindly—he is ferocious. His heart is full of tenderness for the weak and little. Give him orphans to watch over, he will watch over them, and clasp them to his heart;' but set him at defiance, touch a person or a principle which he honours, and he glares on you with an eye like Apollo's on Marsyas, 'watching how the whetting sped.' Professor Masson has remarked that Hugh Miller never engaged in controversial battle without not merely 'slaying, but battering, bruising, and beating out of shape' his antagonist. But the moment his enemy was vanquished, his anger died away. A magistrate of Edinburgh once awakened his wrath. He thought that the civic dignitary had used the power of place to annoy or crush a more honest man than himself, and there was a pompousness in his public behaviour, and a meanness in some of his money-making practices, carefully disguised from the public eye, which gave Miller advantage over him. An article appeared in the *Witness* which made him the laughing-stock of Edinburgh. Next day, when Miller stept into the publishing office, some one made a remark on the severity of his article. 'Ah,' said Miller, in his calmest tone—a very dangerous tone—'I have another shot in the locker for the Baillie.' 'Really, Mr Miller,' replied the first speaker, 'I think you ought to forbear. Baillie —— has had his head shaved.' Miller left the second shot in the locker.

Occasionally he was hounded on his prey by the clerical magnates who took interest in the *Witness*,—never, I believe, by Chalmers; and he has been heard, on becoming acquainted with persons to whom he had administered the lash, to express his regret, and to add that, 'if he had known what manner of man this was, not all the ministers in the Free Church would have persuaded him to inflict the castigation.' That his blows should fall sometimes on the wrong head, or descend with undue momentum, was a necessity of the case. Among the contradictions and limitations of our condition none perhaps is more sad than this, that the very intensity of one's devotion to one's principles tends to incapacitate him for dealing fairly with men who deem it their duty to assail them. 'If we clearly perceive any one thing,' says Coleridge, with fine wisdom and delicate charity, 'to be of vast and infinite importance to ourselves and all mankind, our first feelings impel us to turn with angry contempt from those who doubt and oppose it. Whenever our hearts are warm, and our objects great and excellent, intolerance is the sin that does most easily beset us.' Few controversialists have erred less on the side of severity than Hugh Miller, none perhaps with keener contrition when he found that he had been in the wrong. He told Mrs Miller that he found it necessary to 'abstract' men before he could punish them, and that 'the sight of the human countenance, if it had but a tinge of geniality, so softened and unmanned him,' that he could not shake off the thought of the individual, and had no heart to attack him.

It has been said that Miller, as editor of the *Witness*, felt himself in his place. The stimulus of a strong excitement was useful in rousing his mind to full exertion, and in dispelling the meditative, pensive, almost languid mood in which, in the stillness of Cromarty, he might have indulged. His style, after he came to Edinburgh, compared with that of the *Scenes and Legends*, is improved in energy and fervour. To do his best, he required to be moved; and his most powerful compositions are, I think, his earnest newspaper articles and the Letter to Lord Brougham. Dr Guthrie said with reference to the article on the siege of Acre, that he would rather have written it than taken the fortress. Doubtless, also, Miller was at this time happy. 'He drank delight of battle with his peers.' Fervid emotion bathed the framework of his intellect in flame. The excitement brought its own reward. The additional power and keener sympathetic joy, which a great agitation produces, more than compensate for the daintier pleasures of the intellectual recluse. But of course, in this heroic joy there is a burning which consumes the earthen vessel. While Miller rejoiced in spirit as a strong man to run a race, his body and brain gave unmistakeable evidence that the pace bore hard upon him. Hour after hour he would sit writing, until the letters danced before his eyes and every nerve tingled under the strain. Heedless of exposure, and working deep into the long winter nights, he caught influenza. No matter; he would not pause; he would not lay aside the pen which he had taken up in the cause of his Church and his country. The giddiness of mere exhaustion became the semi-delirium which accompanies inflammatory affections of the lungs and pleura. Had the intense excitement of the conflict been suspended, he would probably have fallen into a state of prostration like that which overtook his father in the sea-fight, who, while the guns continued to roar, did the work of two men, and, when they ceased, fell upon the deck more feeble than a child. Miller grew haggard in the conflict, but he never flinched.

The darkest hour, however, was now past, and streaks of dawn appeared on the horizon. In April, 1840, he was joined by Mrs Miller, who brought with her his infant daughter, Harriet. It was a dark and dreary evening when Mrs Miller saw his figure, in grey suit and plaid, looming through the mist on Granton pier. Her presence, she found, was much needed. Miller looked ill, and his circumstances were comfortless. His sitting-room was dingy with dust and littered with papers. The lodgings were kept by a widow, who, with her family, occupied the rest of the floor. One of the sons took the style of artist, and sat painting in the principal sitting-room. Another cultivated poetry. The only daughter spent a large part of her time in the practice of singing, with a view to appearing before the public. She accompanied herself on the piano, and her shakes and bravuras, on which she industriously lingered, would not have a soothing effect on Hugh's nerves while engaged with his articles. The only servant in the establishment was a 'little maid of all work—a child some twelve years of age who did not look above nine,' —and who 'had all that passed for cleaning in the apartments to accomplish by her own little self.' Mrs Miller did something to alleviate for her husband the horror of all this, but it is obvious that the sole adequate remedy was to get out of it.

So deeply had Miller felt the discomforts of his situation that, before Mrs Miller's arrival, he had taken a small house in Sylvan Place on the southern or country side of the Meadows. 'It was No. 5, the house farthest up the lane. There was a tiny morsel of railed ground in front, a bit somewhat larger under grass at the back. A pleasant market-garden surrounded this enclosure on two sides. A small white-washed dairy stood embowered in trees at the farther end of the lane.' A short walk through lanes and fields led to the gently sloping rise of the Braid hills, from which Scott described and Turner drew Edinburgh. Under these improved aspects Hugh Miller set up his first household in the Scottish capital.

To furnish the house was no easy matter. It had been thought best to dispose of the furniture which had been in the dwelling in Cromarty, but, owing to a continuance of heavy rains, the attendance at the sale was meagre in the extreme, and the household goods, valued at about £150, had yielded a net amount of £40. Such was the sum available for the new furnishing. Miller's salary was £200. He absolutely refused to permit any article to be bought on credit. The difficulty of solving the problem was increased by the Edinburgh custom of requiring the occupant of a house to supply grates and 'fittings' of all kinds. Nevertheless, with activity in attending sales of furniture, contentment with small beginnings, and patient waiting between purchases, Mrs Miller, to whom, in his entire absorption in the work of the paper, her husband left the matter, contrived to make the house first habitable and then comfortable. After a little time she began to assist in the editorial department, first with paste and scissors, then with pen, as contributor of reviews of books, earning thereby some £20 per annum, which went to the furnishing. Once when Miller returned from breakfasting with Chalmers, he told his wife that the great man had complimented him on one of the *Witness* critiques, and that he had never felt so proud in his life as in saying that it was by his wife. The business of furnishing, though difficult, was not without its alleviations.

Capable of existing with perfect convenience in a cave, one stone serving for table, another for seat, and a plate, knife, and fork the whole 'plenishing,' Hugh suffered nothing from the anxiety to put things on a respectable footing which oppressed Mrs Miller. The spring leaves opening round him, the blue sky above, the family of genius with their piteous little drudge at safe distance, his wife beside him and his little daughter on his knee, he could feel once more that he had a home.

A daily visitor from the first was the sub-editor, Mr James Mackenzie, then a student of theology, subsequently minister of Dunfermline. Young, ardent, enthusiastic beyond measure in the cause of spiritual independence, proud of Miller as the popular champion, affectionate in disposition and caressing in manner, sincere, natural, impulsive as a child, he completely won

Miller's heart. 'I scarce ever,' says Mrs Miller, 'saw any one get *so near* my husband as James Mackenzie did. I often remarked that men of a certain *feminine* cast of affection—not intellect—could approach him more closely than others. Mr Mackenzie had that clinging affectionate nature which nestles into a home wherever it finds space. He brought down to the editor all the latest pieces of intelligence, accompanied always with vehement comments of his own. He saved him a vast deal of personal trouble, and was indeed to him as a son or younger brother. Ten to one, when he had run down with some piece of information, and was shouting it out, our little Ha-ha, whom he loved with a most intense fervour, would have climbed into his arms, wound her arms round his neck, prattling her sweet words into his ear, and his awful epithets would be thundered out in the interval between kisses and caresses.'

The Ha-ha here referred to was Harriet, Miller's little child. 'Ha-ha, as she styled herself, was an uncommonly lovely and attractive child. A fair, azure-eyed little thing, with golden curls which reached to her waist. Naturally refined, with great elegance of figure, she glanced about the house like a sunbeam, her childish voice bursting out in a continual ripple of prattle and song. Song more often even than prattle—for she was one of those children richly gifted *as* children, with a certain genius blended with infantile graces which sometimes developes with the growth, sometimes falls away like the blossom of a too early spring. Such was Scott's pet Marjory. Little Ha-ha, however, had one gift which I have never met with in any other child. It was that of natural improvisation. Every little incident, every phase of feeling, was embodied in song and poetry, which she would continue through a long summer's day. The music was always full of melody, the poetry a sort of measured blank verse, sometimes rhyming, sometimes not. It fell on the ear like the ripple of a

> "hidden brook
> In the leafy month of June,
> That to the sleeping woods all night
> Singeth a quiet tune."'

The section of Edinburgh society which Miller would probably have found at heart most congenial was closed to him by the sectarian animosities of the time. Scotch literature had found him out before the clerical world of Scotland cast eyes on him, for it was as a literary man that he had interested Principal Baird, and we have seen proofs of the cordial feeling with which he was regarded by Mr Carruthers, Sir Thomas Dick Lauder, and Mr Robert Chambers. But these, with the large majority of literary men throughout Scotland, viewed the claims of the Church either with positive disapproval or with indifference, and would have told Miller that, in becoming the fighting man of the non-intrusionists, he was throwing himself away. He, on the other hand, was so fervent in his devotion to the cause he had espoused, so confident that the best and dearest interests of his country were at stake, that he could not enjoy any society unless it gave him on this point not only tolerance but sympathy. He did not, therefore, during the early period of his residence in Edinburgh, mingle in literary circles. The clerical leaders, with the exception of Dr Chalmers, with whom he soon grew into warm friendship, did not welcome him to their firesides with the genial hospitality to which he had been accustomed in Cromarty. His only intercourse with them in a social capacity was at an occasional dinner-party given by an evangelical publisher or lawyer.

His editorial contests in capacity of non-intrusion

man-at-arms, were occasionally so severe, the castigation he administered so stinging, and the fury it awakened so conspicuous, that he conceived it possible that he might be personally assaulted, and, with that constitutional timidity or constitutional pugnacity which characterized him, took means to guard against such a contingency. One evening Mr Carruthers had called for him in company with a friend at his house in Sylvan Place, and, not finding him at home, was returning in the thick dusk by the road across the Meadows. Suddenly Miller strode past without recognizing them, and Mr Carruthers turning round exclaimed with mock ferocity, 'There goes that rascally editor of the *Witness*.' Hugh at once faced round and presented a pistol. Another word and glance of course revealed the mistake, and Miller excused himself by saying that it would not surprise him to be attacked any day. As Mr Carruthers and his friend resisted Miller's request to return with him, he accompanied them to their hotel, and they spent one or two pleasant hours together.

Miller never became what printers and newspaper proprietors call a ready editor. To receive at midnight proof-sheets of the report of a long Parliamentary debate, or to sit to a still more advanced hour with attention fixed upon every tedious or wandering speaker who has taken part in it,—an exercise exhaustive and distressing in the last degree to soul and body,—and then to take pen in hand and strike off, as fast as the letters can be traced, an analysis of the discussion, with examination of the principal arguments and judgment upon the whole, in clear and forcible language, garnished with historical allusions, pointed illustrations, and a spicing of eulogy or banter,—this feat, accomplished fifty times a session by practised hands on the London and, *mutatis mutandis*, the provincial press, was never attempted by Hugh Miller. Chalmers remarked of him that, when he did go off, he was a great gun, and the reverberation of his shot was long audible, but he required a deal of time to load.

It was his delight to find the subject of an article in some topic suggested by a friend, the more if old and pleasant associations were recalled by it. When a Danish corvette, for example, appeared in the Bay of Cromarty, and Mrs Allardyce, widow of a Scottish clergyman and one of Miller's valued friends, sent him from his old home a few lines which she had written on the occasion, the lines were at once felt to be too good for the mere poetical corner, and were kept back until they could be presented to the reader in the setting of an editorial article. For days or weeks the thing would dwell in Miller's thoughts, and at last the article, with gossippy speculation as to the relationship between the Danes and the Scotch, and a copious account of the friendship of Cousin Walter and young Wolf, and one more description of the Sutors and the Bay, would charm all those—and they were not few—who took the *Witness* for the sake of the editorial essays of Miller.

The routine work of editing he never so completely mastered as the routine work of banking; rather, perhaps, I should say that he never could feel or affect that transcendent interest in the petty political questions of the day, which beseems the man whose function it is to keep others perpetually interested in them. On Church questions he viewed things in his own way, and even when there was no discrepancy of opinion between him and the leaders, he was apt to surprise them by the time and manner of his discussing questions or advocating causes. During the few months preceding the Disruption, for example, Chalmers was bringing all the energies of his genius to solve the grave economical problems with whose practical solution the Free Church would require to grapple. The religious mind of Scotland, however, was at the moment deeply agitated by what may be called the Railway phase of the Sabbath question. Hugh Miller's enthusiasm for the Sabbath was as great as his enthusiasm for the Church, and he apprehended national disaster on a large scale as the ultimate result of the running of railway trains on Sunday. Accordingly, on the 4th March, 1843, there appeared in the *Witness* a couple of columns from his pen, headed, 'A Vision of the Railroad,' in which the possibilities of the future were delineated in appalling hues. 'Writing a Vision of the Railroad,' growled Chalmers, 'when we want money!' The Vision is so characteristic, and, as a piece of literary workmanship, so striking, that I insert it here. The 'Mr McNeill,' who somewhat spectrally appears amid the imagery of the piece, can be understood to have been some lawyer of the period whose opinions on the first day of the week seemed to Miller to be latitudinarian.

A VISION OF THE RAILROAD.

———, Isle of Skye.

* * * I know not when this may reach you. We are much shut out from the world at this dead season of the year, especially in those wilder solitudes of the island that extend their long slopes of moor to the west. The vast Atlantic spreads out before us, blackened by tempest, a solitary waste, unenlivened by a single sail, and fenced off from the land by an impassable line of breakers. Even from the elevation where I now write,—for my little cottage stands high on the hill-side,—I can hear the measured boom of the waves, swelling like the roar of distant artillery, above the melancholy moanings of the wind among the nearer crags, and the hoarser dash of the stream in the hollow below. We are in a state of siege; the isle is beleaguered on its rugged line of western coast, and all communication with man in that quarter cut off; while in the opposite direction the broken and precarious footways that wind across the hills to our more accessible eastern

shores, are still drifted over in the deeper hollows by the snows of the last great storm. It was only yester-evening that my Cousin Eachen, with whom I share your newspaper, succeeded in bringing me the number published early in the present month, in which you furnish your readers with a report of the great Railway meeting at Glasgow.

My cousin and I live on opposite sides of the island. We met at our tryst among the hills, not half an hour before sunset; and as each had far to walk back, and as a storm seemed brewing,—for the wind had suddenly lowered, and the thick mists came creeping down the hill sides, all dank and chill, and laden with frost-rime, that settled, crisp and white, on our hair,—we deemed it scarce prudent to indulge in our usual long conversation together.

'You will find,' said Eachen, as he handed me the paper, 'that things are looking no better. The old Tories are going on in the old way, bitterer against the gospel than ever. They will not leave us in all Skye a minister that has ever been the means of converting a soul; and what looks as ill, our great Scotch Railway, that broke the Sabbath last year in the vain hope of making money by it, is to break it this year at a dead loss. And this for no other purpose that people can see, than just that an Edinburgh writer may advertise his business by making smart speeches about it. Depend on't, Allister, the country's *fey*.'*

'The old way of advertising among these gentry,' said I, 'before it became necessary that an elder should have at least some show of religion about him, was to get into the General Assembly, and make speeches there. If the crisis comes we shall see the practice in full blow again. We shall see our anti-Sabbatarian gentlemen transmuted into voluble Moderate elders, talking hard for clients without subjecting themselves to the advertisement duty,—and the railway mayhap keeping its Sabbaths.'

'Keeping its Sabbaths!' replied Eachen, 'ay, but the shareholders, perhaps, having little choice in the matter. I wish you heard our Catechist on that. Depend on't, Allister, the country's *fey*.'

'Keeping its Sabbaths? Yes,' said I, catching at his meaning, 'if we are to be visited by a permanent commercial depression,—and there are many things less unlikely at the present time,—the railway *may* keep its Sabbaths, and keep them as the land of Judea did of old. It would be all too easy, in a period of general distress, to touch that line of necessarily high expenditure below which it would be ruin for the returns of the undertaking to fall. Let but the invariably great outlay continue to exceed the income for any considerable time, and the railway *must* keep its Sabbaths.'

'Just the Catechist's idea,' rejoined my cousin. 'He spoke on the subject at our last meeting. "Eachen," he said,—"Eachen, the thing lies so much in the ordinary course of Providence, that our blinded Sabbath-breakers, were it to happen, would recognize only disaster in it,—not judgment. I see at times, with a distinctness that my father

* Marked for early death or overwhelming calamity.

would have called the second sight, that long weary line of rail, with its Sabbath travellers of pleasure and business speeding over it, and a crowd of wretched witnesses raised, all unwittingly and unwillingly on their own parts, to testify against it, and of coming judgment, at both its ends. I see that the walks of the one great city into which it opens are blackened by shoals of unemployed artizans,—and that the lanes and alleys of the other number by thousands and tens of thousands, their pale and hunger-bitten operatives, that cry for work and food. They testify all too surely that judgment needs no miracle here. Let but the evil continue to grow,—nay, let but one of our Scottish capitals, —our great mart of commerce and trade, sink into the circumstances of its manufacturing neighbour Paisley, and the railway *must* keep its Sabbaths. But, alas ! there would be no triumph for party in the case. Great, ere the evil could befall, would the sufferings of the country be, and they would be sufferings that would extend to all." What think you, Allister, of the Catechist's note ? '

'Almost worth throwing into English,' I said. 'But the fog still thickens, and it will be dark night ere we reach home.' And so we parted.

Dark night it was, and the storm had burst out. But it was pleasant, when I had reached my little cottage, to pile high the fire on the hearth, and to hear the blast roaring outside, and shaking the window boards, as if some rude hand were striving in vain to unfasten them. I lighted my little heap of moss-fir on the projecting stone, that serves the poor Highlander for at once lamp and candlestick, and bent me over your fourth page, to scan the Sabbath returns of a Scottish railroad. But my rugged journey and the beating of the storm had induced a degree of lassitude,—the wind outside, too, had forced back the smoke, until it had filled with a drowsy, umbery atmosphere the whole of my dingy little apartment,—Mr M'Neill seemed considerably less smart than usual, and more than ordinarily offensive, and in the middle of his speech I fell fast asleep. The scene changed, and I found myself still engaged in my late journey, coming down over the hill, just as the sun was setting red and lightless through the haze, behind the dark Atlantic. The dreary prospect on which I had looked so shortly before was restored in all its features ;—there was the blank, leaden-coloured sea, that seemed to mix all around with the blank, leaden-coloured sky ;—the moors spread out around me, brown and barren, and studded with rock and stone ;—the fogs, as they crept downwards, were lowering the overtopping screen of hills behind to one dead level. Through the landscape, otherwise so dingy and sombre, there ran a long line of somewhat brighter hue,—it was a long line of breakers tumbling against the coast far as the eye could reach, and it seemed interposed as a sort of selvage between the blank, leaden sea, and the deep, melancholy russet of the land. Through one of those changes so common in dreams, the continuous line of surf seemed, as I looked, to alter its character. It winded no longer around headland and bay, but stretched out through the centre of the landscape, straight as an extended cord, and the bright white saddened down to the fainter hue of decaying vegetation. The entire landscape underwent a change. Under the gloomy sky of a stormy evening, I could mark, on the one hand, the dark-blue of the Pentlands, and on the other, the lower slopes of Corstorphine. Arthur's Seat rose dim in the distance behind ; and in front, the pastoral valley of Wester Lothian stretched away mile beyond mile, with its long rectilinear mound running through the midst,—from where I stood beside one of the massier viaducts that rose an hundred feet overhead, till where the huge bulk seemed diminished to a slender thread on the far edge of the horizon.

It seemed as if years had passed,—many years. I had an indistinct recollection of scenes of terror and of suffering,—of the shouts of maddened multitudes engaged in frightful warfare,—of the cries of famishing women and children,—of streets and lanes flooded with blood,—of raging flames enwrapping whole villages in terrible ruin,—of the flashing of arms, and the roaring of artillery,—but all was dimness and confusion. The recollection was that of a dream remembered in a dream. The solemn text was in my mind,—'Voices, and thunders, and lightnings, and a great earthquake, such as was not since men were upon the earth,—so mighty an earthquake and so great ;' and I now felt as if the convulsion was over, and that its ruins lay scattered around me. The railway, I said, is keeping its Sabbaths! All around was solitary, as in the wastes of Skye. The long rectilinear mound seemed shaggy with gorse and thorn, that rose high against the sides, and intertwisted their prickly branches atop. The sloe-thorn, and the furze, and the bramble choked up the rails. The fox rustled in the brake ; and where his track had opened up a way through the fern, I could see the red and corroded bars stretching idly across. There was a viaduct beside me : the flawed and shattered masonry had exchanged its raw hues for a crust of lichens ; one of the taller piers, undermined by the stream, had drawn two of the arches along with it, and lay a-down the watercourse a shapeless mass of ruin, o'ermasted by flags and rushes. A huge ivy that had taken root under a neighbouring pier, threw up its long pendulous shoots over the summit. I ascended to the top. Half buried in furze and sloe-thorn there rested on the rails what had once been a train of carriages ;—the engine ahead lay scattered in fragments, the effect of some disastrous explosion ; and damp, and mould, and rottenness, had done their work on the vehicles behind. Some had already fallen to pieces, so that their places could be no longer traced in the thicket that had grown up around them,—others stood comparatively entire, but their bleached and shrivelled panels rattled to the wind, and the mushroom and the fungus sprouted from between their joints. The scene bore all too palpably the marks of violence and bloodshed. There was an open space in front, where the shattered fragments of the engine lay scattered ; and here the rails had been torn up by violence, and there stretched across, breast-high, a rudely piled rampart of stone. A human skeleton lay atop, whitened by the winds; there was a broken pike beside it; and, stuck fast in the naked scull, which had rolled to the bottom of the rampart, the rusty fragment of a sword. The space behind resembled the floor of a charnel-house,—bindwood and ground-ivy lay matted over heaps of bones; and on the top of the hugest heap of all, a scull seemed as if grinning at the sky from amid the tattered fragments of a cap of liberty. Bones lay thick around the shattered vehicles,—a trail of skeletons dotted the descending bank, and stretched far into a neighbouring field; and from amid the green rankness that shot up around them, I could see soiled and tattered patches of the British scarlet. A little farther on there was another wide gap in the rails. I marked beside the ruins of a neighbouring hovel, a huge pile of rusty bars, and there lay inside the fragments of an uncouth cannon, marred in the casting.

I wandered on in unhappiness, oppressed by that feeling of terror and disconsolateness so peculiar to one's more frightful dreams. The country seemed everywhere a desert. The fields were roughened with tufts of furze and broom;—hedgerows had shot up into lines of stunted trees, with wide gaps interposed;—cottage and manor-house had alike sunk into ruins;—here the windows still retained their shattered frames, and the roof-tree lay rotting, amid the dank vegetation of the floor,—yonder the blackness of fire had left its mark, and there remained but reddened and mouldering stone. Wild animals and doleful creatures had everywhere increased. The toad puffed out its freckled sides on hearths, whose fires had been long extinguished,—the fox rustled among the bushes,—the masterless dog howled from the thicket, —the hawk screamed shrill and sharp as it fluttered overhead. I passed what had been once the policies of a titled proprietor. The trees lay rotting and blackened among the damp grass,—all except one huge giant of the forest, that girdled by the axe half a man's height from the ground, and scorched by fire, stretched out its long dead arms towards the sky. In the midst of this wilderness of desolation, lay broken masses, widely scattered, of what had been once the mansion-house. A shapeless hollow, half filled with stagnant water, occupied its immediate site; and the earth was all around torn up, as if battered with cannon. The building had too obviously owed its destruction to the irresistible force of gunpowder.

There was a parish church on the neighbouring eminence, and it, too, was roofless and a ruin. 'Alas,' I exclaimed, as I drew aside the rank stalks of nightshade and hemlock that hedged up the breach in the wall through which I passed into the interior,—'alas! have the churches of Scotland also perished!' The inscription of a mutilated tombstone that lay outside caught my eye, and I paused for a moment's space in the gap, to peruse it. It was an old memorial of the times of the Covenant, and the legend was more than half defaced. I succeeded in deciphering merely a few half-sentences,—'killing time,'— 'faithful martyr,'—'bloody Prelates;' and beneath there was a fragmentary portion of the solemn text, 'How long, O Lord, holy and true, dost thou not judge and avenge our blood?' I stepped into the interior :—the scattered remains of an altar rested against the eastern gable. There was a crackling, as of broken glass, under my feet, and stooping down, I picked up a richly-stained fragment ;—it bore a portion of that much-revered sign, the pelican giving her young to eat of her own flesh and blood,—the sign which Puseyism and Popery equally agree in regarding as adequately expressive of their doctrine of the real presence, and which our Scottish Episcopalians have so recently adopted as the characteristic vignette of their Service-Book. The toad and the newt had crept over it, and it had borrowed a new tint of brilliancy from the slime of the snail. Destruction had run riot along the walls of this parish church. There were carvings chipped and mutilated, as if in sport, less, apparently, with the intention of defacing than rendering them contemptible and grotesque. A huge cross of stone had been reared over the altar, but both the top and one of the arms had been struck away, and from the surviving arm there dangled a noose. The cross had been transformed into a gibbet. Nor were there darker indications wanting. In a recess set apart as a cabinet for relics, there were human bones all too fresh to belong to a remote antiquity ; and in a niche under the gibbet lay the tattered remains of a surplice dabbled in blood. I stood amid the ruins, and felt a sense of fear and horror creeping over me,—the air darkened under the scowl of the coming tempest and the closing night, and the wind shrieked more mournfully amid the shattered and dismantled walls.

There came another change over my dream. I found myself wandering in darkness, I knew not whither, among bushes and broken ground ; there was the roar of a large stream in my ear, and the savage howl of the storm. I retain a confused, imperfect recollection of a light streaming upon broken water,—of a hard struggle in a deep ford, —and of at length sharing in the repose and safety of a cottage solitary and humble almost as my own. The vision again strengthened, and I found myself seated beside a fire, and engaged with a few grave and serious men in singing the evening Psalm with which they closed for the time their services of social devotion.

'The period of trial wears fast away,' said one of the number, when all was over,—a grey-haired, patriarchal-looking old man,—'the period of trial is well nigh over,—the storms of our long winter are past, and we have survived them all ; — patience, — a little more patience,—and we shall see the glorious spring-time of the world begin ! The vial is at length exhausted.'

'How very simple,' said one of the others, as if giving expression rather to the reflection that the remark suggested, than speaking in reply,—'how exceedingly simple now it seems, to trace to their causes the decline and fall of Britain. The ignorance and the irreligion of the land have fully avenged themselves, and have been consumed, in turn, in fires of their own kindling. How could even mere men of the world have missed seeing the great moral evil that lay at the root of—'

'Ay,' said a well-known voice that half mingled with my dreaming fancies, half recalled me to consciousness ; 'nothing can be plainer,

Donald. That lawyer-man is evidently not making his smart speeches or writing his clever circulars with an eye to the pecuniary interests of the railroad. No person can know better than he knows, that the company are running their Sabbath trains at a sacrifice of some four or five thousand a-year. Were there not an hundred thousand that took the pledge? and can it be held by any one that knows Scotland, that they aren't worth overhead a shilling a-year to the railway? No, no; depend on't, the man is guiltless of any design of making the shareholders rich by breaking the Sabbath. He is merely supporting a desperate case in the eye of the country, and getting into all the newspapers, that people may see how clever a fellow he is. He is availing himself of the principle that makes men in our great towns go about with placards set up on poles, and with bills printed large stuck round their hats.'

Two of my nearer neighbours, who had travelled a long mile through the storm to see whether I had got my newspaper, had taken their seats beside me, when I was engaged with my dream, and after reading your railway report, they were now busied in discussing the various speeches and their authors. My dream is, I am aware, quite unsuited for your columns,—and yet I send it you. There are none of its pictured calamities that lie beyond the range of possibility,—nay, there are perhaps few of them that at this stage may not actually be feared; but if so, it is at least equally sure that there can be none of them that at this stage might not be averted.

We should hardly have known Miller for what he was without reading this remarkable performance. It will occasion astonishment to not a few. Puritanism so austere, confronting us in the middle of the nineteenth century, shows like a massive boulder, standing, grey and stern, in the midst of a modern meadow. 'One man esteemeth one day above another; another man esteemeth every day alike: let every man be fully persuaded in his own mind.' This is Paul's deliverance on the subject, and Calvin was content to accept it in all its breadth of Christian liberty. Miller knew, however, that in Scotland devout religion had for two centuries been inseparably connected with Sabbath observance, and he naturally looked upon desecration of the day as leading to profanity and licentiousness. For the rest, we are to recollect that, though the concluding words of the 'Vision' prove Miller to have seriously believed that the running of Sunday trains on the Edinburgh and Glasgow Railway might bring calamity on the empire, its colouring is avowedly the colouring of a dream.

In a short preface to a volume of leading articles by Hugh Miller, selected from the file of the *Witness*, I have made some remarks on his capacities as a writer of journalistic essays which I may be permitted to insert here:—'He meditated his articles as an author meditates his books or a poet his verses, conceiving them as wholes, working fully out their trains of thought, enriching them with far-brought treasures of fact, and adorning them with finished and apposite illustration. In the quality of *completeness*, those articles stand, so far as I know, alone in the records of journalism. For rough and hurrying vigour they might be matched, or more, from the columns of the *Times;* in lightness of wit and smart lucidity of statement they might be surpassed by the happiest performances of French journalists,—a Prevost Paradol or a St Marc Girardin; and for occasional brilliancies of imagination, and sudden gleams of piercing thought, neither they nor any other newspaper articles have, I think, been comparable with those of S. T. Coleridge. But as complete journalistic essays, symmetrical in plan, finished in execution, and of sustained and splendid ability, the articles of Hugh Miller are unrivalled.'

In short, he modelled his newspaper essays, as he modelled the chapters of his books, on the productions of his beloved Addison and Goldsmith, rather than on those of the 'eminent hands' whose slashing leaders have made their reputation on the London press. It was his habit to fix upon his subject a few days, or even longer, before the article was to appear, and nothing pleased him better than to have Mrs Miller as volunteer antagonist, to maintain against him, at the supper table, the thesis he proposed to controvert. Supper was his favourite meal. At breakfast he hardly tasted food, a cup of coffee and crumb of bread being the limit of his wants. After working at his desk in the early part of the day, he would walk out, make his way into the country, saunter about the hills of Braid or Arthur-seat, with his eye on the plants and land shells and geological sections, or explore for the thousandth time the Musselburgh shore or the Granton quarries. He never clearly admitted the canonical authority of the dinner hour. He expected something warm to be kept ready for him; but if the day was particularly favourable, or if a storm had strewn the coast with the treasures of the deep sea, or if some new phenomenon struck him in connection with the raised beach of Leith and required interpreting and thinking out, or if he met with a brother naturalist and got into talk, the shades of evening would be falling thick before he again crossed his threshold. Even at that hour he had little appetite. It was not until his brain, obeying what his habits of night-study had made an irresistible law for him, awoke in its fervour about ten o'clock, that he showed a keen inclination for food. Porter or ale, with some kind of dried fish or preserved meat, formed his favourite supper. On these occasions he conversed with great freedom, and found it both pleasant and profitable to have his views and arguments vigorously controverted. There can be no doubt that the extraordinary success of many of his articles,—the repeated case of their being the town-talk and country-talk of the day,—was due, in a considerable degree, to his having beaten over the ground with Mrs Miller.

In illustration of his habits of composition as an editor, I may refer to sundry slips which have been found among his papers, containing what seem the outlines of contemplated essays. He first states a question, and then sets down the heads of an excursive reply to it. Here is one of these essays in skeleton,—I copy *verbatim* from Miller's MS.

'Query—Whether Writing or Painting convey to the mind the more vivid ideas.

'Original method of transmitting history and communicating ideas. Egyptian Hieroglyphics—Mexican ditto—North American. The deaf boy.

'Disadvantages of this method. The operation of mind indescribable,—the achievements of conduct. Causes given for effects. and effects for causes. Winter personified as a Hog. The animals made symbolical. Idolatry.

'1st Conclusion, that the field of Painting is extremely narrow, and that narrow field filled with uncertainty. That of Writing, on the other hand, wide and clear.'

'Painting in a different view. In one particular province of description more clear than Writing. A drawing gives a more minute idea of a person, a building, a landscape, or a machine, than any written description. To men of common minds it gives more vivid ideas of things of a different province, but not to men of genius. The illustrations of the Waverley Novels; of Milton's Paradise Lost; of Shakespeare.—This perhaps a consequence of the painter's inferiority to the writer. Happy illustration involved in Tam O'Shanter by Thom and the sculpture of the ancients. The Italian school—The Flemish.

'Conclusion 2nd. In one narrow field Painting gives more vivid ideas than Writing. In another field it almost disputes the palm;—but the fields of Writing as extended as human thought or the works of the Creator. Nay, infinite; seeing God is a writer.'

We shall take another of these outlined essays or Debating Society speeches,—they would serve equally well for either.

'Query—Whether Luxury be advantageous or disadvantageous to a State.

'A brief view of the history of Luxury from the most remote to the present times. Assyria. The feast of Belshazzar. Persian Luxury in its primitive cast. The savage a great drinker, the Persian so. The boast of Cyrus over his brother Artaxerxes. The men who live by hunting often obliged to fast for a long time, and in the habit of gorging voraciously. The Persians ate only once a day. Consolation of the Abderites. The Greeks similar in this respect to the Persians. Plato's remark on the Sicilians. The Spartans and Athenians luxurious in the latter days of their respective States.

'Roman luxury very peculiar. Cleopatra's pearls. Esopus the

player had one dish that cost nearly £5000—the tongues of singing-birds. Lucullus's suppers amounted to £1600 each. Favourite dishes were the fat paps of a sow—the livers of *scari*—the brains of phenicopters. Apicius sails from the coast of Italy to that of Africa to taste a species of oyster—one of the common dishes of Rome the (word undecipherable) boar.

'Ancient Scandinavian feasts. Intemperate in eating and drinking. As among the Persians, to hold much liquor reputed a heroic virtue. Drinking to the gods and the souls of departed heroes. Their heaven that of the Turks. The old English. William of Malmesbury's account of them—continuing over their cups night and day. Saints' days times of festivity. In Scotland the extreme sumptuousness of the feasts at marriages, christenings, and funerals restrained by Act of Parliament in the reign of Charles II. These great feasts a proof that the people were not habitually luxurious. Our own recollection of Christmas and New Year's day. Luxury of the monks in the Middle Ages. Giraldus Cambrensis' anecdote of the monks of St Swithin who prostrated themselves at the feet of Henry II., complaining of their Bishop. His decision regarding their suit. Angels, patriarchs, apostles, prophets in poetry. Luxury of the present age. Its effects on small farmers, operative manufacturers. Rack-rented tenants. People becoming extremely poor or oppressively rich.

'Luxury the cause of decline in almost all nations. The Persians as invaders and invaded. Alexander conquered by luxury. The Greeks, both Athenians and Spartans. The Romans employ mercenaries. Our own country. Goldsmith's apostrophe.'

Hugh Miller conducted the *Witness* for sixteen years, and he cannot have written for the paper fewer than a thousand articles. 'Having surveyed this vast field, I retain the impression of a magnificent expenditure of intellectual energy,—an expenditure of which the world will never estimate the sum. By far the larger portion of what he wrote for the *Witness* is gone for ever. Admirable disquisitions on social and ethical questions, felicities of humour and sportive though trenchant satire, delicate illustration and racy anecdote from an inexhaustible literary erudition, and crystal jets of the purest poetry,—such things will repay the careful student of the *Witness* file, but can never be known to the general public.'

It was a tragic element in Miller's lot as a newspaper

editor that he had no particle of enthusiasm for the press, no confidence in the newspaper as an educating agency. He has put it on record that the mechanics he had known whose culture consisted in life-long familiarity with newspapers were uniformly shallow and frivolous. Of himself he has spoken as doomed to cast off shaving after shaving from his mind, to be caught by the winds, and after whirling lightly for a little time, to be blown into the gulf of oblivion. Perhaps he did not enough take into account the essentially ephemeral nature of human productions, or reflect that the longest-lived book and the newspaper article of the hour are alike covered up one day.

> 'The memory of the withered leaf
> In endless time is scarce more brief
> Than of the garnered autumn sheaf.'

Nay, inasmuch as a powerful newspaper writer lodges his thoughts in the minds of men engaged in affairs, and has them thus woven into the web of events and the fabric of institutions, it might be argued that he least of all toils without result of his labours. Hugh Miller, at any rate, looked with fixed distrust upon journalistic writing, both as culture for a man's own mind and as a means of influencing his fellows. He regarded science as a counteractive to the deteriorating effects of this kind of work upon his intellectual powers.

CHAPTER IV.

THE DISRUPTION.

IN writing the history of any event or series of events an author is inevitably influenced by the point of view which he has taken. It would be possible to draw up an historical account of the French Revolution without mention of the Paris newspapers of the period. These had no official, formal, technical connection with the occurrences of the time, and from the dynastic point of view, or the Parliamentary point of view, or the official point of view, it would be easy to delineate the Revolution without naming them. If a just account, however, of a great historical transaction involves a description of those agencies which, assisting its progress and modifying its character, contributed mainly to make it what it was, the omission of all mention of the French press in a history of the French Revolution, would be a capital mistake. When the movement which resulted in the Disruption of 1843 and the creation of the Free Church of Scotland is contemplated from the ecclesiastical point of view, it is possible, without any intentional injustice, to overlook the importance of the part played in it by Hugh Miller. The Courts of the Presbyterian system, though the congregation is represented in them by the lay elders, are essentially Church Courts, and the clergy constitute their preponderating element. It was in these courts that the battle of the Church, in its formal and technical aspect, was fought; and one who took part in their discussions, and subsequently wrote an account of the movement, might think that, when he had detailed what took place in Presbytery, Synod, and General Assembly, with mention of corresponding occurrences in the Court of Session and the British Parliament, and occasional glances at the chief public meetings and the most notable pamphlets, he had presented a just and adequate view of the historical transaction which he undertook to depict. Such is doubtless the fair explanation of a circumstance which has often been the subject of remark, namely, that, in a well-known history, entitled 'The Ten Years' Conflict,' distinguished on the whole by lucidity, vigour, and animation, the name of Hugh Miller is conspicuous by its absence. An eloquent passage is quoted in that work from the Letter to Lord Brougham, and the author, Dr Robert Buchanan, has never hesitated to acknowledge the importance of the service rendered by the *Witness* and its editor to the Evangelical party; but Miller nevertheless felt that, in a history of the origin of the Free Church, he ought to have been named. The book was put into his hands for review, and with characteristic magnanimity and candour he reviewed it favourably.

At the time when Hugh Miller undertook the editorship of the *Witness*, the sympathies of the people of Scotland had been to but a comparatively slight extent awakened and secured for the contending Church. A vague but potent impression swayed the public mind that the agitation was a mere clerical affair. I can still recollect how this notion would crop up in the small talk of the day.

'The ministers wanted power. They would like to put down patrons with one hand, and to silence with the other every luckless parson who did not vote in the Presbytery and preach in the pulpit as the Evangelical majority were pleased to dictate. Perhaps it was just as well that the Court of Session should keep these petulant little popes in their own place.' No man did so much to dissipate these notions, so perilous to the movement, as Hugh Miller. The Church of Scotland, he proclaimed, was standing once more, as she had so often stood, on the side of the people, and he tore to shreds the flimsy plea that the dogs, valiantly defending the fold, had an eye only to their class interests. On the other hand, his influence was mightily exerted to prevent the mere ecclesiastical element from assuming that predominance which many alleged to be the object of the whole struggle. Hugh Miller felt with a depth and solemnity of conviction which converted the feeling into a sentiment of duty, that the *Witness* was to be the organ of no clerical party, the sounding-board of no Church Court, but was to represent the movement in all the breadth and independence of its national characteristics.

He took no vulgarly practical, no merely political or secular, view of the interests at stake in the Ten Years' Conflict. His appreciation of the position and claims of the Church was, in a strictly theological point of view, masterly. He recurred, as we saw, to the first Reformed Assembly of the Church of Scotland, its Book of Discipline and Confession of Faith; nor did he stop there, but pressed onward until the principles which he found in these instruments led him to 'the fundamental idea of all Revelation,' that, man's supreme allegiance being due to God, the ultimate formula of freedom in relation to his fellows, a formula which he is bound to enunciate and maintain, is liberty to do his duty towards God. The ancient standards of the Church of Scotland 'embodied,' he said, 'in all their breadth the Redeemer's rights of prerogative as sole Head and King of His Church, and, with these, all those duties and privileges of the Church's members which *His* rights necessarily involve and originate.' The Church, therefore, owed it to Christ that she should not surrender either her spiritual independence or the rights of congregations.

The lofty and ideal conception of the cause he championed, which animated Hugh Miller, was entertained also by the people of Scotland. A thrill of spiritual enthusiam passed through the heart of the nation more ardent and elevating than has been experienced in the United Kingdom since the seventeenth century. I consider it one of the greatest advantages of my life to have witnessed the excitement of the time. So steadily do the forces at work in modern society tend to make the possession of money the all-absorbing ambition of life,—so pallid and ineffectual, in contest with this tremendous power, are the æsthetic and political influences which act upon cultivated men,—that it is a strengthening and priceless consciousness to have lived in a society which in very deed esteemed principle more highly than gold. More highly than gold; yes, and more highly than what to finely-toned minds is of greater value than gold, to wit, the distinction belonging to an Established Church, the deference, the social consideration, yielded to the representatives of a legally recognized and honoured institution.

Within the present century no day has dawned on Scotland when the heart of the nation was so profoundly agitated as on that on which the majority in the General Assembly of 1843 left St Andrew's Church and proceeded to Canonmills Hall. There had been no intention on the part of the protesting ministers and members to form a procession, but when the first exultant shout with which the emerging figures were greeted had subsided, the crowd fell back on either side in spontaneous reverence, and formed a lane through which the procession moved. Dr Welsh, the Moderator of the preceding year, with Chalmers and Gordon, two men whose appearance, the one for its massive and leonine manhood, the other for its severe intellectual majesty, would have attracted notice in any assemblage in Christendom, led the way. Cunningham, Candlish, MacDonald of Ferintosh, Campbell of Monzie, Murray Dunlop, men whose names had become household words in Scotland, followed. As they headed the long column on its way down the broad swell of undulation on which the new town of Edinburgh is built, the Frith of Forth before them, the Bass Rock far on the right, the blue hills of the north closing in the distance, there were lookers-on who, although they had opposed the movement, felt their eyes moisten with proud joy that they had seen such a day. It was the old land yet,—the stuff of immortality, the asbestos thread of incorruptible national character, the light that struggled in Falkirk's wood, and beamed out at Bannockburn, and played in fitful gleams upon the storm-tost banner of the Covenant, survived in Scotland still.

What to most Englishmen—to those who have derived their idea of clerical character from the recent history of the Anglican Establishment—will seem specially surprising in this event, is that the Scottish clergy who, in 1843, abandoned their livings, did not find some plausible pretext, some ingenious sophism, some 'softly-spoken and glistening lie,' wherewith to lull their consciences and enable them to evade their duty. Such plausibilities lay round in abundance,—the raw material of them, that is, requiring only to be woven into webs and veils. The protesters might, for example, have continued their protest *within* the pale of the Establishment. They might have raised the flag of rebellion, have told the civil authorities that, let them do what they chose, the majority would exercise the spiritual government of the Church, and have declared that, whenever a minister called himself the pastor of a parish against the will of the people and in defiance of the Church, they would let him consume the loaves and the fishes, but would depose him from the office of the ministry, appoint another man in his stead, and direct the people to recognize the latter as their pastor. This course of denouncing and defying the State is the highest pitch of devotion to Church principle to which the typical Anglican attains or aspires. Henry, Bishop of Exeter, overruled by the Privy Council in his dealings with one whom he believed to be a heretic, excommunicated all men, the Primate of the Church of England included, who should dare to interfere with his ecclesiastical jurisdiction. When his gesticulations and exclamations, his protests and pronouncements and anathemas, were treated as so much amusing pantomime, Henry crossed his legs in the House of Lords and was content. The gesticulating and exclaiming, though it had produced no manner of effect on Mr Gorham, or on the Archbishop of Canterbury, or on the Judicial Committee of Privy Council, or on any other earthly man or thing, possessed the singular property of calming Henry's agitated bosom, and setting free his energies for assault and battery upon all 'schismatics' from the Anglican Church. But in all ages of the world save ours, and in all countries of the world save that Eden which lies between the Cheviot hills and the English Channel, the proceedings of the Bishop of Exeter, when beaten down and baffled in the Gorham case, would have been considered a mixture of unseemly insolence and atrocious humbug. Mr Bennett, of Frome, refuses to plead in what are called the ecclesiastical courts, regarding them and their doings as mere imaginary entities in relation to the Church of which he holds himself to be a member. How he would name that Church I cannot tell, but he stands before the public as a minister of the State-Church of England. The Rev. Mr Mossman of Wragsby, rejoicing that 'the lurid, murky flame of Protestantism, enkindled in the sixteenth century, is rapidly becoming quenched, and the true light of the gospel, which twice before came to England from Rome, is once more beginning to beam upon us from the Eternal City,' hoping for the day when 'it will come to pass that Anglicans will see that it is God's will that they should submit to the Holy Apostolic See, and that it is their duty as well as their privilege to be in communion with that Bishop who alone is the true successor of Peter, and, by Divine Providence, the Primate of the Catholic Church,' nevertheless continues an official of the Anglican Establishment. By the simple process of regarding the doctrinal formularies of the Church as meaningless and the ecclesiastical courts as imaginary—by merely ignoring the little ceremony of subscription, the preliminary fib which has to be uttered, the passing of conscience under the Caudine Forks which has to be undergone—unlimited freedom is obtained to expatiate in the sweetness and light of the English State-Church. Thank God, the clergy of the protesting Church of Scotland in 1843 did not content themselves with rendering it an insoluble or even an intricate problem to make them out to be honest men. They had distinctly defined their position, and the popular intelligence apprehended it; they said what they would do, and they did it. The morality of all ages—the instinct of veracity in the human breast—responds to the declaration of Chalmers, perhaps the noblest which ever came from his lips, that, if the majority in the Church had belied their position and remained in the Establishment, they would have made it a moral nuisance. A Church of Christ which would have perplexed the human sense of truthfulness, which would have shuffled out of its promises and nestled comfortably into its benefices, which would have given sanctimonious phrases and slippery evasions for plain deeds, would, God wot, have been a moral nuisance.

Strange to say, this declaration of Chalmers's has not only been misrepresented, but has been turned into something which, while retaining a resemblance to it in sound, is, as nearly as possible, its opposite. It has been alleged that he pronounced the Scottish Establishment, after it was left by the Free Church, a moral nuisance. Happily he had an opportunity, before the Parliamentary Committee on the refusal of sites to Free Church congregations, for exposing this malignant falsehood, and stating what he actually said. In point of fact, the mere circumstance that the battle had been bravely and honestly fought on both sides enabled the two sections of the dissevered Church of Scotland to regard each other, though with inevitable irritation, yet with a manly and soldierly respect. Dr Cunningham of Edinburgh and Dr Robertson of Ellon, the foremost clerical athletes in the controversial fray, were in their latter years intimate personal friends. No candid historian of the contest will refuse to admit that the genuine Moderate party, the inheritors of the policy of Robertson, occupied a position which upright men, of good intellectual parts, could occupy. They would not grant that they yielded the fundamental principle of spiritual independence as maintained by the ancient Scottish Church. Even in respect of the right of the people to have no ministers intruded on them, they objected, not to the claims of congregations in themselves, nor to the sympathy and co-operation of the Church with the popular efforts to give those claims effect, but to the means which the Church had taken to express her sympathy and enforce her co-operation,—to the want, to say the least, of deference to the civil law evinced in the enactment of the Veto and in the admission of the chapel ministers to full Presbyterial powers. There was, they argued, no intention on the part of the State to deprive the Church of the privileges recognized as hers in the Treaty of Union; but, in definitely bestowing, by the Veto, a power upon congregations to set aside rights of patrons which the State legally countenanced, and which could not practically be distinguished from rights of property, and in giving chapel ministers the status and authority of parish ministers, the Church, they maintained, had dealt arrogantly and domineeringly with the State, and the natural consequence was that the State put its back up against the Church. Accordingly the course prescribed alike by sound policy and by Christian temper was, they said, that the Church should rescind her obnoxious legislation, and take counsel with the allied State, in a friendly spirit, as to how the fundamental principles of her constitution could be carried out.

The reply to these arguments might be satisfactory, but it is impossible to deny that honest and able men could use them. The majority might say that, since the principles involved were confessedly fundamental, since the Church had done no more than guard them in what she deemed an effectual manner, and since she was willing to let the Civil Courts do what they pleased with the temporalities, the State was bound to ratify her proceedings; that persistence in the course she had taken was no rebellion, inasmuch as appeal was made to fundamental principles of her constitution, but that, if the Church retraced her steps and did the bidding of the Civil authorities, she would be held by all the world to have surrendered at discretion, and would, at best, possess her freedom in future by sufferance of the State. I humbly hold, with Chalmers and Cunningham and Hugh Miller, that the reply of the majority was conclusive; but the scales of argument were rather nicely balanced, and perfect respect could be accorded to every man of either party who was 'firmly persuaded in his own mind.' The essential point was that word should correspond to belief and that action should correspond to word. This is the foundation canon of all morals. Whether a man or a Church stands morally on a foundation of adamant or on a foundation of sand is determined by the answer to be rendered to the question whether said man or said Church conforms to this canon or does not. The founders of the Free Church did not 'palter with eternal God.' Whatever was the persuasion of their opponents, *they* believed that it was their duty to sacrifice emolument and position rather than submit to the Civil Courts. They conferred not, therefore, with flesh and blood, but did their duty. Their act was the noblest thing done in Europe in their time. This, I understand, is Mr Carlyle's estimate of the Disruption.

Among those who were assembled in Canonmills Hall to welcome the Free Church, the stalwart form and great shaggy head, and earnest, thoughtful features of

Hugh Miller, were particularly noticed. One can fancy how the fire would glitter in his moist eye, and the enthusiasm glow on his face, as he listened to words like these in the address of Chalmers:—'We read in the Scriptures, and I believe it will be found true in the history and experience of God's people, that there is a certain light, and joyfulness, and elevation of spirit, consequent upon a moral achievement such as this. There is a certain felt triumph, like that of victory over conflict, attending upon a practical vindication, which conscience has made of her own supremacy, when she has been plied by many and strong temptations to degrade or to dethrone her. Apart from Christianity altogether, there has been realized a joyfulness of heart, a proud swelling of conscious integrity, when a conquest has been effected by the higher over the inferior powers of our nature: and so, among Christians too, there is a legitimate glorying, as when the disciples of old gloried in the midst of their tribulations, when the spirit of glory and of God rested on them, when they were made partakers of the Divine nature and escaped the corruption that is in the world; or as when the apostle Paul rejoiced in the testimony of his conscience. But let us not forget, in the midst of this rejoicing, the deep humility that pervaded their songs of exultation; the trembling which these holy men mixed with their mirth: trembling arising from a sense of their own weakness: and then courage, inspired by the thought of that aid and strength which were to be obtained out of His fulness who formed all their boasting and all their defence.'

It is worthy of mention that the name 'Free Church of Scotland' appears to owe its origin to Hugh Miller. He had made use of it in articles in the *Witness* months before the Disruption. Calmly foreseeing that event, he had meditated profoundly on the position which the protesting Church would occupy, and the course she ought to pursue. Loving the Church of Scotland with passionate attachment, his grand anxiety was that the Free Church should be, in all respects save that of formal alliance with the State, the old Scottish Church. He dwelt earnestly on the necessity of separating her cause from that of political party, and letting it be known that she was to do the work neither of Whigs, Tories, nor Radicals. The strength of her foundation was to be the personal godliness of the Scottish people, and she was to seek no other basis. The counsel of Hugh Miller to the Church on this point was as wise as it was pious, and there can be no doubt that it contributed largely to secure for him that confidence of the devout, quiet, solid-minded laymen of the Church, which constituted a power rendering him independent of every political section as well as of its clerical partisans. 'There is a call in Providence,' he said, 'to the Church, that she dissipate not her powers in the political field.'

At great length, also, he insisted upon the importance of Free Church maintenance of 'the Establishment principle.' He looked upon that principle as an integral part of the British Constitution, and his reverence for the Constitution was such as would have been sympathized with by Burke or by Johnson rather than by modern Whigs. 'It would be as impossible,' he said finely, 'for mere politicians to build up such a Constitution by contract, as it would be for them to build up an oak, the growth of a thousand summers.' And in the Constitution he believed that 'the Establishment principle' was embodied, whereas 'the Voluntary principle' was not. The testimony of the Church of Scotland had, he declared, been 'for the Headship of Christ, not only over the Church, but over States and Nations in their character as such.' His experience had made him acquainted with secular voluntaryism rather than with the spiritual voluntaryism represented by such men as Dr Heugh of Glasgow and Dr Cairns of Berwick. According to these, the principle of a free Church in a free State, with clergy sustained by the free-will offerings of their flocks on the apostolic model, affords amplest scope for the recognition and performance, both by States and Churches, of their duty to Christ as King, and is therefore in strictest consonance with the spirit, if not with the letter, of the old Scottish testimony. Be this as it may, the fact must be clearly stated in a biography of Hugh Miller, that, at the moment when he saw what he considered the infatuated policy of the State rending the Church of Scotland asunder, he 'would fain press on every member of the Free Church the great importance of the Establishment principle.'

By joining the Free Church, however, he declared in the most forcible manner that the only Church, established or disestablished, to which he could adhere was a Church exercising every right of self-government. Unless 'the Establishment principle' meant that the State was to recognize and endow the Church, and the Church to arrange her affairs and exercise her discipline in subjection to the law of Scripture alone, he practically discarded it. All the rights of the Church —all that constituted her vitality as an institution—had, he held, been bought by Christ and had descended from heaven. Speaking as a theologian and a Churchman, John Henry Newman concentrated his rhetoric of condemnation, applied to the State-Church of England, into the phrase, 'a mere national institution;' and Hugh Miller, as he looked upon the spectacle of the Disruption, could say nothing more sternly sad of the State-Church of Scotland than that she had become a 'State-institution.' The difference between the Oxford doctor and the Free Church champion was this, that, whereas Anglican Bishops have since Henry VIII.'s time ordained *vice regis*, in the *person* of the king, thus recognizing the sovereign as sole fountain of spiritual authority for the establishment, the Scottish layman could point to the Treaty of Union, expressly exempting the Church of Scotland from the spiritual supremacy vested in the English Crown, and guaranteeing to her all those rights and privileges which a century and a half of terrible contending had secured her.

The importance of this distinction became evident in the sequel. Dr Newman left the Anglican Church more conspicuously a 'mere national institution' than before: the Free Church won not only her own battle, but the battle of the Scottish Establishment also. The dust of the conflict having cleared away, the men of character and power in the State-Church of Scotland, the Macleods, the Cairds, the Tullochs, and men of like stamp, took up the position of the old Presbyterian fathers, claiming spiritual independence for the Church, and doing their best to secure for congregations the election of their ministers. Nay more, the Court of Session has, by a recent decision, treated the right of the Church to exercise spiritual discipline in the suspension and deposition of ministers as unquestionable. Whether the law lords would have been thus generous, if the Established Church of Scotland had not become so modest, unassuming, and inoffensive that a wigged gentleman might safely pat her on the head, were perhaps an ungracious subject of inquiry.

That an event shall become of world-historical

significance,—that it shall be a grain cast into the seed-field of Providence to grow and bring forth fruit,—depends not upon its consequences being foreseen by those who take part in it, but upon its being, at the moment of its occurrence, true and right. It is the branch which sheds its leaves timeously in autumn that puts forth buds vigorously in spring. Sufficient unto every day is the work thereof, and no generation ever trod this world which could exactly measure the shoes of the succeeding generation. It always happens that the future is an

> 'Offspring strange
> Of the fond Present, that with mother-prayers
> And mother-fancies looks for championship
> Of all her loved beliefs and old-world ways
> From that young Time she bears within her womb.'

We have already ceased to be particularly moved by the watchwords of the Voluntary Controversy, or the Non-Intrusion Controversy, and the Constitution seems to have survived the disrespect shown to the Establishment principle in recent legislation on the Irish Episcopal Church: but the Disruption was a brave, resolute, and honest piece of work,—a triumph of veracity and pious courage,—and its fruits are following it. The principle of the headship of Christ over His Church, which the whole world heard it proclaim, may be slightly regarded in its theological aspect, but reasonable men acknowledge that, when interpreted into the right of Churches and religious denominations to manage their own affairs, it is one which liberates the energies of statesmen and lawyers for their proper duties, and is eminently conducive to the welfare of the State. Not one man in ten thousand knows or cares to know the exact difference between the Voluntary principle and the Establishment principle; but the Free Church has afforded the noblest demonstration the world ever saw of the power of a Christian Church to maintain an educated ministry throughout the entire extent of a country, and common sense drives home the conviction that this is the obvious, the natural, the expedient, and, from the statesman's point of view, the economical way of managing the business. There is not an institution of recent times from which the world has learned so much as from the Free Church of Scotland.

It is on the lines of principle laid down by the Free Church, that the remarkable union which has been proceeding for many years among the branches of the Reformed Communion throughout the world has taken place. The Reformed Church of the sixteenth century, —justly so-called because, in contradistinction to the Lutheran and Anglican Churches, which contented themselves with washing the dirty clothes of Rome and putting them on again, she went direct for her Constitution to Christ and His apostles,—has reunited her broken ranks in Australia and in America, and there are, as we have seen, none but circumstantial hindrances to her reunion in Scotland. Disunion on Christian principles —disunion at the bidding of conscience and duty—is the natural prelude to union on Christian principles.

In how far Hugh Miller would have sympathized with the recent movements for union in the Reformed Church, I shall not pretend to say. At the time when he wrote his articles extolling the Establishment principle, his views were those of Dr Cunningham and of others who subsequently shared in the impulse towards union. Hugh Miller was a man of definite opinions, and held them tenaciously; but he was not devoid of that capacity of growth, which is perhaps the ultimate characteristic of great minds. Such an intellect as his could not become a fossil, however exquisitely coloured and definitely traced might be the markings on it. As a man of science he kept the gates of his soul grandly open, and as a churchman and theologian he would assuredly have seen 'what main currents draw the years' and permitted the wind of Providence to fill his sail. More I cannot say.

There are some who think that, in devoting himself to the Free Church, Hugh Miller abandoned a higher career, which might have awaited him if he had given his undivided energies to science. To this view I cannot subscribe. To take part in founding an important national institution is a higher and more honourable service to humanity than to make discoveries in science or to write eloquent books. So long as the Free Church endures, she will bear upon her the image and superscription of Hugh Miller. She is his best monument,—a sound piece of building, creditable to the journeyman mason. All things considered, there was no one of her founders who played a more important part in rearing her than he. It was at his voice that the people of Scotland awoke to a perception of what they had to lose or gain in the Church controversy. 'There does not,' he said, 'exist a tenderer or more enduring tie among all the various relationships which knit together the human family, than that which binds the Gospel minister to his people.' No one proclaimed this during the conflict so powerfully as Hugh Miller, and it was his proclamation of this, with accompanying exhibition of the fact that all the tenderness and all the sacredness of the relation between pastor and people were at stake in the conflict, which tended more than aught else to fill the courts of the Free Church with the acclaiming voices of the people of Scotland.

CHAPTER V.

DIFFERENCE WITH DR CANDLISH.

NO company of earnest men in this world,—whether good or bad, whether as imaginatively delineated by the true poet, or as described by the historian,—from Achilles and Agamemnon to the Apostles Peter and Paul, —have ever been able to agree. An agreeable man— the definition is Mr Disraeli's—is one who agrees with you, and the talent of agreeing all round is precisely that in which men of original and decisive character, keeping a conscience, are apt to be defective. Hugh Miller, as we have had occasion to see, had his own convictions as to what the Free Church, and the *Witness* as the principal representative of the Free Church in the press, ought to be. He had not derived these from man, and there was no man living at whose beck he would change them. On two points his mind was made up with an impassioned emphasis of determination, first that the *Witness* should be kept independent of party, and, secondly, that there should be no entering into league with 'Voluntaries' for the purpose of assailing the

Established Churches. He was resolved, also, that the *Witness*, under his management, should have a high character as an intellectual newspaper, by no means confining itself to ecclesiastical topics, but making wide incursions into the realms of literature and still more into those of science.

We have seen the active part taken by Dr Candlish in bringing Mr Miller to Edinburgh. In his sketches of the General Assembly of 1841, the latter expressed enthusiastic admiration for Dr Candlish, and for some time the relations between them were most cordial. At what date they began to differ it were bootless to inquire, but no serious discrepancy seems to have occurred in their views of principle and policy until after the crisis of the struggle and the formation of the Free Church. Three years after the Disruption, Miller felt so deeply aggrieved by the proceedings of Dr Candlish, that he addressed a letter to the Committee of gentlemen who had from the first interested themselves in the paper, repelling the attacks of which he believed himself to be the object. The document contains every particular with which it is necessary for the reader to be acquainted, and the only preface which it may require is a statement of the reasons which have induced Mrs Hugh Miller, her family, and myself to publish it.

I. Although it was formally private and confidential, Mrs Miller has a distinct recollection that Mr Miller always held himself at liberty to make it public, and did not look upon that contingency as improbable.

II. Mrs Miller and her family are convinced that, since the vindication of Hugh Miller can be trusted to no pen but his own, the publication of this letter is necessary in order that justice may be done to his memory.

III. It is widely known that such a letter was written, printed, and sent; and were it either withheld, or tampered with in any way, this biography would not secure or deserve the confidence of the public.

IV. Several copies of the letter are certainly known to be in existence, besides those possessed by Mr Miller's family. Its publication, sooner or later, therefore, may be considered as certain; and it seems best that it should be published by those who are specially bound to guard the good name of Hugh Miller.

V. As rumour magnifies all things, the fact that an important document written by Mr Miller against Dr Candlish was withheld in this biography would lead suspicious and scandal-mongering persons to suggest that the disagreement between the two men was of some specially dark and terrible character. It is, therefore, best that the simple truth should be stated.

The copy of the letter from which it is now printed is that which was sent by Hugh Miller to Dr Chalmers. Many of its paragraphs have vehemently assenting pencil marks on the margin, and Chalmers adds one very brief but characteristic note, which will be found in its place. The 'send back the money business,' alluded to by Miller, was the noise raised by sundry enemies of the Free Church because, when the Churches of the Southern States of America sent her a civil greeting and a contribution of money, she did not requite the former with a snarl and throw back the latter in their faces.

[*Strictly Private and Confidential.*]

LETTER

FROM

HUGH MILLER,

EDITOR OF THE 'WITNESS,'

TO

DR THOMAS CHALMERS; DR ROBERT GORDON; DR R. S. CANDLISH; DR WILLIAM CUNNINGHAM; DR JAMES BUCHANAN; THE REV. THOMAS GUTHRIE; THE REV. JOHN BRUCE; THE REV. JAMES BEGG; THE REV. ANDREW GRAY OF PERTH; THE REV. ALEXANDER STEWART OF CROMARTY; ALEXANDER DUNLOP, ESQ.; ROBERT PAUL, ESQ.; J. G. WOOD, ESQ.; AND JAMES BLACKADDER, ESQ.

'GENTLEMEN,

'IT is with much pain, and not without much serious deliberation, that I avail myself at this time of this peculiar mode of addressing you. My communication, however, though a printed one, is strictly confidential: the gentlemen named above are the only individuals who receive copies; and such has been the amount of precaution taken in passing it through the press, that it is even more, not less, private in its present form, than if it had been simply engrossed by a clerk.

'In one certain quarter the *Witness* Newspaper has of late given extreme dissatisfaction. Influential gentlemen of the Parliament House, and gentlemen whom the influential gentlemen influence, have indignantly thrown it up; and such has been its effect on the nerves of Dr Candlish, that he never opens it now "without a feeling of dread lest something untoward should be in it." The Papers of the editor evince, the Doctor is clear, an entire want of delicacy,—there is no "taste or tact" shown in the handling of public questions,—and "the damage done by certain recent editorial articles to the Sabbath cause is," he holds, "incalculable." In short, a great crisis has arrived in the history of the Paper; and Dr Candlish, to prevent its rapid sinking, can bethink himself of but one expedient: the individual who now addresses you must retire forthwith from his place as Editor of the *Witness*, to make way for an Editor of taste, tact, and delicacy,—a gentleman connected with the Parliament House.

'Now, while I entirely agree with Dr Candlish that a crisis has arrived in the history of the Newspaper with which I have been so long connected,—may I not add, identified?—I differ from him very considerably regarding the means best suited for obviating the danger. On the policy of my removal from the Editorship I am, of course, incapacitated from deciding. It may be the best possible policy in the circumstances, and yet, from the disturbing influence of personal feeling, may seem bad policy to me. But on some of the other points my judgment is, I think, less open to bias. The Doctor communicated his scheme to my co-partner in the Paper, Mr Fairly, whom it took somewhat aback; and who, not exactly seeing at the moment how much it involved, or perhaps with some tacit reference to my standing as a *bonâ fide* proprietor in the concern, asked him whether it should be the new Editor or the old whose decision should be the final one, were we to come to differ on some point of principle or policy? The matter, replied Dr Candlish, might be referred to a Committee. There are thus, obviously, two elements in the Doctor's scheme,—the element of a Parliament-House Editor, prepared, doubtless, by the peculiar practices of his profession, to take his clue from his clients; and the element of a controlling Committee; and either of them would, I am convinced, be fatal to the *Witness.* It is with great diffidence that I in any case oppose my judgment to that of Dr Candlish; but in this matter I possess peculiar opportunities of information. The *Witness* has been not merely an organ through which I have communicated with my readers, but it has also led my readers to communicate much with me. I have received many a hint and suggestion in many a note and letter, signed, initialed, and anonymous; and have thus felt, as with a stethoscope, how the heart of the Free Church has been beating outside the disturbing influences of Edinburgh society. And it is in a large measure from the knowledge thus acquired—though I am not without other sources of information—that I derive my convictions regarding the ineligibility of the scheme of improvement suggested by Dr Candlish. The Doctor seems convinced that the *Witness,* as at present conducted, is in imminent danger of sinking. I will stake its life any day against that of a *Witness* redolent of the peculiarly unwholesome atmosphere of the Parliament House, and under the direct control of a Managing Committee!

'I must speak my mind on this point with the full freedom which the urgency of the crisis demands. And, first, let me indulge in a few general observations, which may, I trust, be found, ere I conclude, not much beside the mark.

'I remark, then, in the first place, that the destruction, or even very considerable diminution, of the Sustentation Fund, would be one of the gravest evils which could befall the Free Church. The Fund constitutes the great heart of her material frame,—the all-important muscle, if I may so express myself, whose propelling energies preserve, as in the body natural, the distant extremities from torpor, gangrene, and death. And whatever threatens to injure it from without, or to affect it as with disease from within, is to be guarded against with the most sedulous care. The central habitat of this institution in Edinburgh must of necessity connect, in an especial manner, with its management and control, the Edinburgh Free Church leaders; and the soundness of its condition, and its consequent measure of prosperity, must in the future very much depend on two classes of circumstances,—on the conduct of these men, and their degree of independence of suspicious secular influence; and on the amount of confidence reposed in them, and of cordiality entertained for them, by the ministers and people that occupy the Church's great outer area,—all of Scotland that is not the capital.

'There is one important respect in which the circumstances of the Free Church differ most materially from those in which the members and ministers that compose it were placed previous to the Disruption. There has ever existed in the true Church of Scotland,—that Church which has ceased to exist as an Establishment,—a sort of Presbyterian jealousy, not unwholesome when under the influence of right principle, of what our country clergy used to term the 'big wig' influence, ecclesiastical and lay. And how often has not the perfect parity of Presbyterian ministers been insisted upon by the rasher spirits, even in a sense in which parity does not exist! In the most important sense possible Presbyterian ministers are *not* equal. Their equality is merely an equality of the *letter*, appointed, it would seem, in order that the true episcopacy, which is of the *spirit*, may have room to develop itself, and that the Church may be benefited by the graces and talents of her true bishops, the good and great men whom God has given her. But even the jealousy which clung thus to the dead letter of equality was, when confined within due bounds, a thing not to be greatly regretted. Our Church is very differently circumstanced now from what it was when the frame-work of the Establishment sheltered its ministers, and rendered them independent in their secular affairs, each of all the others. The institution of the Sustentation Fund is an institution new to a Presbyterian Church; and there is the profoundest prudence and discretion demanded, in order to accommodate it to the Presbyterian genius and character. And, further, the Free Church has, what no Dissenters ever had, the Highlands of Scotland, where there are not a few parishes entirely with us which will too certainly never be self-sustaining ones. There is thus a centre of comparative wealth in the body, and a large circumference of dependency. And to the influence of central wealth there is, besides, added a centre of official management, with a certain amount of patronage attached. Wesleyism has also its centre of official management, and its outer circumference of poor non-sustaining Churches. But the Wesleyism of England—an offshoot from English Episcopacy, and still, to a considerable extent, marked by the Episcopal character—is essentially different in its genius from old Scottish Presbytery. Even among the Wesleyans, however, a reaction on the centralization system has produced the New Connexion. A reaction on the analogous tendency in the Free Church would be a very different affair indeed. Among the Wesleyans the reactionists are persons infected by modern notions and feelings, and, in many cases, singularly unsound in their creed; while among us the reactionists would, on the contrary, be men of the long-derived historic type and sternly orthodox school. The reaction in the one case is that of new liberalism against *old* centralization; in the other it would be a reaction of old Presbyterianism against *new* centralization.

'Now, I need scarce remark to any of the gentlemen whom I have now the honour of addressing, that the development of a spirit of this character would be one of the gravest evils which could befall the Free Church. It would be found mainly operative in those large and most important tracts of country in which the congregations are self-sustaining and a little more, and which lie between the outer circumference of necessity and the inner focus of influence. And if once developed, it is all too obvious that it might come to operate with most disastrous effect on the Sustentation Fund, and, by consequence, on the very existence of the Church in the non-sustaining localities. And that this spirit exists in a semi-latent state, I have the most direct means of knowing. Sorley's five-edition pamphlet was a mere straw thrown up by a man of infirm temper, in a moment of excitement; but, straw as it was, it indicated the direction of a current. And be it remembered, that we are tending, in the natural course of events, to a state of things that must be greatly more favourable to the development of the jealous spirit than the present. The great and good man who now imparts lustre to the Free Church by the splendour of his name, and ballasts it by the moral weight of his character, is far advanced in years; nor can the eye rest on a successor who will be to it what he has been. The men of the Disruption will be dropping away, and leaving their places to others less under the influence of the kindly feeling of a soldiery who have fought side by side in the same battle. And as they are removed and disappear, the scheme of unequal incomes will be coming into effect. But above all, in an age in which there are great religious questions arising among the governed classes, and a singularly temporizing spirit manifested with regard to them by the classes that govern,—a suspicion on the part of the Church that the central body had in any degree lent itself to party purposes, or was desirous, for party ends, either to misdirect or neutralize the Church's proper influence, would be of a nature peculiarly perilous, reactive, and disorganizing. Nothing can be more certain than that, unless all the greater care be taken, the Parliament House may prove a most unneighbourly neighbour to the Sustentation Fund. There is much in prospect, of a character fitted to foster the natural Presbyterian jealousy of centralization; and I must be permitted to say, that the various schemes suggested during the last three years for the improvement of the *Witness* have been all of a kind suited to add to their effect. The several courses recommended of steering the Paper have all been courses that bore direct upon the rock.

'It is now about two years and a half or so since Mr M'Cosh of the *Northern Warder* suggested to Dr Candlish a scheme of centralizing the Free Church Press. I am unacquainted with the details of the suggestion, but it at least involved the control of the *Witness* by a central Edinburgh body. The scheme met, I am informed, Dr Candlish's approval; and, with the intention, apparently, of carrying it into effect with regard to the *Witness*, I was waited on, at the time, by Mr J. G. Wood, and sounded as to whether it would be agreeable to me that Mr M'Cosh should have a share given him in the Proprietorship of the Paper, and to have him at the same time as my Sub-Editor. I may mention, that I had not yet become one of the proprietors of the *Witness*.

The proposal did not take my fancy at all. I did not like the centralization idea; nor could I realize what my position as Editor would be with a Sub-Editor who, as a Proprietor of the paper, would, of course, be one of my employers, and be perhaps backed, to boot, in his centralization views, by the influence of Dr Candlish. And as the time had not yet come for proposing that I "*ought* not to be Editor" of the *Witness*, the design was suffered to drop.

'What, however, I most dreaded in the scheme was, I must say, the direct control of some Edinburgh Committee, and the inevitable effect of the arrangement on the centralization jealousy outside. It is of great advantage to a purely political newspaper, that it should be the accredited organ of the Government of the day. The Ministry, as such, is a standing part of the Constitution,—it is the Monarch in his responsible servants; and its choice of a newspaper, as an organ through which to speak, imparts to the Paper selected not a little of that weight which the Treasury must always possess. But it is not according to the constitution of Presbytery that it should have a central government, analogous to a King's Ministry, and armed with a sort of secular Treasury influence. Nor, were the peculiar necessities of some Presbyterian Church to demand that a body of the kind should be called into existence, would it be desirable or prudent to put into its hands any such control over the newspaper press of the body, as that which a King's Ministry possesses, and ought to possess, over its organ. And, further, were the true Presbyterian pulses to beat vigorously in the body, and were its Presbyterianism to be of the old earnest school, that man might be well deemed over-sanguine who would calculate largely on the influence and success of a press so controlled and directed. Mr M'Cosh's centralization scheme would be fraught with danger, not only to the Free Church from its inevitable tendency to foment and exaggerate a feeling to a certain extent inevitable in the Church's peculiar circumstances, and which every means should be taken to sooth and allay, but with danger also to the press which it would centralize. It is peculiarly unsafe to fasten an undecked skiff to a central pier in a tideway, against which it is the tendency of the current to run strong. Let the stream but reach a certain momentum, and the skiff is sure always to founder at its moorings.

'I may remark in the passing, that as the *Witness* is already tied down to principle by legal document, the arrangement suggested by Mr M'Cosh could have but the effect of tying it down to policy also. So long as the Paper rests on its present independent basis, the Editor stands alone. He is an insulated individual, possessed of no other influence than what he owes to the degree of confidence reposed by his readers in the integrity of his principles or the soundness of his judgment. He is not the mere convenient pendicle of a centralization scheme; he is a felt check on it rather; and, considering the nature of the danger, not wholly useless in that capacity. If he take a false step, the error is that of an individual; and its effects, if it have any, are easily neutralized. But a false step taken by a newspaper which represented a central body of management would be in reality the false step of the body itself, and, in consequence, a greatly more serious affair. What, for instance, would be the position of the *Witness*, if thus tied down, were some such controversy as that of the year 1837 to arise now?—a controversy in which a great majority of the central Edinburgh body would take decidedly the one side, and at least nineteen-twentieths of the entire Church take as decidedly the other.

'The scheme next originated for improving the *Witness*,—a scheme of uniting into one Edinburgh paper both the *Witness* and *Guardian*,—was still more directly fraught with danger; and the fact that it should be at all entertained shows how thoroughly the disturbing influence of the sort of cross eddy which Edinburgh society in reality forms, prevents some of our leading Free Churchmen, who live much in the whirl, from taking note of the current outside. What would be the first effect of an union of the two Papers? At present the proceedings of the Edinburgh Presbytery are reported at great length in the *Witness*, and those of the Glasgow Presbytery in the *Guardian*; but in no single paper could both be published in a manner equally full. Even had the Edinburgh Paper a reporter at Glasgow, for the special purpose of giving the Glasgow Presbytery as much attention as it now receives, no set of newspaper readers would bear, after they had perused some three or four columns of the proceedings of the one Presbytery, to set themselves to peruse some three or four columns more of the proceedings of the other. The Glasgow Presbytery would assume, in consequence of the curtailment in its reports, unavoidable in the circumstances, a secondary and provincial character; and one mighty step towards the dreaded centralization would be accomplished. The Church would come to have its Archbishop Presbytery, associated with its centre of secular influence. Such was the view which I could not avoid taking of the intended union of the *Witness* and *Guardian*. I had received very direct information regarding the matter, but it had not been formally proposed to me; an union of the *Scottish Re-*

cord and *Edinburgh Evening Post* was opportunely taking place at the time; and I wrote such an article on the sort of marriages through which expiring newspapers get into a state of coverture and disappear, as rendered it impossible that, without the most ludicrous inconsistency, the *Witness* and *Guardian* should ever be united.

'I may mention, that I received about this time two several visits from Mr Sheriff Monteith. What the Sheriff had to suggest on these occasions was a good deal veiled in that peculiarly delicate kind of phraseology, somewhat professional, which has to a simple man the effect of casting a misty obscurity over the subjects which the words envelop. On thinking, however, with all my might, over his proposals, I came to discover that they embodied a scheme of salaried censorship for the *Witness*, devised, it would seem, for the express purpose of keeping the editorial articles right; and I straightway intimated, in a note to Mr J. G. Wood, that what I had succeeded in seeing of the business I did not at all like. The judgment of the proposed Censor might, I thought, be scarce worth a hundred a-year to the *Witness:* it might be even a judgment not a great deal stronger or more honestly directed than my own, save and except, of course, in the matter of Whig elections: and, besides, if the Censor and I came to differ, we might quarrel. After communicating with Mr Wood, I heard no more of the design, nor did I receive, though I have experienced some kindness at the Sheriff's hands, any after-visits from the Sheriff. It is perhaps scarce worth while adverting to this circumstance; but it may probably have some bearing on the present state of matters, and there can be no doubt that it indicated at least one incipient attempt more to improve the *Witness*.

'The next scheme of improving the *Witness* was

originated a few months later, and it was on this occasion that I became one of the Proprietors of the Paper. I undertook to repay, by instalments, the thousand pounds originally advanced by the subscribers to Mr Johnstone, with the interest, year by year, of the unpaid portion in hand, until the entire debt should be extinguished. It had at first been proposed to make the thousand pounds over to me as a boon ; but I was startled by some of the conditions attached, and insisted that the terms of the transaction should be those of ordinary business. Strange enough, however, though the wish was acceded to, the terms were not softened ; and the document drawn up for the signatures of the Proprietors by Mr J. G. Wood appeared so strange, even to my unpractised eyes, that I requested my partner, Mr Fairly, to submit it to his man of business. I found I had not mistaken its character ; —the man of business had never seen such a document before. And so a somewhat different sort of document had to be drawn up by another hand ; and it is this second document that defines the existing relations of the *Witness* Proprietors and the Committee. Nothing certainly could be more strangely despotic than the terms of the first. To one of these I must specially refer : it would have so stringently fastened down upon the Proprietors, Mr Somers of the *Scottish Herald*, in the capacity of Sub-Editor, that were they subsequently to become dissatisfied with him (and he was an untried man at the time), they could not have shaken him off. His position was to be rendered legally independent, both of the Editor and the Proprietors of the *Witness*. I shall not here inquire whether Mr Somers's after-management of the Paper quite justified the wisdom of the position which Mr Wood, in the behalf of his clients, would so fain have assigned to him ; but I may surely take the liberty of asking, whether a standing so independent of the parties within the *Witness* establishment did not bear a too obvious reference to the exercise of an influence from without, and to the favourite centralization scheme? I may be permitted to ask further, as the "tact, taste, and delicacy" of Mr Somers did not prove quite what had been expected, whether Dr Candlish's proposed Parliament-House Editor be a man known in the republic of letters, or one who has yet to give his first proof of literary ability? I must speak my mind freely;—I am on my trial; and, be it remembered, that I address myself not to the public at large, *a la* Sorley, but to a select few of the wisest heads and soundest hearts of the Free Church. I injured no one by the stand which I made on this occasion; had I not made it, I would have greatly injured both my partner and myself, and degraded myself to a position which no man of a right spirit could for a moment submit to occupy. I injured no one; but I must say that the part I acted seems never to have been pardoned me. Are there men among us who, though they have perhaps principle enough to forgive an enemy, have not magnanimity enough to forget a disappointment? *

* 'The following extract from an outline, by our man of business, of Mr Wood's document, may serve to illustrate its character:—

'"Clause third provides that arrangements shall be made for the copartnery publishing the *Scottish Herald*, or some similar weekly newspaper in connection with the *Witness*, at the present price of the *Herald*, viz. threepence, and to engage the present Editor of the *Herald* as its Editor, and Sub-Editor of the *Witness;* and from this obligation the Company cannot be relieved, until they shall have satisfied the Committee that the undertaking cannot be continued without certain loss to the Proprietors of the *Witness*, after a trial. The Committee's consent is also declared to be necessary in dispensing with the services of the Editor of the *Herald* as Sub-Editor of the *Witness;* and the copartnery are again taken bound, *at the sight of* the Committee, to allow a good and sufficient salary for the Sub-Editor of the *Witness;*

'I became, I have said, a Proprietor of the *Witness* on ordinary business terms. The printing material, in

thereby investing the Committee with the absolute power of determining what is 'a good and sufficient salary,' without regard to the Company's idea of the matter. Another obligation laid upon the Company is, that in the event of their obtaining the consent of the Committee to discontinue the *Herald*, they shall be bound to permit the Editor of that paper to carry it on upon his own account, if he shall be disposed to do so. [At page 14] it is declared that, for the better regulation and carrying on of the said business (of printing and publishing the *Witness* and *Scottish Herald* newspapers), the Company 'bind and oblige themselves to implement and fulfil the whole conditions of the above-quoted minute, and to make payment of the penalty of £500, in the event of their not implementing the whole or any one of said conditions.'" And the following is the opinion given us for our guidance by our man of business, regarding the legal consequences of the entire document :—" The effect of such a contract would be to divest the copartnery of every vestige of control over their own business, and to throw it wholly into the hands of the Committee ; whilst the only conditions attached to the loan of the £1000 to Mr Johnstone were, those which provided for the maintenance of the principles of the Paper and for the repayment of the loan."

'I must further mention here, that to these strange terms I had strongly objected by letter, in Mr Fairly's name and my own, ere they had yet been embodied in the document, and existed but in a Minute, said to be that of the Committee ; and that Mr Wood at once sustained my objections as just and reasonable, but proceeded, notwithstanding, to frame the document founded on the Minute, without intimating to me his intentions on the subject, as if no such objections had ever been made. The following paragraphs form part of my letter :—

'" And, first, with regard to the *Scottish Herald :* We find it stated absolutely that it should be published at its present price. Now, it occurs to us that it could be only necessary to do so, so long as there is a competing paper in the field at the same price, or so long as the competitor does not advance *his* price. Suppose Harthill's paper were to come down, just because its price proved not remunerative, the clause binding us not to advance ours might, in such circumstances, have but the effect of bringing the *Herald* down too ; and this at a time when the removal of our rival gave us a chance of keeping it up by an advance of price.

'" What we desiderate, in the next place, is, that we should not be held bound to keep up the *Herald* unless it proved a fair speculation, in a business point of view. Profit we do not expect, but an unvaried return of loss we cannot bind ourselves to incur. A simple fact may place the matter in a strong light. Within the last quarter of a century no fewer than eleven papers have been started in Edinburgh, and which Mr Johnstone had invested the £1000 originally advanced him by the subscribers for the purpose of starting the Paper, had so far deteriorated, that its value, at the time I entered on the proprietorship, was estimated, on the principle usually adopted in such cases, at some-

of these eleven only one—the *Witness*—has succeeded. The attempts to establish the others have all ended in disappointment, and have entailed on the proprietors losses in each instance of from a thousand to fifteen hundred pounds. But how much greater would the losses have been had the parties been taken bound to maintain their papers, whatever the amount of loss sustained! Some of them could not continue to be published at a defalcation of less than from eight to ten pounds per week. Now, the *Herald* may possibly turn out such a speculation as the one Edinburgh paper that during the last twenty-five years has succeeded, but it is as possible it may turn out such a speculation as the ten that have not; and what Mr Fairly and myself ask, is simply not to be bound to keep it afloat perforce, should it prove unable to float itself. We promise to give it a fair trial; but should it refuse to live on its own means, as an honest paper ought, it must just be permitted to die.

'" In the third place, we are not sure that Mr Somers's name should have been included in the Minute. He is yet in a considerable degree an untried man. He may turn out to be all we wish him, or he may turn out something else. Suppose a great majority of the subscribers were by-and-by to be dissatisfied with his mode of conducting the sub-editorial department of the *Witness*, or the editorial department of the *Herald!* Or suppose we were to find him too much a Radical, and impracticable to boot! His name in the Minute, in these circumstances, might possibly be found a rock a-head. There could be little danger of our breaking with him so long as we found the arrangement a good and beneficial one, though his name were not in the Minute; and the introducing it there may have merely the effect of embarrassing, should the arrangement prove not good or beneficial."

' Such were some of the objections which were held reasonable, and yet suffered to weigh nothing. The Minute on which Mr Wood founded his document was said, as I have stated, to be that of the *Witness* Committee; and the Committee have now an opportunity of knowing whether it was in reality such. The members are, Dr Cunningham, Dr Candlish, Mr Begg, Mr Dunlop, Mr Blackadder, and Mr Wood. I may mention, that none of the meetings on this business were attended by Dr Cunningham; and Mr Dunlop I saw but at one meeting, and that for only a very few minutes. I may further mention, that the names of both these gentlemen were added to those of the Committee, as reappointed on this occasion, at the express request of Mr Fairly and myself,—Dr Cunningham at Mr Fairly's, and Mr Dunlop at mine.'

what under nine hundred. I undertook, however, on its being made over to me by Mr Johnstone, the whole of the original debt; and of course the money-interest of the *Witness* Committee in the Paper ceased immediately on my becoming bound to pay them the sum invested in it by the subscribers. It was suggested, however, that a certain yearly sum should be put at the disposal of the Committee, for securing occasional contributions to the Paper, in addition to the editorial ones; and, regarding the gentlemen that composed it as thoroughly identified with Free Church interests and feelings, I at once did what the mere man of business would have declined to do,—acquiesced in the proposal. I, however, did not at all like the form which it took in Mr Wood's document; and, in consequence, submitted to him, in the letter already referred to in my note, the following statement respecting it:—

'With regard to the second article in your minute, Mr Fairly and myself feel that we cannot treat with it until it takes a more definite form. It would be desirable, we think, that a sum for occasional contributions should be specified, which the Committee could in no case exceed, and that the said sum should be subject to a sort of sliding-scale reduction, proportioned to any fall that might take place in the profits of the Paper. There are Papers in Edinburgh which, though they have existed for many years, and were at one time highly remunerative, yield no profits now, and can scarce pay an Editor. The ———, for instance. . . . . And yet, at one time the ——— was a piece of more valuable property than the *Witness* is now, and could have afforded more for occasional contributions. But what would be its present condition, were there an outlay nearly equal to the Editor's salary rendered imperative on the Proprietors, without reference to existing circumstances? I have suggested to Mr Fairly a scheme of arrangement in this matter which satisfies him, and which, I trust, you will deem reasonable. He tells me that the profits of the *Witness* for the present year promise to amount to six hundred pounds. Let us assume six hundred as the average profits, and fix the sum to be set apart for occasional contributions as equal to the one-sixth of these, i.e. one hundred. Should the profits sink, let the sum set apart sink in an equal ratio. Let a profit of three hundred pounds per annum yield, for instance, only fifty pounds for contributions; but should the profits sink to two hundred, I trust you will deem it but fair that the claim should be extinguished altogether, and that the Paper, thus reduced to the level of what are the lower orders of Scotch newspapers in a business point of view, should be restricted, like them, to merely the usual newspaper staff of Editor, Sub-Editor, and Reporter. Just one word regarding the occasional contributions themselves. I trust it is unnecessary for me to stipulate, that I must possess over them, in every instance, the proper editorial control. I shall pledge myself to exert it on no occasion, without being prepared to yield a sufficient reason to the Committee, whether for curtailment or rejection.'

'I regarded this proposal as fair; the mere man of business would have perhaps deemed it something more; Mr Wood at once admitted it to be at least nothing less. The next time, however, that his document was submitted to me as altered and modified, I found it did not embody the proposal which he had apparently deemed so reasonable; but that, without having made a single objection to the suggested proportion of a one *sixth* part of the profits being set aside for the occasional contributions, he had quietly inserted in his document a one *fifth*. And then finding myself treated, not as a man who would fain have dealt with friends whom he trusted, in a generous and liberal spirit, but as a mere simpleton, who, having already yielded much, might be expected to yield more, I indignantly broke off all negotiation in the business, resolved that, if the terms of Mr Wood should turn out to be those of his clients the Committee, to receive at the hands of such a body no share in the proprietorship of the *Witness* on any conditions whatever.

'Some short time after I had arrived at this determination, I waited on Dr Candlish at the Doctor's own request, and we talked over the business alone together in his study. At this distance of time I can give but the purport of the conversation,—not the words. I immediately opened to him my intention of demanding of the Committee whether the terms of Mr Wood's document were in reality theirs, and if so, to have done with them at once. I was taking, said the Doctor, much too serious a view of the matter. Wood was my friend, and a good fellow; but, of course, a lawyer, who did things in the stiff, strict, professional style. His conditions were really nothing,—it was with friends that I had to deal. I stated to the Doctor the terms which Mr Wood had in so strange a style rejected, and asked him if they were not fair. Oh, certainly, he replied; nothing can be more fair; but the difference between you and Wood is a mere misunderstanding, and it will never do to let it come to a serious quarrel, by laying it formally before the Committee. Now, of course, in all this Dr Candlish may have acted simply the part of a peace-maker between Mr Wood and me, irrespective of the merits of the point at issue. It may have been a perfectly indifferent matter to him, on at least his own account, whether I submitted the matter to the Committee or no. He perhaps sent for me on this occasion, and borrowed time from his sermon to converse with me (for 'twas on a Saturday), on Mr Wood's account alone. At all events, in consequence of what he recommended, I did not appeal to the Committee; I merely submitted, as I have already said, Mr Wood's document, through Mr Fairly, to a man of business, instructing him at the same time to draw out another, on what I deemed more reasonable terms. The man of business did so; and his paper, on being submitted by Mr Wood to Mr Dunlop for his professional opinion, was pronounced by the latter gentleman to be a not unfair embodiment of the intentions of the Committee. It is with great pain that I have thus adverted to a transaction, the remembrance of which still lies heavy upon me; but circumstances demand that the whole truth should be known to the select few whom, in now addressing, I constitute my judges. I sincerely trust there are no such arts employed elsewhere in the management of the more secular concerns of the Free Church, as those which I found all too palpable in this intended settlement of the *Witness*. They are arts that can lead to but doubtful advantages and damaging exposures; they form but short-lived substitutes—hurtful in their life and offensive in their death—for that broad and massive simplicity, profound in its wisdom and immortal in its nature, which leads in a way everlasting. And, further, they too strongly savour of those wily diplomatists—not of the Church, but of the world—whose achievements and character one of the English satirists has so pointedly described,—

> "Men in their loose unguarded hours they take;
> Not that themselves are wise, but others weak."

To one very agreeable incident connected with this affair I cannot deny myself the pleasure of referring. A report had gone abroad, I know not how, that I had differed on business matters with certain individuals connected with the *Witness*; and a gentleman whom I scarce at all knew, and whose sympathies had been awakened in my behalf on purely literary and scientific grounds, came at once forward, and generously offered, through the medium of a common friend, to place a thousand pounds at my disposal.

'I may just mention, ere I pass on to something else, that there has been no call made on the part of the Committee for money with which to enter the literary market and purchase articles.

'The fifth and last scheme of improving the *Witness* is that which has proved the occasion of this letter. For the special benefit of the Paper I am to retire, it seems, from the Editorship, and a Parliament-House Editor (with perhaps a Committee of Control in the distance), is to take my place. A word or two here on the Parliament House, and the peculiar position of the Free Church portion of it. In all my remarks, however, I must be held to except at least one Edinburgh lawyer. I need not name him. It may be enough for me to say, that of almost all the men I ever knew, he possesses a mind the most exquisitely disinterested and honourable. The step which I have now taken will, I fear, greatly grieve, mayhap offend him,—a consideration which would have so weighed with me against it, could any consideration not founded in absolute principle have done so, that I would have adopted, for the sake of his friendship alone, a more timid and submissive course.

'The adherents of the Free Church may be safely divided into two classes, the people who belong to the Parliament House, and the people who do not; and in at least matters political, when Whiggism is in office, there is, I am afraid, no such thing as urging the duties of the one class without running counter to the interests of the other. I need scarce say, that there exists no body of men among whom political feeling is more vivacious than among our Edinburgh lawyers; nor that peculiarly on the legal profession in Scotland the dew of Government patronage descends, like the dews of heaven on the fleece of Gideon, when all often is dry around. For almost all the higher civil appointments of the country, gentlemen of the law alone are eligible; and it must be admitted at least as a fact, that of these our Parliament-House friends have, in proportion to their number, got a competent share. Now, be it remembered, that these legal Free Churchmen occupy at the present time, with reference to the Whig Government, a singularly important position. The Free Church, ever since the Disruption, has been giving significant indications of her strength; and, fairly broken loose from the moorings of party, she is in circumstances either considerably to damage or considerably to strengthen the cause politic which she may oppose or adopt. And so, if the Whig lawyers possess much influence with the Free Church, they must of necessity, on that account, possess much influence with the Government also. For through them exclusively, and some one or two Free Churchmen high in rank, can the Whigs alone expect to manage the Free Church, either by rendering it subservient to Government designs, or by neutralizing its influence against Government measures. The only other mode of conciliating our Church is a mode which the present Ministry are obviously not to adopt. They might endear themselves to us on the broad ground of principle; but the withdrawal of the Sites Bill,—which certainly would not have been withdrawn were the Whigs, as a party, disposed to support it,—demonstrates too surely that that course they are not prepared to try. I can well understand how, with reference to these grounds, the *Witness* should be an exceedingly obnoxious paper to our Whig Free Churchmen of the Parliament House, and how it happens that some of the more influential among them have of late indignantly thrown it up. Whigs as they are,—and to Whiggism Toryism is naturally opposed,—it must be more decidedly a rock a-head than all the Tory newspapers of the kingdom put together. It comes between them and the Free Church people. It abides firm by the Old Whiggism, but it warns its constituency against the designs of the New. It prefers Protestantism to Macaulay; and might be found exerting a damaging influence on the coming elections. Hence a large amount of Parliament-House criticism on a want of "tact and taste" in the Paper; its views are narrow and its spirit and policy bad. The real cause of condemnation cannot be assigned, but condemned it must be.

'There are no doubt ingenious gentlemen among the Free Church lawyers of the Parliament House, and it is peculiarly their vocation to plead causes; but in this special cause, should the annoyance of insinuation and inuendo continue, I shall not hesitate to encounter them one and all. It is, I trust, in no proud or boastful spirit that I say so; and that my confidence is rather in the character of the men whom I shall have to address, than in any peculiar ability of my own. I shall have the ministers and the people of the Free Church of Scotland for judges and jury in the case,—a class not particularly remarkable, mayhap, for "taste, tact, or delicacy," but certainly not wanting in the characteristic shrewdness of their country, and most assuredly not devoid of solid principle. If compelled to lay open the policy of our Parliament-House friends to the *landward* readers of the *Witness*, the good men will, I fear not, with all their bluntness and rusticity, be quite intelligent enough to understand it. Nor do I despair even of Edinburgh. There has been a strong disposition evinced of late to count heads in our metropolitan Free Kirk-Sessions, and no high satisfaction evinced at finding the lawyerly heads so numerous.

'It is not by mere tact and management that differences consequent on the position of Parliament-House Free Churchmen with regard to the Government, and the peculiar duties of the *Witness* with regard to the ministers and people of the Free Church, are to be reconciled. So long as the Whigs continue what they are, and our Free Church lawyers persist at all hazards in supporting them, the difference must, in many points, be irreconcileable. The Parliament-House Editor who produced a *Witness* suited to satisfy them, would fail most egregiously in his duty to the Free Church and its people. I may here state, as not quite out of place, that I entertain no factious dislike of our Whig members of Parliament. I am not Tory, Conservative, Radical, Chartist. I am a Whig, and have been so ever since I was able to form an opinion. When many of my respected friends in the north country, alarmed, some ten or twelve years ago, by the Whig measures of the day, were casting their influence into the Conservative scale, I remained a Whig still; nor have I yet laid down any one of the political principles which I then held. That I am now no thorough supporter of the Whig Ministry, is a consequence, not of my having changed my Whiggism, but of my being more a Protestant than a politician, and that my real party principles are those of the Free Church of Scotland.

'I have of late heard much regarding the faults of the *Witness*, and I am, I trust, not blind to them. It may be well known, however,—for on scarce any other topic do newspaper Editors so frequently insist,—that what one reader deems a fault, another does not deem a fault at all; and that a Paper thoroughly to the mind of one class, would be in utter discordance with the taste of another. The *Witness* has many faults which Dr Candlish has observed,—it has perhaps some others which he has not discovered yet. There are occasional complaints made at the *Witness* Office, which at certain times bear on a particular point; and the following extract from a letter dated the 21st November, which I received in Cromarty from my partner Mr Fairly, ere the present storm arose, indicates its nature very directly:—" You will observe," says Mr F., "from the state of the Paper to-day, that it was impossible to find room for the twentieth chapter of your *Impressions*. I am sorry that our efforts to please, by giving a full report of the Commission, have had a very different effect. There have been a good many grumblers about the Publishing Office to-day, who say they would like an outline of the business much better than a full report, and would willingly wait the publication of the *Monthly Statement* for such documents as Dr Candlish's speech [Report] on Education. I suspect that now, when the great excitement is over, a succinct outline of ecclesiastical affairs, in a well-selected general newspaper, would be more acceptable to the greater part of our readers, than a Paper almost entirely filled with heavy reports. A question may occur occasionally to fill the public mind, and it would be well to give full scope to all such; but I have found that our very full reports of the ordinary Church business have, for some time back, attracted no notice except to be grumbled at for their length, and have failed to defray the extra expense of getting them up."

'Now, there is a meaning in this little bit of fact, which the proprietors of the *Witness* cannot afford to disregard. Dr Candlish, in his present scheme for improving the Paper, suggests that "occasional Supplements ought to be given, if required." Certainly the necessities of the Paper require them: in the course of one fortnight I reckoned up no fewer than nineteen columns of ecclesiastical reports and intelligence, which appeared in no other Edinburgh Paper.* And the

* 'In the two last *Witnesses* alone, that of Saturday the 9th and that of Wednesday the 13th instant, there are upwards of fourteen columns of Free-Church-speech reports, which are to be found in no other Edinburgh newspaper.

other Edinburgh Papers, possessed of an equal breadth of surface, had, of course, instead, their nineteen columns, filled with the staple news of the day. But though in this sense the *Witness* very much requires Supplements, the public, it seems, does not require them at all; a single half-sheet Supplement costs £16, and £16 worth of speaking, given often away, for which we, the Proprietors, would not receive the thanks of those who got it gratis, and which those who did not would not buy, though it might be sport to the speakers, would be death to us. Free Churchmen, in Edinburgh and elsewhere, must not fall into the mistake of reckoning on what is by what has been. The series of events which terminated in the Disruption formed a great and intensely exciting drama, and the whole empire looked on with an interest much more than commensurate with the ability of the men who spoke or the men who wrote. The speeches and motions were pregnant with a transaction of vast historic importance. But we are now thrown more on our own resources, and are simply speakers and writers of a certain definite calibre. The curtain has fallen on the master drama, and the afterpiece resembles, in comparison, that unfinished play of Bayes in the "Rehearsal," which was complete, said the poet, in all the dialogues, and wanted only the plot.

'There is obviously some grave difficulty in the way of rendering the *Witness* a good general newspaper. It has now had in succession no fewer than six Sub-Editors, and all of them have failed to satisfy. And for eight months it had the incessant attention of Mr Fyfe, a gentleman for many years connected with the *Scotsman*, one of the best sub-edited newspapers in Scotland,—and who, thoroughly a master of his profession, had found no difficulty in giving order, interest, and an air of lightness and intelligence to the columns of that Paper. The very superior ability of Mr Fyfe attracted the notice of the *Times*, and he was lost to us, through an arrangement with that journal which no Scotch Paper could afford to make. I requested him to state, when leaving us, why it was he could not make the *Witness* what he had previously made the *Scotsman*, and he did so, assigning three distinct reasons,—first, our long and heavy reports, the weight of which the *Scotsman* and every other Edinburgh newspaper escape; second, the insertion of documents, such as the Reports of Committees, &c., which, though highly important in themselves, are unsuited to the columns of a newspaper; and, third, disturbances from without, of a kind which he had never before experienced:—time after time, at a late hour, when the order of the paper had been completed for publication, Mr Fyfe found that some document, or article, or report, had to be inserted, agreeably to instructions, and that the news of the day had in consequence to give place,—to be inserted in some succeeding number, when no longer fresh. Such was the decision of a practical man, long conversant with the getting up of a newspaper; and I submitted it, among others, to Dr Candlish. And the Doctor now tells the Proprietors that "on passing topics" the *Witness* "is often silent or tardy," and that he "attaches no weight to the excuse of long reports having to be inserted." With regard to what Dr Candlish deems the still more serious faults of the Paper,—those of the writer of this letter,—such as "a want of taste and tact in the handling of public questions,"—faults of so grave and hopeless a character, that the Doctor is convinced "Mr Miller ought not to be Editor" any longer, but reduced to the standard of a mere subordinate contributor of articles,—I shall go direct to the point at once. The scheme in view is simply that of a censorship; and the Censor, assisted by the nice taste and tact of the Parliament-House Editor, is to be Dr Candlish.

'There are important questions on which the Doctor and I entirely differ; and one of these I must be permitted to adduce, that it may be at once seen how such a censorship would work. The opening article in *Lowe's Magazine* for the present January,—that on the "Duty of Electors,"—is, I have the most direct means of knowing, from the pen of Dr Candlish; and it contains an advice, entirely practical in its nature, to the Free Churchmen of Scotland. It earnestly urges them, if they wish effectually to prevent the endowment of Popery, to join the Voluntaries in a vigorous attack on all the existing religious Establishments of the empire.

'Now, this advice I deem so singularly unwise and dangerous, that on no account could I be induced to second it in the *Witness*. Whether right or wrong in my convictions, I am at least thoroughly convinced that it would have the effect, if acted upon, of placing the great Protestant front of the empire in a fatally false position, and would, besides, be peculiarly injurious to the Free Church. In the first place, a war with Establishments would have the inevitable effect of cutting off the earnest Protestantism of our own body from the earnest Protestantism of the English and Irish Churches; and, be it remembered, that in these Churches, notwithstanding their signal faultiness as Establishments, there are very many truly excellent men who are doing what the Voluntaries have long since forgotten to do,—honestly maintaining a consistent protest against Popery. In the second place, it would have the effect of arraying on the side of Popish endowment a numerous class who now stand neutral,—men who, like the ministers of the Scottish Establishment, and the men, clerical and lay, of a corresponding character in England and Ireland, have a large secular stake involved in the question, and who, much rather than that Establishments should fall altogether, were they to find them assailed in the mass, would unite in calling in to their assistance and support the seven millions of Popish Ireland. In the third place, we have, I think, direct evidence, that though the war against Popery is, in its effects on those who prosecute it, an eminently safe war, the war against Establishments is not;—never, at least, was the Church more spiritual than when she was warring against Popery;—never did any Church, in any controversy, become more thoroughly secular than the Voluntaries of Scotland when warring against Establishments. In the fourth place, while the war against Popery would be strictly constitutional, the war against Establishments would not: it would of necessity endanger, with the assailed institutions, not a few precious remnants of the Revolution Settlement, in which no class have a larger stake than we; and it is, besides, a grave question, whether the Free Church would not lose immensely more by forfeiting the esteem of all solid men, hostile to a position so decidedly revolutionary, than she could possibly gain through any consequent accession to the number of her allies, from the ranks of the restless and the dissatisfied. In the fifth place, such a war as the Doctor desiderates would justly lay our Church open, if waged ere the present generation had passed away, to a charge of gross and suspicious inconsistency; seeing he recommends it should be a war carried on not simply on mere consideration of expediency, which may lead at various times to various courses of conduct, but also on a strictly religious principle,—a principle which, if sound at all, must have been as much so ten years ago, during the heat of the Voluntary controversy, as it is now; and seeing, further, that though since that time the Scottish Establishment has been revolutionized and degraded, no change has taken place in the constitution and government of the English and Irish Establishments, both of which we in the course of the controversy strenuously defended. And, in the sixth and last place, I deem it all too obvious, that the Voluntaries would not, as a body, be heartily our allies in the conflict. A vast majority of their number in Scotland lost their hearty dislike of the Scottish Establishment when we left it; and it is to us,—as manifested in the "Send back the money" business, and several other transactions of resembling character,—that the dislike now points. When the vast majority of the old Dissenters of the country put the slave-having Churches of America so emphatically under the ban, it was a clear proof, to my mind at least, that the Voluntary controversy was fully over. A State never dismisses its auxiliaries until its wars are ended. Dr Candlish's scheme, if promulgated from the platform in the Music Hall, would, I doubt not, elicit the hollow hosannahs of Edinburgh Voluntaries; but I am equally clear that it would do infinite mischief to the cause of Protestantism, by lowering its spirit and dividing its forces. It seems all too characteristic of a mind greatly more adroit in preliminaries and details, than sagacious in the master businesses to which the preliminaries should lead and the details belong,—a mind skilful in coasting along shallow bays and miniature promontories, but that, when it launches into the open sea, lacks both compass and rudder, and carries with it no nicely graduated sextant, by which to determine lines of bearing or points of position.*

'Nay, even on considerations of expediency, in the lowest sense of the term, I entirely differ from Dr Candlish regarding the policy of what he recommends. His proposed means of check and counteraction are ludicrously inadequate. Great bodies of men can be moved by but the lever of great interests or great principles; and there is no possibility of awakening a nation to enact a stratagem. "Let the demand," says the Doctor, "for the removal of all existing Establishments, on the ground of those two principles, be as clear and unequivocal as the cry for new Popish or miscellaneous endowments. Let statesmen of all parties know, that they have to deal with a firm and compact body of men, opposed to all Establishments as now constituted, on religious considerations, and determined, on every occasion, when new endowments are talked of, to move the previous question of confiscating of the old for public exigencies of the State." Ingenious, but surely not very wise. The people are to be made earnest and energetic, it would seem, on the strength of splitting hairs and drawing distinctions, and a weak-minded Ministry to be appalled and arrested by the manœuvre of "moving the previous question." Does Dr Candlish in reality hold, that a country is to be aroused, or a Government overawed, by the exercise of the same kind of arts through which a Kirk-Session is to be managed, or a Deacons' Court held in check?

'Now, here is exactly one of the sort of matters on

* 'In the entire article in *Lowe's Magazine* for January,—very unworthy of its November precursor,—the trumpet gives an uncertain sound. Independently of the fatal advice which it tenders, it is of a character suited to give the Protestant cause no help, and the policy of the most secular-minded Whig that walks the Parliament House, be he who he may, no hindrance.'

which the proposed censorship would be brought to bear. But though it seems natural enough that on so difficult a question Dr Candlish and I should differ, it is perhaps not quite so obvious on what principle his judgment should invariably control and over-ride mine. Such, however, is the principle which a censorship involves. It might be all quite well that Dr Candlish should keep me right; but, pray, who is to keep right Dr Candlish?

'I must speak my mind freely;—much better now, in secret conference, ere matters have yet gone too far, than from the house-top. I complain that the most justifiable stand which I made regarding Mr Wood's document has never been forgiven me by the Doctor. "How comes it that you have offended Dr Candlish?" I was asked, a considerable time since, by a friend; "have you quarrelled?" "Not at all," I said; "what can have led you to think so?" "The marked change in his mode of addressing you in the *Witness*," was the reply; "you once used to be his dear Sir; you are now simply Sir." "Oh," I rejoined, "the Doctor never thinks of his mode of addressing people; he has something else to do." "I beg your pardon," was the reply; "in matters of that kind Dr Candlish is always wide awake." Let me not be misunderstood: I solicit no man's friendship; I merely advert to a circumstance that has attracted notice. I complain, further, that disadvantageous criticisms have been dropped into circulation regarding the *Witness*, which, had the real interests of the Paper been consulted, should have been addressed direct to myself. A remark let fall in the hearing of the pliant officials of No. 38, York Place, flees far without losing its original character. I have been cognizant of what was going on in certain quarters, in this way, for a considerable time past.

> "For wherever he went, he was 'ware of a sound,
> Now heard in the distance, now gathering round,
> And it irked him much that that sound should be,
> But the *soul* of the cause of it well guessed he."

I yet further complain of the part acted by Dr Candlish in the Edinburgh Presbytery, when he stood up to say he had detected a sophism in an article of mine on communion with the slave-having Churches. Did the Doctor really imagine that, by casting the *Witness* as a tub to the whale on that occasion, he could neutralize the bitter hostility evinced by the Establishment and the Voluntaries on the slave-having question, to the Free Church? Alas! the hostility lies too deep. I build nothing on the fact that, for the article to which Dr Candlish so pointedly and so publicly referred, I had already received, in an exceedingly kind note, the thanks of Dr Chalmers; and that he, in consequence of seeing either worse or better than the critic, had failed to detect the sophism. Nor did it in the least move me that the criticism should have furnished Mr M'Beth of Glasgow with a text on which to deliver himself, in a pamphlet, of a little of his superfluous nonsense. The article *did*, perhaps, contain a sophism, though I have yet failed to see or Dr Candlish to show that it did. I shall, however, take the liberty of asking the Doctor whether it was wise,—I shall not say friendly,—to strike thus, as if from behind, at a humble but devoted man-at-arms, engaged in fighting in the same front of battle with himself against the same common enemy? I ask him, further, if he thinks it was from a sense of fear, or from a lack of materials for criticism in *his* productions, spoken or written, or from any other cause than simply a sincere regard to the interests of the Free Church, that I made no reprisals? This I shall dare say,—that were

I disposed to make sport to the Philistines, I could find in the recorded thinking of Dr Candlish not a few magnificent immaturities with which to amuse them. There are some minds that, from a certain structural idiosyncrasy, lie peculiarly open to animadversion; and were I as regardless of general as of personal consequences,—of my Church as of myself,—I might perhaps find exercise for ingenuity in the dissection of one special intellect of this character, that, like a summer in the higher latitudes, produces much, but ripens little,—that is content often to acquiesce in its first hasty conclusions, without waiting for what the second cogitations may produce,—and that bears on its incessant stream of thought, ever copious and sparkling, many a fragile air-bell, that, though it reflects the rainbow hues of heaven on its surface, owes all its dancing buoyancy to a lack of weight, and is singularly hollow within.

'Of the more elaborate articles on the Sabbath question which lately appeared in the *Witness*, three were of my writing, and the others by the Assistant-Editor, Mr Wylie. I may mention, that Sir Andrew Agnew, in writing to the office, deemed them at least worthy of thanks; and that immediately on the appearance of my first paper,—the one with the reference to Bishop Gillis,—I was requested by Mr Hanna, son-in-law of Dr Chalmers, to furnish him with an article on Sabbath Observance, "more in its social than its peculiarly religious aspect," for the *North British Review*. Now, on these articles Dr Candlish founds one of his urgent reasons for my retirement from the Editorship of the *Witness*. "The damage done," he says, "to the Sabbath cause by certain recent articles is incalculable;" and I have learned from Edinburgh,* through more

* This letter was written in Cromarty.

than one channel of communication, that he has been unsparing in his condemnation of my assault on Bishop Gillis. I shall not defend myself in this matter. It has perhaps been my misfortune that, somewhat curious in my reading, I have lived too much in the past, when even the best of men were unconscious of fault in taking liberties equally great with Jesuitical bishops, extreme unction, and the wafer; and I have, I am afraid, failed to note that the atmosphere of the present time has been materially changed by exhalations blown aslant, over the face of even the Evangelic Churches, from the bogs and fens of a hollow liberalism, that professes to respect all religions, and believes none. I make no defence. I am convinced, too, that were all the errors to which man is liable of the especial character of mine, Dr Candlish might with a clear conscience cast at the peccant Editor the first stone.

'I have thus stated my views on the general question, in its bearings on the *Witness* newspaper; and I must now be permitted to say a few words regarding myself. I am neither in the mood nor the circumstances to act rashly. My health and strength, though gradually improving, are not what they have been. I have several poor relatives dependent on me for support. I have a wife and family. I have been doing what little I could for the Church; and as I have not yet touched a farthing of the profits of the *Witness*,—and, until the debt to the Committee be honourably discharged, never will,—I am a poor man.* But though not in a condition, and certainly not in a frame, to act rashly, I trust, with God's help, to be enabled to act firmly, and in a manner not unworthy of the Church of the Disruption. I need not

* Should not refuse that money—subscribers are willing to give up.—*Note by Dr Chalmers.*

remind some of the gentlemen whom I now address, that the place which I occupy as Editor of the *Witness* was not of my seeking, and that I entered upon its duties in weakness and great fear, thoroughly convinced of the goodness of the cause, but diffident indeed of my own ability to maintain it as it ought to be maintained, against the hostile assaults of well-nigh the whole newspaper press of the kingdom. Little did I think, after reading Lord Brougham's speech in the Auchterarder case in my house in Cromarty, and then passing a sleepless night, brooding over the state of great peril in which the Church that I loved was placed, that the result of my cogitations should have had the effect of leading me, by a "way which I knew not," into so peculiar a position. Once there, however, I found myself in my true place; and I trust I may be permitted to say, that I have striven to perform its duties not without reference to the Providence whose hand I recognized in the entire transaction. I have been an honest journalist. During the seven years in which I have edited the *Witness*, I have never once given expression to an opinion which I did not conscientiously regard as sound, nor stated a fact which, at the time at least, I did not believe to be true. My faults have no doubt been many; but they have not been faults of principle; nor have they lost me the confidence of that portion of the people of Scotland to which I belong, and which I represent. And possessing their confidence, I do not now feel myself justified in retiring from my post. Dr Candlish and his Parliament-House friends are not the ministers and people of the Free Church of Scotland; nor do I recognize the expression of the Doctor's will in this matter as a call to me in Providence to divest myself of my office, in order that the Free Church Press may be made the subject of a centralizing experiment, or that the Free Church itself may be laid open to the temporizing influences of the Parliament House. I shall at least, ere I yield, carry my appeal from one minister of the Free Church to all the others, and from the *Sheriffs* to the counties.

'I may, however, cease to be Editor of the *Witness*. Dr Candlish and Mr Wood I consider as parties in this business, and so I do not put myself into their hands; but with regard to the other gentlemen whom I address, I do so with the fullest confidence. The *Witness* is property; it belongs *boná fide* to Mr Fairly and me. If, however, the friends whom I now consult think I should cease being its Editor, the course is obvious. Let it be valued by persons mutually chosen by the *Witness* Committee and the Proprietors; and immediately on the Committee's coming under the necessary obligations, it shall be made over into their hands, and I as its Conductor, and Mr Fairly and I as co-Proprietors, shall close our connection with it for ever. Permit me to state further, that if there is to be war between Dr Candlish and me, it must be open war. "Of wiles, more inexpert, I boast not." The true springs of the under-current of "hinted faults" and "hesitated dislikes" must be fairly uncovered, should the stream continue to flow. The difference must either close entirely, or the people of Scotland must be made fully acquainted with the grounds on which it rests.

'I am, Gentlemen,

'With sincere respect,

'Your most humble,

'Most obedient servant,

'HUGH MILLER.'

POSTSCRIPT.

'The above was forwarded from Cromarty in MS. to Mr Fairly, who, with his own hands, assisted by my brother, a printer in the *Witness* Office, set it into type. On my arrival here I found it awaiting me in proof, and, with the proof, a note from Dr Candlish, in which the Doctor states that, having come to learn from a letter he had received from Mr Stewart of Cromarty, that his "well-meant interference in the affairs of the *Witness*" had been wholly misunderstood both by Mr Stewart and me, he now wished, in consequence, all his communications on the subject to be "superseded, set aside, burned, and held as non-existent." The request comes too late; —*I* take the present step; it will of course be the Doctor's turn to take the next. If my views regarding the danger to be apprehended from a system of Free Church centralization, and the influence of secular politics on our Edinburgh leadership, possess any measure of soundness, it would be well, surely, that they should be soberly discussed by our wiser and cooler heads, so long as Chalmers is still with us. Dr Candlish's written communications have, I may here state, been marked "private;" but neither Mr Fairly nor I were the proper depositories of the secret which they embody. The man who has let his neighbour understand in strict secrecy that he intends bleeding him for his benefit, by sending a bullet through him in the evening, has no reason to complain that his neighbour betrays his confidence by blabbing to the police. I understand from various quarters, that there is to be forthwith another Free Church paper started in Edinburgh. Assuredly one special class of Edinburgh Free Churchmen shall by-and-by greatly need such an organ; for so long at least as I edit the *Witness*, it shall never be the representative of their views, or the advocate or apologist of their policy. I would fain advise them to term their vehicle "THE FREE CHURCH CENTRALIZATION JOURNAL, and *Parliament-House Gazette;*" and I shall have very great pleasure indeed in contributing the Prospectus.

'EDINBURGH, 14th January, 1847.'

In a biography of Hugh Miller an opinion on the merits of this quarrel may be expected, but I am not sure that I am qualified to form one with impartiality, because, on the most important matter in dispute, the duty of the Free Church to take her place in line of battle with the Nonconformists of the realm, in order to strike their bonds from the 'State institutions' and change them into Free Churches, my humble opinion coincides with that of Dr Candlish. I have also had the opportunity of taking note, in quite unguarded moments when we happened to meet at Kilcreggan on the estuary of the Clyde, and were sometimes together alone in a little boat on Loch Long, of Dr Candlish's feelings towards Hugh Miller in the last years of his life. Clear acknowledgment of his intellectual power, satisfaction with the part played by himself in bringing him to Edinburgh, and a regretful, not angry, feeling that he could not forget and forgive and let by-gones be by-gones,—these are the impressions which remain with me of what I heard from Dr Candlish in 1856 respecting Hugh Miller.

If the readers of these pages have formed the same conception of Miller as I have, they will not fail to understand how offence, and just offence, might be taken by him at the conduct of Dr Candlish, although the latter was imperfectly conscious, or not conscious at all, of the offence he was giving. Miller was a man of exquisite sensibility, and his life in Cromarty, the admired of all admirers, the delicious incense of Miss Dunbar's worship ever in his nostrils, was not adapted to fit him for those rough collisions of man with man which are every-day occurrences in political and ecclesiastical life. Not only was he personally sensitive, he was filled with a devotion more than knightly to the cause of the Church, and anything like a hint, direct or indirect, that he had not served her efficiently wounded him to the quick. Dr Candlish has lived in the eye of the public for more than thirty years, and every one who has either watched him or worked with him knows that, in carrying out a public object, in giving effect to an idea which has taken possession of him, he consults no sensibilities, is reckless of minute proprieties, and hurries to his mark in conspicuous disregard of the ribs he may elbow or the toes he may crush. There can, also, I think, be no doubt that the conception formed by Dr Candlish as to what the *Witness* ought to be and to do differed in essential particulars from Hugh Miller's. Dr Candlish may never have said as much in words, or even brought the thought to distinct consciousness, but fundamentally his notion was that it should be more exclusively a Church paper, and more emphatically a Whig paper, than its editor could or would make it. Miller felt that he had a right to put his own image and superscription on the *Witness*. Throughout Scotland he was identified with it; the acclamation of Scotland justified him in the belief that the splendid success which had attended it was due to him. Surely this was no more than that honest pride, that manly independence, which prompts a colonel to resent interference with him in fighting his regiment or a captain in sailing his ship.

It was, of course, in no sense or degree incumbent on Dr Candlish to think that the *Witness* was faultlessly conducted;. but, presuming Miller to be correct in believing the Doctor to have considered his positive removal from the editorship necessary, most of us probably will think the decision surprising, and no one will be astonished that Hugh Miller should regard it with indignation. For the rest, opinion will be divided upon the views taken by him and by Dr Candlish respectively on many of the questions which crop up, somewhat in the manner of geological strata through fields bearing the corn and the weeds of to-day, in this pamphlet. Miller appears as the representative of Free Churchmen 'of the long-derived historic type and sternly orthodox spirit.' Bishop Gillis was simply a Jesuitical bishop, to be unmasked; the Sabbath was to be maintained in all strictness, without regard had to 'exhalations blown aslant, over the face of even the Evangelic Churches, from the bogs and fens of a hollow liberalism, that professes to respect all religions, and believes none;' and the splendour of Macaulay's genius, the energy of his political Whiggism, the brilliancy of his Parliamentary career, were to be made of no account by the citizens of Edinburgh in electing a member of Parliament, if only he adhered to his determination to vote in favour of the Maynooth grant. Exhalations, whether of the fen or of the dawn, have assuredly been dimming the glory of this 'long-derived historic type' of Scottish Presbyterian religion; but it has been the faith of many a brave heart and strong brain, and never of a braver or a stronger than Hugh Miller's.

The triumph of Miller was instant, absolute, final. Dr Chalmers, having perused the letter, lost not a moment in convoking the Committee. He addressed it with great animation on behalf of Miller, dwelling on the absurdity of officious meddling with such a man, and asking, with peremptory, leonine glance, 'Which of you could direct Hugh Miller?'

There seems to have been a kind of understanding among the members of the Committee who were present at the meeting that the copy of the letter which each possessed should be destroyed. Dr Begg tells me that he 'went honestly home and destroyed his;' but the understanding must have been vague, or several of the members of Committee were absent, for there is no doubt that a plurality of copies exist, and Dr Bruce is known to have declared, with characteristic emphasis, that he set far too high a value upon his to think of burning it.

No accurate information as to what had occurred reached the general public; but an indistinct conviction pervaded the Free Church that Hugh Miller had vindicated the independence of the *Witness* against Dr Candlish, and the lay element in the Church against the clerical. It was clearly understood that the opinions of the *Witness* would be those of its editor, and that the weight and worth of his independent judgment were guaranteed against all extraneous influence. After being shaken a little by Parliament-House defection, the paper pursued its career of prosperity.

Though Miller triumphed, the memory of the fray rankled deep in his heart to the last, and its result on the whole, even to the paper, was melancholy. He had imbibed an invincible repugnance to handling the Free Church questions of the day. Resolute to speak no man's mind save his own, and shrinking from the bitter necessity which he knew to be entailed upon the frank utterance of his own sentiments, he became shy of Church matters altogether, made it his aim to afford scope, in the columns of the paper, for an expression of opinion, in speech and letter, by rival parties, rather than to influence either in leading articles, and sought for himself a more congenial sphere in science and literature. In these views his partner, Mr Fairly, entirely acquiesced, and held it, after Miller's death, not only to be due to his memory, but to be advisable on other grounds, to maintain the same policy in the conduct of the paper. The independence of the journal had been vindicated, once and for ever; the clergy never came near the *Witness Office;* but, after all, it was isolation rather than independence that was attained, and there was a gaunt and desolate feeling about it. What it must have been to Hugh Miller is not easy to realize. Mrs Miller thinks that about the sum of it is that it broke his heart. But even the successor of Hugh Miller in the conduct of the *Witness,* for whom I am qualified to speak, found the position sometimes peculiar. A brief illustration will show what is meant.

The liberty of every Free Church in the United Kingdom was directly or indirectly threatened by the attempt of Mr MacMillan of Cardross to induce the Court of Session to punish the Free Church of Scotland for excluding him from her ministry. The most influential Edinburgh newspaper tried to make it appear that Mr MacMillan's plea against the Free Church was strictly analogous to Sir William Dunbar's plea against Bishop Skinner, Primate of the Free Episcopalian Church of Scotland, when the Bishop visited, or attempted to visit, Sir William with ecclesiastical censures. The question was the key of the Free Church position in a legal point of view, for if the Reverend Sir William Dunbar, on the one hand, and the Primate of the Free Episcopalian Church of Scotland, on the other, had consented to plead, and had been accepted as pleaders, at the bar of the Court of Session, how could the Free Presbyterian Church claim exemption from their jurisdiction? A perusal of the report of the Dunbar trial in the records of the Court of Session showed, what any one acquainted with the historic position of the Sir William Dunbar section of Scotch Episcopalians might have guessed beforehand, that the cases were essentially distinct. Sir William Dunbar and his section claimed relations with the Church of England, and repudiated the jurisdiction of the Scottish Episcopal Church. This independence constitutes, in Scottish ecclesiastical history, the very meaning and *ratio essendi* of the little knot of Episcopalian congregations referred to. In relation, therefore, to the Primate of the Scottish Episcopalian Church, Sir William Dunbar was a layman; and the plea of Sir William, when the Bishop flung his ecclesiastical net over him, was substantially that the latter had no more right to meddle with him than with the editor of the *Scotsman*. To determine the question of fact, both parties came, and rightly came, before the Court of Session. Mr MacMillan, appearing in his character as a Free Church minister, demanded that the Court of Session should compel the Free Church to remove the ecclesiastical censures which she had pronounced upon him. On the question whether she owed him money, or the question whether she had masked a private attack under ecclesiastical forms, or any other question involving the civil rights of Mr MacMillan, the Free Church declared herself ready to plead: but she maintained that, if Mr MacMillan admitted her ecclesiastical jurisdiction over him, which he did; and if his claim was not for money, which it was not, and if the proceedings against him were allowed to have been ecclesiastical, which they were, then she could not be compelled to answer to the Court of Session at his complaint, save on grounds which would subject to the Court of Session the internal discipline of every unendowed denomination in Scotland.

The first articles in the *Scotsman* attempting to mystify the public as to the relation between the Sir William Dunbar Case and the Cardross Case may have been contributed, for what I know, by some sleek Edinburgh parson, whose hatred of the Free Church was an illustration of the infinitely great, and whose love of truth was an illustration of the infinitely little. The busy, careless public were ready enough to take the view suggested by the *Scotsman*. The editor of the *Witness*, having read the report of the Dunbar trial, perceived that the matter was one on which it was impossible for men who took this trouble to hold two opinions. He said so in the columns of the *Witness*, stated the evidence, told people to satisfy themselves by turning for an hour into the Signet Library and reading the report, and, with a simplicity quite primeval and lamb-like, argued that the *Scotsman* ought to retract a demonstrated error. The misleading articles continued to appear. Hereupon the virtue of our Nathaniel took fire, and he inveighed against the dishonesty of his controversial opponent. Whether the editor of the *Scotsman* had previously interfered, or had not, he now came upon the field, cudgel in hand, and with the experienced fury of one who had fought many newspaper battles, showered his blows upon the head of the ingenuous young man. Upon the merits of the controversy, the cudgel-bearer said little; what nerved his arm for vengeance was the preposterous idea that, if black is black and white white, and a gentleman of the press, knowing black to be black and white white, calls white black and black white, it is consistent with the usages of society to tell said gentleman of the press that he is not a gentleman in the ordinary sense. What I had to remark, however, was that, all the time the *Witness* was striving, amid the most virulent opposition, to defend a vital point of the Free Church position, not one syllable of recognition of the *Witness* as taking part in the defence, was uttered by any minister, or in any Court of the Free Church. The matter at length came up in the General Assembly. Speaker after speaker, lay and clerical, among the leading Free Churchmen, took the same view of the Sir William Dunbar Case as had been taken by the *Witness*, and the essential importance of disabusing the public mind of the fallacy of comparing it to the Cardross Case, except by way of contrast, was amply acknowledged; but not one syllable betrayed that the *Witness* had done the work beforehand, or that the Free Church had any manner of connection with a paper called the *Witness*.

The experience of the proprietors and the editor on this occasion enabled them to realize with vividness the isolation, which, under circumstances incomparably more agitating, pressed upon the soul of Hugh Miller. In times of danger and of difficulty he had been the eye of the fleet, keen to discern the approach of peril, ready always to draw the fire of the enemy upon himself and his ship; and now, when smoother waters had been reached, all the services of the night, and the tempest, and the fray were forgotten, and uniformed captains of the fleet, waiving him a cold adieu, held on their way without acknowledging that he or his ship were in existence. He never flagged in his service. While he continued to breathe, his name was a tower of strength for the Free Church, and had he seen her in danger, he would have rushed to the rescue as in those days of young and proud enthusiasm when he addressed his Letter to Lord Brougham; but his mind dwelt with brooding anguish upon the isolation into which he had been thrown, and on the whole, in its incidents and its results, we may pronounce this quarrel with Dr Candlish the unhappiest occurrence of his life.

One word must be added, to obviate misconception. Though Hugh Miller belonged to the school of sternly anti-Popish Scottish religionists, and had no mercy for Roman Catholic Bishops, he recoiled from the acrid extravagance and toothless bigotry of those self-elected representatives of Protestant zeal, who band themselves into associations, publish hateful prints, make Protestantism a laughing-stock in Parliament, and on the whole render considerable, though contemptible, service to Archbishop Manning and the Pope. Miller honoured the intention of these poor creatures, and went with them as far as he could. Though few of the anti-Popish articles in the *Witness* were from his pen, he permitted a great number to be inserted by an eloquent and popular contributor, and any one who glances over the old files of the paper must be surprised that even Miller's genius could render it acceptable to readers with those perpetual doses of anti-Popish harangue. Nevertheless the bigots were not satisfied. The *Witness* was pronounced lukewarm in the Protestant cause. A new paper, called the *Rock*,—Miller called it a *trap* rock,—was started, and for a limited number of weeks or months the craziest Protestant champion in Edinburgh could not complain that 'the banner' was not held high enough. Miller had that hatred of Popery which seems constitutional to Scotchmen of strong religious instincts,—it has always characterized Mr Carlyle, for instance,—but he had no fellowship with those bigots who, having lost that fervour which made the intolerance of the seventeenth century respectable, retain that intolerance which cast over the fervour of the seventeenth century a ghastly and ensanguined glare.

And now, not without satisfaction, we can turn the leaf upon this Sturm und Drang—storm and stress—chapter in Hugh Miller's history.

BOOK VI.

MAN OF SCIENCE.

Thus you see we maintain a trade, not for gold, silver, or jewels; nor for silks; nor for spices; nor any other commodity of matter; but only for God's first creature, which was *Light.*

CHAPTER I.

THE OLD RED SANDSTONE—LETTER TO A CHILD—FIRST IMPRESSIONS OF ENGLAND AND ITS PEOPLE.

AMID the stormful enthusiasm of a great popular conflict, Hugh Miller had not forgotten the serener if not more lofty devotion which had inspired him as a servant of science. 'Of Bacon,' he once wrote, 'I never tired;' and often, probably, when the excitement and sword-clashing of polemical battle filled the air around him, would that vessel rise before the eye of his imagination which Bacon saw, speeding on, age after age, across calm ocean spaces in search of light, horizon after horizon opening before her, constellation after constellation kindling in her skies. And now, when he had been for the better part of a year editor of the *Witness*, he ventured to yield to the prompting of his heart, and to recur to those geological studies which had been his delight in the quiet days of Cromarty.

On the 9th of September, 1840, there appeared in the *Witness* the first of a series of articles under the title, *The Old Red Sandstone.* There were seven in all, each occupying two or three columns. The last was published in the *Witness* of October 17th, 1840. The moment was propitious. Hugh Miller could state in the outset that the Old Red Sandstone formation had been hitherto considered as remarkably barren in fossils; that a continental geologist, in tabulating the various formations, had appeared to omit this one altogether; and that Lyell, whose standard work on the Elements of Geology had been issued two years previously, had devoted but two and a half pages to its description. He could add that he had 'a hundred solid proofs' lying close to his elbow, that the fossils of the system are 'remarkably numerous.' Nor were they less strange than they were abundant. 'The figures on a Chinese vase or an Egyptian obelisk are scarce more unlike what now exists in nature than the fossils of the lower Old Red Sandstone.' They seemed to be products of 'nature's apprenticeship.'

The importance of the Old Red Sandstone, as part of the geological record, had begun to be surmised by naturalists; and remarks like these were fitted to awaken the curiosity of the general public. The ear of the world, therefore, was open for the word which Hugh Miller could speak. Before September had closed, his reputation as a geologist was made. On the 23rd day of that month the British Association for the Advancement of Science held its annual meeting. In the Geological Section, Mr Lyell in the chair, Miller's discoveries were brought under the attention of leading geologists. Mr Murchison, now Sir Roderick, spoke in terms of high eulogy of his perseverance and ingenuity in the geological field, declared that he had raised himself to a position which any man might envy, pointed to the specimens forwarded by him to London, and invited M. Agassiz to describe the class to which they belonged. The distinguished Frenchman followed in a similar strain, and proposed to name one of the most remarkable specimens *Pterichthys Milleri.* Dr Buckland's enthusiasm knew no bounds. He 'had never been so much astonished in his life by the powers of any man as he had been by the geological descriptions of Mr Miller.' That wonderful man described these objects with a felicity which made him ashamed of the comparative meagreness and poverty of his own descriptions in the *Bridgewater Treatise,* which had cost him hours and days of labour. He would give his left hand to possess such powers of description as this man.' It was, in Dr Buckland's view, 'another proof of the value of the meeting of the Association, that it had contributed to bring such a man into notice.'

There is something fine in this spectacle of the magnates of science welcoming with glad acclaim a brother who, coming at one stride from the quarry, makes out his title to rank as one of them. Men of letters were almost equally astonished at the performance of the Cromarty stone-mason. The benefit which Miller had derived from his long discipline in literary composition was now evident. Into the description of bare and rigid organisms, he could throw a fascination which charmed every lover of literary form. Here was a self-educated man who had educated himself not to mere copiousness of glittering words, but to the chastened strength, the subtle modulation, the placid-beaming clearness, of a classic. Every page spoke of ripe thought and confirmed intellectual habits.

The *Old Red Sandstone,* which, in the following year, appeared in the form of a book, was the first literary work executed by Miller in the maturity of his power. It stands at the head of a series of unique and remarkable books with which he permanently enriched English literature; books in which the results of face to face inspection of nature in the quarried hill-side or on the ribbed sea-shore, are interwoven with fresh, racy, sagacious judgments on men and manners; books in which observations distinguished by exquisite scientific accuracy furnish the ground-plan for landscapes over which is poured the softest, ruddiest glow of imaginative colouring. The nature of Hugh Miller's imaginative power is characteristically exhibited in this work. His imagination is bold, yet its audacity is always restrained by reference to ascertained fact. Its pictures are never vague. If, as one critic remarked, his fossil fishes 'swim and gambol,' they do so as the mind's eye of Hugh Miller, after severe inspection and long gaze into the past, had seen them swim and gambol in primeval seas. If the stone branch budded like a rod of Aaron on his page, and forests, breaking from their sepulchres in the rock, grew green again in the sunlight and rustled in the wind, it was not that an oriental fancy delighted in clothing phantom hills with visionary foliage, but because the science of the West had put into his hand a lamp which lighted for him the long vistas of bygone time. This species of imagination is the most valuable which a scientific man can possess, and without it no man, however accurate his observation, however just his conception of individual facts, can be great in science. True workers in science are of three kinds, in ascending order of excellence,—the accurate observer and compiler; the sound generalizer; and the seer of nature, who first observes, then generalizes, and lastly illuminates his generalizations so that they become visions. There were many geologists of his time who, having devoted themselves exclusively to science for nearly as many years as Hugh Miller lived, traversed wider fields of observation and attained a greater acquaintance with fact;

but not the most distinguished of his contemporaries surpassed him in those august operations of the mind, which may be claimed indifferently for science and for poetry.

Viewed in relation to the dimensions of the field it covers, the *Old Red Sandstone* of Hugh Miller is a comprehensive account of the formation as it appears in Scotland. The journeyman mason presented to the world of science a monograph on one of the chief rock-systems of his country, and it proved to be an imperishable masterpiece. On that division of the Old Red which is exhibited at Cromarty, and to which Miller had special access, it was almost exhaustive. He added little afterwards to his discoveries in the Cromarty beds, and no other eye has been keen enough to detect in them anything else of importance. Subsequent research has proved that what he regarded as the Lower Old Red is the Middle division of the system. In relation to the Old Red Sandstones of the southern shores of the Moray Frith and of Fife, Perth, and Forfar, the book was necessarily less complete at its first appearance, but even of these it presented a distinct, and, in the main outlines, a correct account, and when Miller put his finishing touches to it in later editions, it could claim to be the standard work on the Scottish Old Red. Sir Roderick Murchison has since demonstrated that the geological position of the conglomerate of the Western Highlands is in the Silurian system. But this would not have surprised Hugh Miller, who entertained doubts upon the subject.

With every year of his life his skill and care as a geological observer increased, but the keen and exquisite discernment he had already attained may be illustrated by a fact which can be stated in his own words. 'After carefully examining many specimens,' he wrote in 1855, 'I published a restoration of both the upper and under side of pterichthys fully fifteen years ago. The greatest of living ichthyologists, however, misled by a series of specimens much less complete than mine, differed from me in my conclusions; and what I had represented as the creature's under or abdominal side, he represented as its upper or dorsal side; while its actual upper side he regarded as belonging to another, though closely allied, genus. I had no opportunity, as he resided on the Continent at the time, of submitting to him the specimens on which I had founded; though, at once certain of his thorough candour and love of truth, and of the solidity of my data, I felt confident that, in order to alter his decision, it was but necessary that I should submit to him my evidence. Meanwhile, however, the case was regarded as settled against me; and I found at least one popular and very ingenious writer on geölogy, after referring to my description of the pterichthys, going on to say that, though graphic, it was not correct, and that he himself could describe it at least more truthfully, if not more vividly, than I had done. And then there followed a description identical with that by which mine had been supplanted. Five years had passed, when one day our greatest British authority on fossil fishes, Sir Philip Egerton, was struck, when passing an hour among the ichthyic organisms of his princely collection, by the appearance presented by a central plate in the cuirass of the pterichthys. It is of a lozenge form; and, occupying exactly such a place in the nether armature of the creature as that occupied by the lozenge-shaped spot on the ace of diamonds, it comes in contact with four other plates that lie around it, and represent, so to speak, the white portions of the card. And Sir Philip now found, that instead of lying over, it lay under, the four contiguous plates; they overlapped it, instead of being overlapped by it. This, he at once said, on ascertaining the fact, cannot be the *upper* side of the Pterichthys. A plate so arranged would have formed no proper protection to the exposed dorsal surface of the creature's body, as a slight blow would have at once sent it in upon the interior framework; but a proper enough one to the under side of a heavy swimmer, that, like the flat fishes, kept close to the bottom;—a character which, as shown by the massive bulk of its body, and its small spread of fin, must have belonged to the Pterichthys. Sir Philip followed up his observations on the central plate by a minute examination of the other parts of the creature's armature; and the survey terminated in a recognition of the earlier restoration,—set aside so long before,—as virtually the true one;—a recognition in which Agassiz, when made acquainted with the nature of the evidence, at once acquiesced.' This is the kind of fact which proves consummate practical ability of that character which cannot be derived from books.

We may here take in a letter written by Miller, about the time when the *Old Red Sandstone* was getting into circulation as a book, to a little boy whom he had known in Cromarty. It contains a few details as to his history at the time which cannot be presented to the reader so pleasantly in any other way, and it adds one other illustration to the many we have already had of his gentle, playful, sympathetic manner with children.

'Edinburgh, 5, Sylvan Place, Sept. 8, 1841.

'My dear Alie Munro,

'I will tell you how it was that I did not reply to your kind letter last spring. It was all in consequence of another letter which I received only two days after I received it, and which entirely put it out of my mind for the time. This other letter was a Cromarty letter; and it informed me that my poor mother was very, very ill, and that unless I hurried north to see her, I might never see her more. And so I did hurry away north, with the very first coach that set out, and was one day and two nights on the road. I found my mother much better than my fears had anticipated, for the disease that threatened her life had taken a favourable turn; and ere I parted from her, which was in about a week after, she was well-nigh recovered. Meanwhile, however, I had forgotten your letter. Now I know, my dear Alie, you will forgive me when you take all this into your grave consideration. My journey was a very unpleasant and sad one, so sad and unpleasant that one might almost make an agreeable story out of it. Wrecks and battles, you know, make good subjects for stories; and the worse and more unpleasant the wreck or battle, just all the better is the story. So long as I was with the coach I had nothing worse than sad thoughts and very bad weather to annoy me. From Fortrose to Cromarty, however, I had to grope my way as if I had been playing all the way at blind man's buff. Never yet have I been out in so dark a night. I had to feel for the road with my staff, and I discovered on two or three occasions that I had got off it only by tumbling into the ditch. It was at least three hours after midnight ere I reached my journey's end.

'I saw Aunt and Uncle Ross in Cromarty, and Cousin Mora. Cousin Mora is a smart, pretty little girl. But I dare say you will deem my news of the north somewhat old, and there is no denying that it is less new now than it was six months ago. It is not so old, however, by a great deal, as the news you gave me about the battle of Hastings; and I of late have been giving much older news to the public in a book on the Old Red Sandstone. You remind me, dear Alie, of the stones and fossils which I used to point out to you on the shore of Cromarty. I have written a whole book about them, with curious-looking prints in it,—the portraits of fish that lived so very long ago that there were no men in the world at the time to give them names. But they have all got names now, stiff-looking Greek names, which only scholars can understand. One of them has been named after me, *Pterichthys Milleri*, which means *Miller's winged fish*, and I send you prints of it that you may see what a strange-looking creature it was. By the way, have you not great chalk cliffs at Hastings? There are very curious fossils found in chalk, sea-eggs, of a kind no longer found alive, spindle-shaped stones, called *Belemnites*, other stones called *Ammonites*, that resemble coiled snakes, cockle-like shells with spines on their backs, and a great many other curious things besides that were once living creatures.

'I trust you will remember me to Aunt Munro, whose kindness to me in Cromarty I very often remember, and who has since been very kind to my sister Jane. I dare say that by answering your letter, though at this late time, I have made her lose her wager. You know lose it she must, if there was no particular time specified. I am very, very busy in these days, thinking, reading, writing, beating one day, beaten the next; called a blockhead at one time without believing it, believing it at another without being called it; living, in short, a hurried, bustling, fighting sort of life. It is very seldom I can command leisure enough to write letters, and sometimes when I have the leisure I want the will. But you see I have at length written to you, and had it not been for one circumstance, of which I have already told you, I would have written you six months ago. I have a little daughter who helps me at times in putting wrong my papers, books, and fossils. She has got language enough to call a dog *bowwow,* and a cat *mew ;* and when she sees a fossil she points to it, and calls it "papa's fish." She had a philosophical desire long ago to ascertain whether the flame of a candle might not taste and feel as pleasantly as it looked; but she is no longer curious on this head. I sometimes sing to her, and she seems much pleased with my music, a thing no one ever was before. I am afraid, however, that her mother will spoil her just and simple taste in this matter; but know not how to prevent it. And now, my dear Alie, I have come to the close of my letter. It is not long, you see, but there is a good deal of nonsense in it, for all that. I would like very much to be a little boy once more, but, alas, I am a big man, and cannot play myself so much or so often as I could wish. Some of my reddish-brown hair is actually getting grey. I am, my dear boy,

'Your affectionate friend,

'HUGH MILLER.'

There is perhaps no work of Miller's by which the general reader can better judge him than his *First Impressions of England and its People.* The subject has the look of being hackneyed beyond all chance of effective writing; yet the book, I venture to say, is one of the most fresh and charming in the language. The materials were gathered in eight weeks of autumnal wandering through England in 1845, and the composition occupied the leisure hours of a hard-worked editor for six months. Yet how admirable is the style! With what subtle felicity does it combine the dignity of elaborate literary form with perfect ease and freedom! And how completely do we feel, as we read, that we are in converse with a cultivated mind! The treasures which Miller had been accumulating since he was six years old,—the impressions, facts, reflections, fancies of life-long observation and study,—flow out upon the page in stintless yet chaste abundance, absolutely without straining or parade. There is no gaudy metaphoric daubing, no wearisome drawing out of similitudes, but the right illustration, brief and happy, always comes in at the right place, and the nice bright word of metaphor, like the honey-touch on the lip of Jonathan when he was weary, never fails.

The skill in which Miller has to this day no rival, the skill of bringing poetic hues and musical tones out of the stony tombs of geology, is in this book exquisitely exhibited. In Dudley glen, for example, he finds himself surrounded with rocks containing Silurian fossils of remote antiquity amid and upon which stand the ordinary trees of English woodland. This is enough to awaken his fancy. 'I scarcely know,' he says, 'on what principle it should have occurred; but certainly never before, even when considerably less familiar with the wonders of geology, was I so impressed by the appearance of marine fossils in an inland district, as among these wooded solitudes. Perhaps the peculiarity of their setting, if I may so speak, by heightening the contrast between their present circumstances and their original habitat, gave increased effect to their appeals to the imagination. The green ocean depths in which they must have lived and died associate strangely in the mind with the forest retreats, a full hundred miles from the sea-shore, in which their remains now lie deposited. Taken with their accompaniments, they serve to remind one of that style of artificial grotto-work in which corals and shells are made to mingle with flowers and mosses. The massy cyathophyllum sticks out of the sides of grey lichened rocks, enclasped by sprigs of ivy, or overhung by twigs of thorn and hazel; deep-sea terebratulæ project in bold relief from amid patches of the delicate wood sorrel; here a macerated oak leaf, with all its skeleton fibres open as a net, lies glued by the damps beside some still more delicately reticulated festinella; there a tuft of graceful harebells projects over some prostrate orthoceratite; yonder there peeps out from amid a drapery of green liver-wort, like a heraldic helmet from a mantling, the armed head of some trilobite: the deep-sea productions of the most ancient of creations lie grouped, as with an eye to artistic effect, amid the floral productions of our own times.'

He lies down on an English hill-side, and as he gazes musingly upon the landscape spread below, and thinks of the time when it was the bottom of a broad ocean sound washing the rocky island that was one day to constitute the Highlands of Wales, rises into this serene flight of prose poetry: 'Was it the sound of the distant surf that was in mine ears, or the low moan of the breeze, as it crept through the neighbouring wood? Oh, that hoarse voice of Ocean, never silent since time first begun,—where has it not been uttered! There is stillness amid the calm of the arid and rainless desert, where no spring rises and no streamlet flows, and the long caravan plies its weary march amid the blinding glare of the sand, and the red unshaded rays of the fierce sun. But once and again, and yet again, has the roar of ocean been there. It is *his* sands that the winds heap up; and it is the skeleton remains of his vassals,—shells and fish and the stony coral,—that the rocks underneath enclose. There is silence on the tall mountain peak, with its glittering mantle of snow, where the panting lungs labour to inhale the thin bleak air,—where no insect murmurs and no bird flies, and where the eye wanders over multitudinous hill-tops that lie far beneath, and vast dark forests that sweep on to the distant horizon, and along long hollow valleys where the great rivers begin. And yet once and again, and yet again, has the roar of Ocean been there. The effigies of his more ancient denizens we find sculptured on the crags, where they jut out from beneath the ice into the mist-wreath; and his later beaches, stage beyond stage, terrace the descending slopes. Where has the great destroyer not been,—the devourer of continents,—the blue foaming dragon, whose vocation it is to eat up the land? His ice-floes have alike furrowed the flat steppes of Siberia and the rocky flanks of Schiehallion; and his nummulites and fish lie embedded in great stones of the pyramids, hewn in the times of the old Pharaohs, and in rocky folds of Lebanon still untouched by the tool. So long as Ocean exists there must be disintegration, dilapidation, change; and should the time ever arrive when the elevatory agencies, motionless and chill, shall sleep within their profound depths, to awaken no more, and should the sea still continue to impel its currents and to roll its waves, every continent and island would at length disappear, and again, as of old, when "the fountains of the great deep were broken up,"

"A shoreless ocean tumble round the globe."'

Of Miller's power as a critic, the passage in this book recounting his visit to the birth-place of Shakspeare, as well as several others, furnish ample proof, and the sketch of the younger Littleton is managed with great adroitness and lightness of touch. There is too much of Shenstone and the Leasowes. Comparing the notes Miller gives us of his visit to Stratford-on-Avon with those on Shenstone's landscape-gardening, we cannot but regret that his modest avoidance of fields whence the 'originally luxuriant swathe' had been 'cut down and carried away,' prevented him from journeying to the shrines of England's great men and detained him so long beside the work of one of her ingenious versifiers.

The Scotchman is, of course, seen peeping from beneath the plaid of Miller as he journeys through England. 'To my eye,' he says, 'my countrymen,—and I have now seen them in almost every district of Scotland,—present an appearance of rugged strength which the English, though they take their place among the more robust European nations, do not exhibit; and I find the carefully-constructed tables of Professor Forbes, based on a large amount of actual experiment, corroborative of the impression. As tested by the *dynamometer*, the average strength of the full-grown Scot exceeds that of the full-grown Englishman by about one-twentieth,—to be sure, no very great difference, but quite enough in a prolonged contest, hand to hand, and man to man, with equal skill and courage on both sides, decidedly to turn the scale. The result of the conflict at Bannockburn, where, according to Barbour, steel rung upon armour in close fight for hours, and at Otterburn, where, according to Froissart, the English fought with the most obstinate bravery, may have a good deal hinged on this purely physical difference.' But if he dearly loves to put in a good word for Scotland, he can do justice to England. 'Scotland has produced no Shakspeare; Burns and Sir Walter Scott united would fall short of the stature of the giant of Avon. Of Milton we have not even a representative. A Scotch poet has been injudiciously named as not greatly inferior; but I shall not do wrong to the memory of an ingenious young man, cut off just as he had mastered his powers, by naming him again in a connection so perilous. *He*, at least, was guiltless of the comparison; and it would be cruel to involve him in the ridicule which it is suited to excite. Bacon is as exclusively unique as Milton, and as exclusively English; and though the grandfather of Newton was a Scotchman, we have certainly no Scotch Sir Isaac. I question, indeed, whether any Scotchman attains to the power of Locke; there is as much solid thinking in his Essay on the Human Understanding, greatly as it has become the fashion of the age to depreciate it, and notwithstanding its fundamental error, as in the works of all our Scotch metaphysicians put together.' Few people, to whatever school of philosophy they belong, would now agree with Miller in setting Locke above Hume; but we cannot read these words without feeling that, though he loved Scotland much, he loved truth more.

An apology may, perhaps, be thought necessary for quoting from a book which you see, beside the popular novels of the day, at London Railway Stations; but Miller's acknowledgment of what he owed to the literature of England, suggested by his visit to Poet's Corner in Westminster Abbey, is so beautiful and so strictly autobiographic, that I must take liberty to insert it. 'I had no strong emotions,' he says, with signal honesty, 'to exhibit when pacing along the pavement in this celebrated place, nor would I have exhibited them if I had.' But the reader will feel that deep and true emotion pervades the words which follow. 'There was poor Goldsmith; he had been my companion for thirty years; I had been first introduced to him through the medium of a common school collection, when a little boy in the humblest English class of a parish school; and I had kept up the acquaintance ever since. There, too, was Addison, whom I had known so long, and, in his true poems, his prose ones, had loved so much; and there were Gay, and Prior, and Cowley, and Thomson, and Chaucer, and Spenser, and Milton; and there, too, on a slab on the floor, with the freshness of recent interment still palpable about it, as if to indicate the race at least not *long* extinct, was the name of Thomas Campbell. I had got fairly among my patrons and benefactors. How often, shut out for months and years together from all literary converse with the living, had they been almost my only companions,—my unseen associates, who, in the rude work-shed, lightened my labours by the music of their numbers, and who, in my evening walks, that would have been so solitary save for them, expanded my intellect by the solid bulk of their thinking, and gave me eyes, by their exquisite descriptions, to look at nature! How thoroughly, too, had they served to break down in my mind at least the narrow and more illiberal partialities of country, leaving untouched, however, all that was worthy of being cherished in my attachment to poor old Scotland! I learned to deem the English poet not less my countryman than the Scot, if I but felt the true human heart beating in his bosom; and the intense prejudices which I had imbibed, when almost a child, from the fiery narratives of Blind Harry and of Barbour, melted away, like snow-wreaths from before the sun, under the genial influences of the glowing poesy of England. It is not the harp of Orpheus that will effectually tame the wild beast which lies ambushing in human nature, and is ever and anon breaking forth on the nations, in cruel, desolating war. The work of giving peace to the earth awaits those Divine harmonies which breathe from the Lyre of Inspiration, when swept by the Spirit of God. And yet the harp of Orpheus does exert an auxiliary power. It is of the nature of its songs,—so rich in the human sympathies, so charged with the thoughts, the imaginings, the hopes, the wishes, which it is the constitution of humanity to conceive and entertain,—it is of their nature to make us feel that the nations are all of one blood,—that man is our brother, and the world our country.'

Though this book is less geological than most of his other works, there is none, I think, in which his power as a geologist is more characteristically displayed. As a naturalist, as a palæontologist, he has been often excelled; but as a geologist, in the strict sense of the term, as one whose aim it is to see the rocky bones of a country beneath its robe of broidered green, he will not easily be matched. For this his work in the quarry, his fifteen years' journeying through various geological districts, had been an education, potent and precious, before he thought of formally studying geology. England was new to him, and he had but a few weeks in which to make himself acquainted with its geological aspects; yet he apprehended every essential feature of English landscape, and sets before us, in brief, rapid, vivid outline, the geological anatomy of the country. The breadth of his handling is superb, revealing the artist and the master. And when he deals with intricate and complex matters, when he gets among the coal-measures and the salt-works, his lucidity is as remarkable as his breadth. The light never is so dazzling as to obscure the object on which it falls; and yet we feel that the light itself is pleasant to the eyes, and that its touch lends beauty to that which it illumines.

'Sir, you have an eye,' Pope's friends in need, when

they wanted to get something out of him, used to observe. Thanks, by the way, to them, for so doing, since they would not, even for purposes of adulation, have ventured on the remark, if there had not been conspicuous splendour in the orb. 'Sir, you have an eye,' was what English critics said when Miller's chapters, detailing his impressions, were published in the columns of the *Witness*. A Birmingham editor, ignorant as to Miller, and fancying, for what reason I know not, that he was a 'dominie,' quoted passages descriptive of Birmingham and its district, and said that the writer must be a very remarkable man, since he had seen a great deal which had escaped the observation of the natives, but which, on its being pointed out to them, they also could see. Mr William Drummond, then sub-editor of the *Witness* and an esteemed friend of Miller's, showed him the article in the Birmingham paper, and he had a hearty laugh at it. Truth to speak, he had an eye that was worth its place in a man's head; searching, inevitable, keen, swift, sure; which gathered information at every moment and in all places, to be hoarded up in the cells of a memory which seems never to have lost an atom of the store.

His judgment on those two grand English institutions, the Church of England and Dissent, is shrewd. He discerns the incommensurable strength of the one, and the ineradicable vitality of the other, representing as each does a separate phase of the individualism of English character. While pronouncing the Englishman distinctly superior to the Scot as an asserter of civil independence, he recognizes that, for all religious ends and aims, the Scot is gregarious, the Englishman individual. The Anglican Church does not, for all its Bishops and Convocations, hide from him the fact that it is essentially a huge, amorphous, loosely-rolling mass of congregational units, the rector king in his own castle, the parish a little realm in itself only vaguely related to other similar realms. Religious Dissent in England he believed to have sunk wholly into political Dissidence. Alliance between religion and politics always excited his jealousy, and the time when he visited England was perhaps unfavourable for understanding the character and estimating the prospects of English Nonconformity. The Puritan watchwords had fossilized; but they had hardly yet visibly begun to be formed into a concrete for the foundation of Free Churches.

It is impossible not to see that Miller's heart warmed to England every day he continued within her borders. He was surprised with the frankness, the ready hospitality and generosity of the people. The fact clearly is, broad as was his accent and plain his garb, that he had in England as elsewhere an irresistible charm for every one who came near him. The old lady who, after a chat in a railway carriage, astonished him by an invitation to visit her, had perceived in that little time that he was gentle and good. Readers of the *First Impressions* will recollect the Temperance Coffee-House of Dudley, with its eight-year-old orator, who savagely denounced every one that tasted wine or strong drink. No preface, therefore, is required to the following, with which I have been favoured by an intelligent correspondent: 'It was about 1849 that I arrived one afternoon in Dudley, and took up my abode in the Temperance Coffee-House there. It was a very unpretending hostelry, but clean and comfortable. The thing that soon arrested me was that though I had never before been in this town, yet the tile-paved kitchen, the elderly hostess, her daughter, the little boy drawing at table, in fact, the whole establishment and its *dramatis personæ* seemed known to me. What kind of new optical or mental illusion was this? At last the idea dawned upon me that my memory was reproducing one of Hugh Miller's graphic descriptions in his *First Impressions of England and its People.* The story of the spinet-player, the account of the boy before me who spoke at Temperance Meetings, and other living touches came fast upon me. On my asking the head of the house whether a Scotchman had once lived there, she eagerly took up the theme, and in one almost incessant flow ran on thus: "Oh, he was such a nice man. But then he had odd tastes. Eliza there used to take him to quarries and out-of-the-way places all round the town, and there he chopped and chopped away with a hammer till he filled a bag with stones. I never knew such a man. What could he ever do with all these bits of stone? But he was so kind to us all. Then after he had gone away who should come to our door one day but the schoolmaster, with a book in his hand. 'Here's a fine description of you all,' said he. 'The Scotchman has been writing about you.' So there, to be sure, was such a nice account about my son playing upon the spinet, and Eliza, and Tommy, and me. Who'd 'a thought that he would write about us? Do you know him, sir, and can you tell me what he does in Scotland?" "I can't say," I replied, "that he is an acquaintance of mine, but, like many people in the North, I know a good deal about him. He has raised himself from being a workman to be a writer of remarkable books, and he is Editor at present of the *Witness*, the organ of a body called the Free Church of Scotland." "La, now, sir, do you say so?" was the overpowering reply. "Well, do you know, *my* eldest son plays the organ at Penkridge?" More followed that need not be repeated, about the patronage which followed the son thither, but my efforts to conceal the tendency to laughter produced by the two organs may easily be imagined.'

CHAPTER II.

SCIENCE AND RELIGION.

'SUCH is the state of progression in geological science that the geologist who stands still but for a very little must be content to find himself left behind.' The words are Hugh Miller's,—they occur in his preface to the first edition of the *Old Red Sandstone.* Their application is of course peculiarly forcible to the geologist whose activity has been arrested by death. No man can do more than his own piece of work in science, and the question on which our estimate of the merit of a scientific worker must depend is not whether he penetrated to the limits of any one province in nature, or uttered the final and absolute truth as to any one of nature's laws and processes, but whether he did the work he professed to do faithfully, honestly, and, to the point to which he carried it, thoroughly.

As science continues to advance, the several positions taken up by Hugh Miller, in prosecuting the sublime enterprise of proving the existence, and illustrating the character, of God from His works, may or may not prove tenable. Without question some of them would now be abandoned by the most eminent geologists. On taking the chair as President of the Royal Physical Society of Edinburgh, in 1852, he enunciated the following proposition: 'There is no truth more thoroughly ascertained than that the great Tertiary, Secondary, and Palæozoic divisions represent, in the history of the globe, periods as definitely distinct and separate from each other as the modern from the ancient history of Europe, or the events which took place previous to the Christian era from those that date in the subsequent centuries which we reckon from it. All over the globe, too, in the great Palæozoic division, the Carboniferous system is found to overlie the system of the Old Red Sandstone, and that, in turn, the widely developed Silurian system.' Having spoken thus without evoking one dissentient symptom in his audience, he could expect an affirmative answer to the question which he proceeded to put: 'I would ask such of the gentlemen whom I now address as have studied the subject most thoroughly, whether,—at those grand lines of division between the Palæozoic and Secondary, and again between the Secondary and Tertiary periods, at which the entire type of organic being alters, so that all on the one side of the gap belongs to one fashion, and all on the other to another and wholly different fashion,—whether they have not been as thoroughly impressed with the conviction that there existed a Creative Agent, to whom the sudden change was owing, as if they themselves had witnessed the miracle of Creation?' Professor Huxley now declares without contradiction that it is 'admitted by all the best authorities that neither similarity of mineral composition, nor of physical character, nor even direct continuity of stratum, are *absolute* proofs of the synchronism of even approximated sedimentary strata: while, for distant deposits, there seems to be no kind of physical evidence attainable of a nature competent to decide whether such deposits were formed simultaneously, or whether they possess any given difference of antiquity.' In the same historical tone, as of one referring to what is no longer open to dispute, Mr Huxley continues:—'For anything that geology or palæontology are able to show to the contrary, a Devonian fauna and flora in the British Islands may have been contemporaneous with Silurian life in North America, and with a Carboniferous fauna and flora in Africa. Geographical provinces and zones may have been as distinctly marked in the Palæozoic epoch as at present, and those seemingly sudden appearances of new genera and species, which we ascribe to new creation, may be simple results of migration.'

The tendency of scientific research throughout every province of nature has been to obliterate lines of demarcation, and to show, stretching beyond us into the infinitude both of time and space, immeasurable curves and undulations of unity. The definite proof afforded by spectrum analysis of the sameness of matter throughout the solar and stellar expanses marked a stage of sublime advancement in our conception of the harmony of things; and correspondences, indubitable though mysterious, between terrestrial magnetism, the spots of the sun, and those systems of aerolites which have recently attracted so much of the attention of philosophers, suggest that the unities of nature are as intimate and as wonderful as her diversities. I venture to throw out the suggestion that a key lately put into our hands by Professor Tyndall may possibly unlock for us the secret that there is unity of life, as well as unity of matter, throughout space. Germs of life, Professor Tyndall has taught us, are of what may be called infinitesimal smallness; and what proof have we that, if aerolites can traverse space, life germs cannot traverse space likewise?

The unity of life throughout even the solar system may be no more than a speculative guess, but something not far short of authoritative unanimity in the scientific world supports Professor Huxley in maintaining the unity of succession in the vital organisms of our planet, against the theory of an indefinite number of creations, each preceded by a cataclysm destructive of all the life upon earth. There is no paradox, however, in the assertion that, at the time Miller wrote, and with the facts Miller possessed, the genuine service which it was his part to render to science implied and required that he should controvert the popular theories of development. To refuse to believe on insufficient evidence,—to point out in what respect the evidence is insufficient,—this is the most effective preparation that can be made for the discovery of truth upon any subject. If philosophers had been content, twenty-five years ago, to accept as conclusively established the theories of Lamarck, Demaillet, and the 'Vestiges,' those researches by which more light has been thrown upon the history of living things in our planet than had been cast in ten preceding centuries, might never have been undertaken. The highest service that can be rendered to science is the discovery of new truth; the next highest is the exposure of false pretensions to the discovery of new truth. In pointing out the unsoundness of the development theories in vogue in his day, Miller was in the same line of battle with his ablest scientific contemporaries. How he would have comported himself if he had lived to see the publication of Mr Darwin's work on the Origin of Species, to witness its reception throughout the world of science, to follow the lines of research and speculation to which it pointed the way, we need not inquire. This, however, I will venture to say: first, that he would have distinctly declared, as, indeed, he did with reference to the old theory of development, that Mr Darwin's doctrine has no necessary affinity with atheism; secondly, that he would have subjected the facts and reasonings of Mr Darwin and his followers to a scrutiny more searching than they have yet received; and thirdly, that, if he had found them incontrovertible, he would, without a moment's hesitation, have proclaimed his assent to them. His reverence for God's truth was infinitely deeper than his regard for his own conceptions of it. That truth he would accept, howsoever and whensoever it was revealed, conscious that the wilful misreading of nature is a sin against Him whose ordinance nature is. Strange imagination, that the Ineffable One is less honoured by reverent caution and hesitation,—by child-like fingering among the letters of His name and child-like diffidence in spelling it out,—than by vociferous dogmatism on the subject! Hugh Miller dared not force his conscience to lie *to* God by bribing his intellect to lie *for* God. His writings on those high questions which belong to the border-land between science and theology have a perennial value, not because of the finality of their matter, but because of the rightness of their manner. With true reverence and sterling integrity, he discoursed of the relations between physical and moral law in this universe, and the reciprocal bearings of God's revelation of Himself in matter and by matter—His Works; and His revelation of Himself in mind and by mind—His Word.

It is to be noted that Miller, while contending against the inadequate and erroneous theories of development which presented themselves to his notice, did not propound a cut-and-dry rival theory, or profess to have formed a scheme of his own upon which to map out and explain every mystery of organized being. With no fixed principle except that the Earth is the Lord's and the fulness thereof, that the laws of life and of death, of matter and of mind, are His laws, and that, on the whole, the procession of existence, moving at times through breadths of gloom and masses of shadow, tends onwards and upwards, he recognized all facts, admitted all difficulties, and preferred to state them honestly and contemplate them deliberately rather than bind them together with logical formulas. The fact of the degradation of races, for example,—of the startling delight which nature appears occasionally to take in eccentric and asymmetrical types,—of the production of what, to human apprehension and sensibility, look mere masterpieces of the horrible and repulsive, 'worm-like creatures, without eyes, without moveable jaws, without vertebral joints, without scales, always enveloped in slime,' like the 'glutinous hag' of the Moray Frith,—had evidently awakened the deepest feelings of wonder in Hugh Miller. 'An ingenious theorist,' he said, 'not much disposed to distinguish between the minor and the master laws of organized being,' could get up 'quite as unexceptionable a theory of degradation as of development.' But the relation of the 'minor laws,' to which presumably we owe the glutinous hag, and the 'master laws,' to which we owe the gazelle and the nightingale, the rose and the wheat-plant, he did not attempt to define, contenting himself with the denial that either are mere processes of physical nature, and with the assertion that both are parts of His government who is wonderful in working, and whose ways are past finding out. Familiar as Miller was with the tremendous reasonings of Hume's Dialogues on Natural Religion, he was not one to take refuge in the amiable platitudes of the Rosa Matilda school, or to wrap himself from the lightnings in garlands of flowers. He would not have shrunk from admitting, in all that width of extension and precision of application which Mr Darwin and his school have shown to belong to it, the law of pain throughout the world of life. Survival of the stronger, with extermination of the weaker by famine and anguish, he would, I think, have allowed to be the law of physical nature. He would, at the same time, have maintained that 'a gradual progress towards perfection,' though, as Mr Huxley points out, it forms no necessary part of the Darwinian creed, is revealed in nature. He would have dwelt with Goethe on the fact that death itself is but a subtle contrivance by which more life is obtained. Above all, he would have insisted, as neither Hume, Goethe, nor Darwin have insisted, upon sin as an explanation of misery, and upon the promise of redemption and immortality through Christ, as sending a stream of celestial radiance far up into the hollow of the terrestrial night.

It was a heart-rending experience for Miller that the erect, fearless, friendly attitude which he felt ought to be that of religion towards scientific truth gave offence to religious people. He would have been so glad, in his visit to England, if he could have convicted the Anglo-Catholics, and them alone, of distrust and alarm with regard to science, and could have left them to nurse their childish Mediævalism in all departments, while the sturdier sons of the Reformers welcomed every revelation of science as Divine. His Evangelical Religion was of that early and vigorous type which, having complete faith in truth, had no fear that there might turn out to be heresy in science. He belonged to the great old Evangelical school, and was perhaps its last representative to whom the title great can be accorded. Whatever that school may have become, it is a fact of history that no religious school of recent times has been so great a power among men. It can lay claim to Cowper; and Cowper is acknowledged to have been one of the quickeners of English poetry from formal correctness and frosty elegance into the freshness and luxuriance of its recent period. It was before its fervid inspirations that the thin, plausible, infinitely clever and witty scepticism of the eighteenth century gave way. To it, directly or indirectly, that missionary movement of recent times, which seems destined to exert a mighty influence on the history of the world, owed its origin. It made itself felt in politics in the form of an intense, indefatigable, sympathetic humanity, and passed, in the same spirit, into all modern philanthropic enterprise. It showed its vitality and strength in Scotland by bringing about what David Hume would certainly have pronounced an impossibility in the nature of things, the relinquishment, by a large body of clergy, of the privileges and pay of a State-Church. Its express contributions, of a high class, to English literature, have not been numerous. The works of Hugh Miller are, on the whole, its most important literary performances. But its indirect influences on English literature have been remarkable. To it, without question, is due that lofty and earnest spirit which has rendered much of the doubt of the present century more religious than much of the faith of the last. All Macaulay's critics have taken note of those touches of old Hebrew grandeur which sublime his noblest passages, nor has the observation failed that they came from that Hebrew Bible which he read in the house of his Evangelical father. In Mr Ruskin's earliest, which many consider also his greatest books,—for their morning softness and depth of glistening blue may be thought to have more enduring loveliness than the perpetual sword-gleam of his later writings,—passage after passage, inferior in eloquence to none in the English language, suggest the idea of an Evangelical Bossuet or Jeremy Taylor. And has it been enough considered that it was from the cottage of a Scottish Old Light Seceder that Thomas Carlyle went forth to write those books which, for good or for evil, have changed the current of English and American, possibly also of European, literature?

Hugh Miller saw with grief inexpressible that the Evangelical party in England, with its *Record* newspaper and its Dean Cockburns, was taking the fatally wrong turn in the matter of science and religion. In Scotland, during his lifetime, there was not much cause for alarm. While Fleming, as Professor of Geology in the Free Church College, sent out clergymen to teach and preach that death and pain existed myriads of ages before Adam, that the starry heavens and the earth are of an antiquity to be measured by millions of years, that the Noachian deluge was local, and while the leading religionists of Scotland recognized with gratitude and approval the importance of Miller's own services in the cause of religious truth, he had no occasion to fear. But the Evangelicals of England never showed the courage and faithfulness in this matter of the Evangelicals of Scotland, and the lamentable exhibitions we have recently had have painfully demonstrated that Hugh Miller's expostulations, printed in his *Impressions of England and its*

People, have been of none effect. It is too probable that if he had lived twenty years longer, he would have found himself looked on askance by some Evangelical magnates of his own country. He had never hesitated to abandon an old position when he found it to be untenable, and he had not been long in his grave when there were unmistakeable signs that part of clerical Scotland was disposed to fall back upon at least one of the lines of defence which he had given up, to wit, the hypothesis of a cataclysm or chaos, embracing the whole world, immediately antecedent to the human period, and separating the entire succession of geological formations from the six days' work of Genesis. Instead of this view he adopted that which assigns an immensely protracted duration to the successive Mosaic days. In his latest work, the *Testimony of the Rocks*, he expounded the Age theory of Mosaic geology with admirable breadth and lucidity; and it is generally admitted that this is the sole hypothesis which can now be maintained with any show of plausibility by those who hold Christian theologians bound to furnish a scheme of reconciliation between geology and Genesis.

Hugh Miller maintained the entire independence of science. 'No scientific question,' he says, 'was ever yet settled dogmatically, or ever will. If the question be one in the science of numbers, it must be settled arithmetically; if in the science of geometry, it must be settled mathematically; if in the science of chemistry, it must be settled experimentally. . . . As men have yielded to astronomy the right of decision in all astronomical questions, so must they resign to geology the settlement of all geological ones.' Again: 'The geologist, as certainly as the theologian, has a province exclusively his own; and were the theologian ever to remember that the Scriptures could not possibly have been given to us as revelations of scientific truth, seeing that a single scientific truth they never yet revealed, and the geologist that it must be in vain to seek in science those truths which lead to salvation, seeing that in science these truths were never yet found, there would be little danger even of difference among them, and none of collision.' This is from the *Testimony of the Rocks.* So also is the following, which proves that if he had once leaned too strongly towards catastrophism, the balance was coming right: 'What more natural to expect, or rational to hold, than that the Unchangeable One should have wrought in all time after one general type and pattern, or than that we may seek, in the hope of finding, meet correspondences and striking analogies between His revealed workings during the human period, and His previous workings of old during the geologic periods?'

An erroneous idea, however, would be conveyed of the general scope and pleasantness of Hugh Miller's writings if they were regarded as constantly, or mainly, controversial. The exquisite accuracy, combined with imaginative beauty, of their descriptions of natural scenes and objects; the scientific fitness and poetic felicity of the illustrations they present of the Divine mind acting in matter; the ethical force of their exhibition of the mutual adaptation of man and his world, the lofty conception they present of man as the tiller of the Earth-garden, the fellow-worker with God in subduing and beautifying the planet and in raising to the height of spiritual and physical perfection the human family; these, and many other excellences which will occur to all acquainted with those books, will render them precious and fascinating when the particular controversies on which they touch have assumed new aspects or have given place to others.

Having been a lover of nature all his life, Miller deliberately chose, at the time when his powers were reaching their maturity, to enroll himself among the votaries of science. For about thirty years the activity in which his mind chiefly delighted was that of scientific research and reflection. Other labour might be strenuous, but it was hard; here he obeyed his ruling impulse, and to obey was to enjoy. Science never failed him. Year after year, as his frame became unstrung, and his nerves shaken by toil at the editorial desk, he returned to the mountains, and in long pedestrian expeditions sought that health which was sure to come with the pure air, and the bracing exercise, and the glow of faculties at their congenial work.

What he could have done as a man of science can never be fully known. Almost all the scientific works he has left are more or less marred by the controversial ends he kept in view, or by imperfect unity arising from publication in form of lectures. The *Cruise of the Betsy,* in which he gives an account of his geological investigations among the Hebrides, has been preferred by some critics to his most elaborate controversial works. Had he lived to realize the ambition of his life, to write a comprehensive book on the geology of Scotland, the world would have been able to take by sight, what must now, to some considerable extent, be taken by faith, the measure of his greatness. If, after an eight weeks' ramble, he made so much of the geology of England as is made in the *First Impressions*, what might we not have expected from the observation and study of thirty years in his beloved Scotland? Records, no doubt, exist, and have been published, of the geological tours in which he prepared himself for his high undertaking, but these are no more than *disjecta membra*, valuable and interesting apart, yet offering no hint by which imagination itself can picture the living whole into which they would have been moulded by the hand of Miller. No man ever knew the geology of Scotland as he knew it. Others may have grappled as vigorously with a particular problem suggested by it here and there, but who ever studied it so long and so lovingly as a whole? And not only would the book have been a masterpiece of geological literature; it would have been unique as a manual of Scottish landscape, doing in prose for Scottish scenery what Scott has done for it in poetry; and it would assuredly have displayed, in soft idealizing gleams across the rugged features of its general framework, that proud and tender affection for Scotland, that reverent acquaintance with all that is noblest in Scottish history and character, which fitted Miller so admirably to have become the historian of his country. As I think of what a work the Geology of Scotland by Hugh Miller might have been, I cannot but recall those words, more fanciful, perhaps, than Thackeray commonly permitted to his masculine genius, in which he referred to an unfinished picture, containing a few figures dimly sketched, by his friend Leslie. 'The darkling forest,' he said, would have grown around them, with the stars glittering from the midsummer sky: the flowers at the queen's feet, and the boughs and foliage about her. They were dwelling in the artist's mind, no doubt, and would have been developed by that patient, faithful, admirable genius: but the busy brain stopped working, the skilful hand fell lifeless, the loving, honest heart ceased to beat. What was she to have been—that fair Titania—when perfected by the patient skill of the poet? Is there record kept anywhere of fancies conceived, beautiful, unborn? Some day will they assume form in some yet undeveloped light? They say our words, once out of our lips, go travelling *in omne ævum*, reverberating for ever and ever. If our words, why not our thoughts? If the Has Been, why not the Might Have Been? Some day our spirits may be permitted to walk in galleries of fancies more wondrous and beautiful than any achieved works which at present we see, and our minds to behold and delight in masterpieces which poets' and artists' minds have fathered and conceived only.' One has difficulty in realizing that it was the giant hand of Thackeray which traced these tender pencillings of fancy. It is no fancy, however, that, in the future to which Christians look forward, neither will the energies of the soul be paralyzed, nor the definite lines of personality be effaced. So long as Hugh Miller's personality retains its mould, he will think with love of Scotland, and be wafted in memory to her pine-clad hills and craggy shores; and if the converse of friends beyond the bourne has any analogy to that of friends on earth, he may vary the note of celestial felicity by dwelling on the wonders and beauties of that great work which he was not permitted to achieve on earth.

CHAPTER III.

THE LAST EDINBURGH PERIOD.

THE life of Hugh Miller, so varied and eventful in its early period, formed no exception, in the ten years preceding its close, to that placid uniformity which proverbially characterizes the lives of literary men, and which precludes detailed description.

During those years he conducted the *Witness* with steady and ever-broadening success, speaking his weighty word on every important question as it arose, and widely accepted as a guide of opinion. mis views on the subject of education were in advance of those of the body of his contemporaries, especially his ecclesiastical contemporaries. He maintained them with his usual courage and frankness, careless of the bitter obloquy to which they exposed him in some quarters. The state of public opinion at this moment proves that the conclusions at which he arrived have become in substance those of the nation. He advocated the exclusion of denominationalism from the machinery of popular education, and the method by which he proposed to effect this object was that of intrusting the power in connection with education to the body of the people. It is scarce necessary to remark that the very idea of excluding the Bible from the National Schools would have been abhorrent to him; but no limit can be set to the decision with which he would have forbidden the inculcation of distinctive denominational tenets by National schoolmasters.

Socially his position was fully established. The words of Miss Dunbar, that the day was coming when his country's greatest would court his acquaintance, had been literally fulfilled, and there was no circle in Edinburgh or in London which would not have felt itself honoured by his presence. In the communications addressed to him by men of rank or reputation, it was assumed as a matter of course that his place was among the intellectual aristocracy of his time, and that he was one of those whose acquaintance conferred distinction. Again and again did the Duke of Argyll solicit the honour of a visit from Miller, resting his hope of a favourable reply, not on his own aristocratic birth, but on community of scientific interests and pursuits. The difficulty was to overcome that feeling on the part of Miller which we found himself describing as diffidence, but which is, perhaps, insufficiently characterized by the term; a feeling which partook little of self-distrust, and still less of haughty coldness, but consisted principally in a shy and sensitive reserve, a consciousness that his mental instruments could work perfectly only in their own placid atmosphere. He was totally devoid of ambition to shine in mixed and fashionable society. On the whole I should say that the word 'shyness' most correctly describes the quality in Hugh Miller which led him inexorably though courteously to decline invitations like that of the Duke of Argyll.

He was not a man to have many intimate friends, and few indeed of those whom he knew subsequently to coming to Edinburgh did he take to his heart with that impassioned ardour of affection which marked his Cromarty friendships. One of the few, however, respecting his regard for whom such language would hardly be out of place, was Mr Maitland Macgill Crichton, of Rankeillour. The rugged, simple-hearted country gentleman, intrepid in thought and word, sincere beyond the tolerance of guile, impatient of generalship and suspicious of expediency, made his way at once to Hugh Miller's heart. There was that difference and that agreement between them which are said to go to the making of the happiest friendships,—Crichton all impulse and impetuosity, Miller all calmness and reflection, both intensely devoted to the cause of Scotland and her Church, both averse to the sly prudences of party leadership, both too strong to be capable of cunning. They perfectly understood and perfectly trusted each other. When Miller determined to leave the town of Edinburgh and locate himself in the suburbs, his friend accompanied him in long roamings for weeks together, the main purpose of house-hunting being doubtless combined with minute exploration of the environs of a city which Miller dearly loved. The word used by him in estimating Mr Macgill Crichton's services to the Free Church was 'gigantic.' Crichton, for his part, threw all the vehemence of his nature into his love for his friend, and insisted, almost with petulance, on his right to have at least an annual visit from Miller. The stated time for the trip to Rankeillour was at Christmas. Sir David Brewster was generally of the party. In this circle Hugh felt himself as much at home as at Cromarty. The completeness and tenderness of Crichton's trust in Miller were evinced by his naming him in his will,—the sole addition to his own brothers,—as tutor and curator to his three youngest children.

For seven years of his last residence in Edinburgh, Chalmers was still in life. The relation between the men was not so much personal friendship as what I may call comprehensive, unwavering alliance on all points of sentiment and opinion. Chalmers had the large and masculine sagacity,—a thing of the heart as well as the brain,—to accept Miller for what he was, a strong-featured, strong-charactered, original man, whose services to a cause he loved were sure to be of incalculable importance, but who was not to be expected to pull in party harness, or to tame down his natural port and action to the step of party drill. Such a man, Chalmers felt, ought not to be fretted by little criticisms. A word, now and then, of frank sympathy, was what he deserved, and such a word Chalmers was always ready to speak. 'You are getting on bravely with your education battle,' it might be, or 'Many, many thanks for your noble article on the West Port;' flashes of radiance, slight but precious, which would kindle a gleam of proud encouragement in the eye of Miller. Chalmers frequently looked in upon the family circle in Archibald Place, and Mrs Miller recollects the 'apostolic fervour,' softened no doubt with fatherly kindness, with which he 'put his hand on Harriet's golden head and blessed her.' A generation has arisen which knew not Chalmers in the body, and already, when the defects of his theological erudition, and the lack of searching and exhaustive power in much of his thinking, are considered, there begin to be whispered inquiries as to wherein his greatness consisted. Even in his books, with their glowing earnestness and massive force, there is enough, I think, to rank him with the sons of the mighty; but it was evidently in himself, in the influence of his personal weight and worth, in the majestic nobleness of one for whom virtue was 'not the price of heaven but heaven itself,' that lay the spell which constrained men like Thomas Carlyle and Hugh Miller to pay him kingly honours. We may read Chalmers's books and think him overrated; but we cannot read Lockhart's description of him, or observe how profoundly his contemporaries felt the contagion of his moral intensity, without recognizing him for a man of great genius. One of his characteristics was an exquisite and joyful appreciation of excellence, however different it might be from his own; and I can well understand his enthusiastic delight in the compositions of Miller, whose sense of literary form was much finer than his own.

In endeavouring to place before the reader a just picture of Miller's Edinburgh life, I am happy to be able to avail myself of the memoranda of a lady who knew him intimately during the whole of the period, and whose delineation will be perceived by all who even partially shared that privilege to be delicately true.

PERSONAL RECOLLECTIONS BY A LADY.

'I first met Hugh Miller in the beginning of 1840, when he came to Edinburgh to edit the *Witness*. He dined at our house with some of the gentlemen who had been the means of setting up that paper. I had the good fortune to sit next to him at dinner, and so had a good deal of conversation with him. We somehow got on the poets, and found several of them to be intimate mutual friends. The delight of that conversation is still fresh in my remembrance. His appearance then was that of a superior working-man in his Sunday dress. His head was bent forward as he sat, but when he spoke he looked one full in the face with his sagacious and thoughtful eye. There was directness in all he said; to have spoken without having something to say would never have occurred to him. He had not the light, easy, inaccurate manner of speech one usually meets with,—every word was deliberate, and might have been printed. There was a total want of self-assertion about him, but at the same time a dignified simplicity in the way he *placed his mind alongside* that of the person with whom he conversed. There is no doubt he was somewhat shy and proud and jealous of his independence, and some found him inaccessible from this cause; but I happened somehow to overleap these barriers the first day I met him, and they were never interposed during our further intercourse.

'When his family came to town, it was with no common pleasure that I recognized in Mrs Miller a young lady who had been a classfellow of mine. This led me frequently to her house, and gave me the opportunity of seeing her husband, and my admiration of and interest in him increased every time I met him. There could be no greater or more exciting pleasure than to converse with Hugh Miller. He did not harangue but conversed, and raised those with whom he did so for the time to his own level. One felt amazed to hear one's own trifling remarks made the means of bringing out his stores of observation and thought; and if by good fortune one brought to his notice some to him yet unknown fact or quotation from the poets, to whose "terrible sagacity" he loved so often to refer, it was indeed gratifying to see the look of pleased attention with which he listened. It seemed as if he could not but be thinking, and that everything brought grist to his mill, set agoing some new train of thought, or confirmed some old one. Then, of course, there were all the exciting subjects of the time,—a time never to be forgotten by those who lived in it, and shared the principles and emotions of those engaged in the struggle which ended in the formation of the Free Church. If one happened to see him the day before the publication of the *Witness*, then one was sure to see in the next day's article some of the observations that had fallen from him the day before. Some, not all, for surely every one who knew Hugh Miller must have felt, as I do, that his mind was deeper, richer, more far-sighted, than any of his written utterances showed it to be.

'How one wishes one could recall some of those pleasant meetings, but

> "The path we came by, thorn and flower,
> Is shadowed by the growing hour,
> Lest life should fail in looking back."

Among the gleams of the past, seen through that shadowed vista, is one of a summer evening when, after dining with the Millers at Sylvan Place, we walked to Blackford Hill. In the valley behind the hill he and the other gentlemen of the party tried who could throw furthest some of the large stones lying about. It was soon seen whose arm had had the advantage of being strengthened by labour, so greatly did his throw exceed the others. I think it was at that time that, referring to the change from labour with the chisel in the open air to labour at the desk, he said the last was far harder than the first, and that the change had come too late to him. "It was too late ere I was caught," was his expression.

'One evening the conversation turned on ghosts, *à propos* of Mrs Crowe's "Night Side of Nature," which had just been published. It was said that that lady had sought out every one who could tell her a ghost-story. He said he was glad she had not found him out, as he happened to have a store of such stories. On being asked to tell us some, he complied most readily, and told us several. I never heard such stories so told. I happened to have made the request, and he "held me with his eye" as he told them, an eye not "glittering," but pale and sad,—the saddest I ever knew. He seemed to see the scenes he described, and compelled one to see them too. It was evident he had been nurtured in the belief of these superstitions, and that in early life they must have had complete sway over his mind,—a sway that might be resumed in hours of weakness. Then, however, he disclaimed all belief in them; and in the conversation which preceded the stories, had made some forcible remarks on the frequent combination in the same person of scepticism and credulity, and on the difference between a real faith in revealed truth, and the ready belief in lying wonders, then beginning to be common.

'Whoever has read that chapter in Hugh Miller's autobiography which relates his religious history will be aware that the inner life and conflict of the true follower of Christ would not in him come to the surface in conversation. But no one could have intercourse with him without plainly perceiving his deep religious reverence and faith. I have heard the opinion expressed, that doubts must have beset him to the end of his life. The very contrary was my impression. He had laid these "spectres of the mind" in earlier life, and they did not rise to haunt him again. He saw the rising tide of scepticism, and foresaw that it was bringing misery to the world and trial to the Church, and to raise some bulwarks against it was his earnest desire. His attempts to harmonize Scripture and geology were more for others than himself. All intercourse with him gave the assurance of his thorough belief that the book of revelation and that of nature were from the same hand, and that this would be manifest one day, though now the last was but half-read, and the first, perhaps, misread. His delight in tracing the proofs of the creation of man's mind in the likeness of his Maker's, as shown in the works of both, was evident to every one who had looked over his museum under his guidance, long before his lectures were published. When he referred to our Lord, it was usually as the *Adorable* Redeemer, an epithet which may seem somewhat formal in the increasing preference for the Saxon element in our language, but which fitly expressed the deep reverence with which he ever named Him. He delighted to observe or hear anything which led to the inference that those whose genius he admired had been led, sooner or later, to listen to the voice of that Adorable Redeemer, and learn of Him. An instance of this may be seen in his remarks on Shakspeare, when relating his visit to Stratford-on-Avon in *First Impressions of England.* I remember discussing with him the various passages in Shakspeare which allude to religious thought or feeling, especially that in Richard the Second, in which the poor king in prison, speaking of the multitude of thoughts fighting within him, says,

> "The better sort,—
> As thoughts of things divine,—are intermix'd
> With scruples, and do set the Word itself
> Against the Word:
> As thus,—Come, little ones; and then again,—
> It is as hard to come, as for a camel
> To thread the postern of a needle's eye."

I do not, however, remember Hugh Miller's remarks with sufficient accuracy to set them down.

'In the end of 1848 I happened, when calling on Mrs Miller in Stuart Street, to hear that her husband was giving some little lectures on geology to a few lady friends, and I was most kindly invited to join the party. We met on Saturday forenoons, and sat round a table on which he had arranged some specimens to illustrate what he was going to tell us about. These lectures were the germs of those he afterwards delivered before the Philosophical Institution, and which have been published under the title of the *Sketch-book of Popular Geology*. Any one who reads that volume may see how pleasant as well as instructive they were. But it cannot convey the interest of being taught by such a teacher, who thought no question too trivial to be answered, and explained himself by all manner of illustrations, homely or otherwise, but always comprehensible and distinct; while over everything his imagination threw an endless charm, and his earnest faith a deeper interest. His manner to women I always thought particularly good,—wholly wanting in flattery but full of gentle deference. Our meetings frequently ended by our enjoying Mrs Miller's hospitality and society at luncheon, when we witnessed the same gentle manner in his own family, and various little incidents which showed his strong parental love. As spring came on our lectures took place in the open air instead of in Mrs Miller's drawing-room, and we had some charming walks to shores and quarries in the neighbourhood of Portobello, and to Salisbury Crags and Arthur Seat, where the Queen's Drive had been lately opened, and afforded us many illustrations of what we had learnt from him during the winter. Never was geology more pleasantly studied.

'We spent a few weeks that autumn in the neighbourhood of Melrose, and having asked him to visit us on the banks of Tweed, I had the following note from him.

"Edinburgh, 2, Stuart Street, 21st August, 1849.

"Your kind note reached me as I was engaged on the last article for the *Witness* which I shall write for at least several weeks. The completion of my book* has at length set me at liberty; on Thursday I leave by the Wick Steamer for the extreme north; and on the Saturday, to which your kind invitation refers, I shall be sauntering, if the voyage be a prosperous one, not along the soft wooded banks of the Tweed, but along the bleak crags that overlord the Pentland Firth. *Your* river has all the beauty on its side, but the broad Pentland with its roaring eddies is by far the more magnificent *river* of the two. My book will not be fairly published until Saturday first; but on Saturday last I did myself the pleasure of forwarding copies to all my *lady-pupils*, though I am not sure that your copy has got further than C— Street. You will, I suspect, find the pure geology of it rather dry; but in the concluding chapters, more especially in the last taken in connection with the chapter on the *degradation principle*, you will, I think, find some thoughts that will interest you. A man who merely refutes an error, if it be an ingenious one and suited to fill the imagination, does only half his work. The void created ought to be filled with something as novel and curious as that which has been taken away; and in the chapters to which I refer I attempt embodying a theory compensating for the development one. I had a note yesterday from Mrs Miller. She was well when it was written, and in spirits, and just emerging from the bustle of a Free-Church-Manse marriage."

* Footprints of the Creator.

'He seemed to feel an increase of kindness to us, his "lady-pupils," after that pleasant season.

'My after-recollections of him at Stuart Street are fragmentary;—of sitting with him and Mrs Miller at the fireside late one dark winter afternoon discussing murderesses, from Lady Macbeth down to Mrs Manning, whose trial had just taken place, and reading the observations I had heard from him in the next *Witness*. Again,—looking over Johnston's Physical Atlas with him, and, on my observing that Cromarty was coloured with the Scandinavian tint, his assuring me with evident satisfaction that most of his ancestors were of that descent. Again, after his return from some visit to the north, his speaking of Nature being an increasing source of pleasure as we go on in life, unlike most other things. At the same time he referred to an anecdote of Thorwaldsen, who said to a friend, that his genius was decaying, for now he felt satisfied with his works, while formerly his idea had always gone far beyond what he could execute. Hugh Miller said he had felt the same, and felt, for the same reason, that he would never now write anything better than he had done.

'In the summer of 1856 there was an archæological collection exhibited in Edinburgh. In the beginning of August I went there early the last day it was open, and came upon Hugh Miller looking at the ethnographical department. He told me he had been ill, and was then on his way home after a week's recreation. I expressed surprise at this, as I thought I had noticed his hand in the *Witness* during the past month. He said I was right, but that he had written these articles during the time of the Assembly. After looking with him for a little at the stone and bronze weapons, I went further to look at some portraits of Mary Stuart, and at a print to which he directed my attention—the print which was the subject of conversation when the boy Walter Scott met Robert Burns. As I returned I found Hugh Miller standing by Robert Bruce's sword, which had been placed in an upright position in the centre of the gallery. He measured it, and then, turning to me, recited from the "Lord of the Isles" the lines referring to it. Lord Elgin had sent the sword, and we were speaking of his being of the same family as the King, when Hugh Miller told the story of the old Scotch lady of the name, who, on being asked if she was of King Robert Bruce's family, answered that the King was of *her* family.

'On one of the first days of December my sister and I went down to Portobello to call for Mrs Miller. We met him at the gate of Shrubmount, and he came into the drawing-room and remained while we were there. We thought him looking well, and remarked to each other afterwards that there was an increase of dignified ease in his manner, as of a man assured of his position.

'He showed us a stag's horn, found in the bank of a burn that flows into the Solway, which a friend had sent him. The burn, swollen by the rains in October, had carried away some of the soil from the bank, and the horn was seen sticking out from it. Though it was not whole, from what remained it was evident that it had belonged to a stag of at least sixteen tynes. This confirmed an opinion he had for some time formed, that before the British deer had been driven by man to the higher and poorer districts of the country, they had been a larger race. Lord Kinnaird, to whom he had shown it some days before, had mentioned another confirmatory fact. A friend of his in England, who had very much enlarged his deer-park, found that his deer were developing larger horns than usual. We spoke of

Thackeray's Lectures on the Georges, which he had just given in Edinburgh. Hugh Miller had not heard them. Thought Thackeray too like Swift in his estimate of human nature. Spoke of the admirable imitation of the style of the writers of Queen Anne's time in Esmond. *A propos* of Thackeray's comments on the "First Gentleman of Europe," he referred to the complete change in the use of the word "*genteel*," formerly used by the well-bred to signify good-breeding—and now only vulgar or ironical in its use.

'He showed us some newspapers sent to him from America, with an attack on him, in the American fashion, for an Article on "Dred" which had appeared in the *Witness*. He had not written the article (which was a laudatory one), and said he would not have so expressed himself. I said I had been sure that article was not written by him, but that I had been much struck by one on the immoral doctrines brought forward by American divines on behalf of slavery. *It* was written by him. He said it was most remarkable how these old doctrines, striking at the root of all morals, were brought up again in a new dress to serve a new cause. The conversation then turned on the chances of the continuance of the White supremacy in the Southern States. He thought the chances all in favour of the Whites from the superiority of race. From all he could learn there was no man of mark among the Blacks or Half-castes. Frederick Douglas was their most remarkable man. He had met him, and thought but little of him. He compared the supremacy of the Whites to that of the English over the Irish in Cromwell's time and afterwards—with this difference, that Cromwell's despotic supremacy was exercised with a good conscience, however sternly, while the immoral doctrines to which he had referred showed that this could not be altogether the case with the Southern Americans. He thought the religious parts of "Dred" very inferior to "Uncle Tom,"—rather sentimental and sickly. He referred to the opinions of Mrs Stowe's brother, Dr Beecher, on Sin and Eternal Punishments. "He would make us out to be ticket-of-leave men," he said with scorn.

'Ere the new year had come he had passed away from amongst us in that paroxysm of bodily and mental illness in which his strong will gave way, and he lost his self-control.

'Edinburgh streets have seemed emptier since I could no longer hope to meet there his rugged stalwart form, and see his pale abstracted eye brighten with the smile of kindly recognition.

'There is a city which the streams of the River of Life make glad. May we meet as fellow-citizens there!

'M. W.'

At Cromarty and Linlithgow Miller, as we saw, was an indefatigable and voluminous correspondent. In Edinburgh, penning two or three leading articles per week and with a book generally on hand, he required no further vent for his literary productiveness, and wrote no such letters as those which it formerly had been his delight to pour forth. He told his friends that they must consider the *Witness* a bi-weekly letter from him, and confined his epistolary performances to notes of reply to invitations, brief answers to geological querists, and the like. In this rapid fashion he corresponded—if correspondence it can be called—with a very large proportion of the most eminent scientific men of his time. Professors Owen, Agassiz, Sedgwick, Sir Charles Lyell, Sir David Brewster, Mr Mantell occur to me at the moment as among those who compared notes with him as a scientific peer and fellow-worker. But it is evident that, how distinguished soever these and others may be, and apt and seasonable as were Miller's brief scientific replies, these last were of an occasional nature, and, the rather that they touch almost invariably on matters treated of in his printed works, are not generally suited for publication. The course I have adopted is to glean from the memorials of those years such letters and jottings by Miller as illustrate his manner of life and bring into view the several occurrences by which it was diversified, adding, as formerly, such notes or letters from his correspondents as tend to elucidate the narrative or otherwise interest the reader.

We shall begin with a batch of letters addressed to Mrs Miller during a tour undertaken by him in the North of Scotland—Sutherland, Cromarty, Inverness, the Caledonian Canal—in the summer of 1843.

'Port Gower, Sutherlandshire, 8th July, 1843.

'The boat anchored off Aberdeen for about an hour, and I went ashore. As I had to restrict myself to sauntering about the harbour, I saw only tide-waiters, sailors, and boatmen. I picked up, however, a little story for Harriet, which you may read to her as from papa. Just as we had gone out from the pier on our return, a small white dog came down, whining and pawing to be taken on board, but the boatmen rowed on. The mole runs a long way out, and as we were scudding alongside of it below, the dog was keeping pace with us above, and stopping every moment to make the most pitiful appeals to our compassion. The poor animal had, it would seem, lost his master, and expected to find him in the steamboat. At length the mole terminated, and on its outer sides there was a heavy sea breaking, but the little dog, after looking doubtfully on the broken water for a moment, dashed in and swam hard after us. The poor animal won on the regards of even the surly boatmen, and after he had made way for about fifty yards against a strong flowing tide, they rested on their oars and took him in. He lay cowering in a corner of the boat, showing that it was evidently not among us but in the steam vessel he hoped to find his master; the moment we reached the vessel's side he climbed up the ladder, explored every corner from stem to stern above and below, and not finding him whom he sought, took leave of us by leaping overboard. As there was a strong shore-going tide in his favour he would, I doubt not, safely have made the land. I dare say after reading my story to Harriet you must translate it for her into what Swift used to call the "little language." The perseverance and solicitude of the creature served to show me how dogs who have lost their masters many hundred miles from home have contrived to trace them out, making good their passage across wide seas by working on the feelings of seamen. The appeals of a devotedness like that of the poor little doggie the rudest natures cannot withstand.

'On landing in Wick I made my way to the inn, and after waiting to take a draught of lemonade, the only thing my strained stomach could bear at the time, I opened my bag to get out my hammer and chisels, that I might spend two hours ere the coming in of the south mail in geologizing. My walk, pleasant in the main, though the morning was rather close, ran along the Wick, a small stream which flows through a flat valley remarkable chiefly for being rich and green with grass and corn, and yet totally bare of bushes and trees. In the fence of a field formed of upright flags I saw numerous scales, occipital-plates, and gill-covers of some of the better known fish of the Old Red, but nothing worth bringing away.

'On the journey south I saw from the top of the coach several melancholy marks of the Disruption,—preaching tents on green hill-sides with a few rude forms around them, and not far distant parish churches locked up. There is no preaching in the vacated churches in Caithness; Moderatism is sitting down helplessly and attempting nothing. The poor people of Scotland have been grievously injured, but the extent of the injury is not yet felt. The coach passed great numbers going to a Free Church sacrament. As the hour was comparatively early, it was chiefly elderly men and women we saw, who, to use their own phrase, were "stepping on before," the more vigorous walkers not yet having taken the road. I passed on through Helmsdale to Port Gower, where I now am in a very pleasant little inn. The scenery in the immediate neighbourhood is not so fine as at Helmsdale, which is a very striking place in the opening of a deep, grand glen; but the situation is more central for geological inquiry, and Helmsdale lies within reach of an ordinary excursion. I have been out for a few hours among the rocks, but they do not seem at all rich in the neighbourhood of the inn. The cliffs on the shore are of a rude and very hard breccia, which, save foı some two or three belemnites that I have detected in it, I would not know to be oolitic at all. You can have no idea how thick the population lies on this narrow border of coast. The bases of the hills are so studded over with miserable cottages surrounded by miserable patches of corn land, not larger than ordinary-sized gardens, that for miles together the appearance is that of one vast straggling village. Depend on it, if the Poor Laws be introduced into Scotland, the extra profits derived to the Duke of Sutherland from his sheep farms will go but a small way in supporting his paupers. He will find his northern injustice taking to itself the form, not of a mere crime, —for that would be nothing,—but of a very terrible blunder.'

'Sunday, 9th July.

'I have just returned from Helmsdale, where I have been hearing a sermon in the open air with the poor Highlanders. The sun was very hot, and the portion of my forehead that used to be best sheltered by my hat is all blotched with red, and will, I dare say, be blistered to-morrow. The congregation was numerous, from six to eight hundred at least, and all seemed serious and attentive. It must have been the mere power of association, but I thought their Gaelic singing, so plaintive at all times, even more melancholy than usual. It rose from the green hill-side like a wail of suffering and complaint. Poor people! There stretched inland in the background a long, deep strath, with a river winding through it. It had once been inhabited for twenty miles from the sea, but the inhabitants were all removed to make way for sheep, and it is now a desert, with no other marks of men save the green square patches still bearing the mark of the plough that lie along the water-side, and the ruined cottages, some of them not unscathed by fire, with which these are studded. But though thus desolate within, the sides of the two hills, in which it terminates at the sea, are so thickly mottled by little dwellings that it would seem as if the people who had been forced out through its entire length had stuck at its mouth. The inmates of these poor dwellings are all adherents of the Free Church. They surrounded the tent by hundreds, wonderfully clean and decent considering their extreme poverty, but with a look of suffering and subdued sadness about them that harmonized but too well with the melancholy tones of their psalms. There is, it is said, a very intense feeling among them. "We were ruined and made beggars before," they say, "and now *they* have taken the gospel from us." I hear that when the Duke last passed through the place the men stood sulkily looking at him, or slunk away into their houses, but that some of the women began to *baa* like sheep. I can get no information at the inns; the Duke is, of course, omnipotent in the county, and there is a general fear of trusting a stranger with what might be unfitted to reach his ears, and might reach them notwithstanding.'

'Monday evening.

'What a very peculiar taste in the ornamental Highland landladies display! My landlady at Brora last year was a stately, lady-looking personage, who, as Dandy Dinmont would say, behaved herself very distinctly. And yet, barbarous to relate, she had, by way of ornament on the chimney-piece of her best room, a child's toy-house, rough with glue, and daubed over with paint. In my present inn, a neat little place, with a good side-board and mahogany chairs in the sitting-parlour, and with a very genteel bed-room, there are square bits of common 2½*d.* per yard room-paper, framed round with a listing of black velvet, that hang on the walls, and supply the place of pictures. My bed-room has got two of these superb ornaments. One of them presents five nondescript brown flowers, with brown leaves on a field of buff. The other boasts of but one flower; but as it is in green and red it has got its superior beauty protected by a cover of glass. Is there nothing akin in Highland gentility to the stately gentility of the savage? The dignified Indian chief, when determined on being very fine, smears his forehead with vermilion, and sticks himself round with glass beads.'

'Monday evening.

'I have been spending the day, from nine o'clock till five, among the cliffs to the north of Helmsdale. Rare sport. Fossil wood and corals may be gathered on the shore by tons, and there are abundance of bivalves, serpulæ, impressions of ammonites, spines of echini, and some fine belemnites. I found a piece of fossil wood, a cross-section of a tree showing the pith and the annual rings, which measures rather more than twenty inches across. It was so weighty that I had to employ three boys to carry it to Helmsdale, where it lies to be shipped for Leith by the first vessel. By being at an expense of a few shillings with it, it may be rendered one of the most splendid specimens of the kind anywhere in existence. If sawn through the centre, and then polished, it will show the growths of at least fifty summers, and the alternating checks of as many winters. The polishing cannot cost me more than six or eight shillings, and it will form an ornament for your mantelpiece splendid to your heart's content. Some of the corals are of great size. I saw one nearly three feet in diameter, and which, to all appearance, would not have attained to so great a bulk in less than a century. At least, the brain-stone coral of the tropics, which it very nearly resembles, takes a full hundred years to attain a size equally great. I could, of course, bring only fragments away with me, but even the smallest are pretty, and when polished will be exceedingly so. The pattern is that of a piece of honeycomb, with a many-rayed star in the centre of each hexagon. Lyell figures it in his Section on the Oolite, and I remember being struck with the beauty of the figure. Both the corals and the fossil wood take the form here of a very rich limestone, and the people gather them on the shores, and burn them in kilns for lime, without dreaming that in some remote age of the world the one has manifested animal, and the other vegetable, life. On passing along the beach I went up to a kiln beside which several tons of limestone had been lately accumulated. Fully two-thirds of the whole was fossil wood, and the half of the remaining third corals. The people have learned to dig the trees out of a shaly conglomerate just as in the centre of the county they once used to dig trees out of the mosses, but though they saw the coaly bark, the pith, and the annual rings, it never once came into their heads to suppose it was trees they were digging. I pointed out to one of the owners of the lime-kiln the true nature of the stones he was employing, and he seemed much surprised. "That," he said, "was never known here before." Much depends on the locality of a marvel. All the world has heard of the fossil-trees of Craig-Leith. Here fossil trees not less in bulk, and quite as finely petrified, are so abundant that the Highlanders dig them out for lime, and yet no one has heard anything of them.

'I had a Highland boy of thirteen carrying my bag all day. He soon became as good a practical geologist as myself, and as expert in finding corals and belemnites. Boys are much less stupid than men. The Highland boy learned to know at the first telling that detached fragments of belemnites were of no value, and that the pieces were only worth keeping when all together made up an entire belemnite. I have taught old —— the same lesson I know not how often, yet he persisted in bringing me the detached fragments notwithstanding. Lamb's joke had some truth in it after all. Boys in the main *are* superior to frivolous members of parliament. "What church," I asked my boy, "does your father go to?" "Oh, the church that's put out," he replied; "none go to the other churches here but the Duke's people." On coming home I made him happy with a shilling and a drink of ale.'

'Brora, Tuesday evening.

'Here I am after a walk of about ten miles along the coast. The day was warm, and my path or rather no-path lay for the greater part of the distance over loose sand, but exercise and a reinvigorated stomach are already beginning to tell in my favour; and so here I am, after an additional spell among the rocks between dinner and tea-time, a fresh active fellow at your service. The Brora, a dark mossy stream, runs at the back of the inn here, between precipitous banks; it abounds in fine small trout that were leaping so thick and looking so happy this evening that I became quite sentimental in looking at them, and formed the romantic notion of eating a few of them with pepper and vinegar just that I might be happy too. Accordingly at tea I got the idea realized into an act, and found it, but you won't believe me, a good workable idea, notwithstanding its mixture of romance.

'Brora is a loose straggling village, with too many people in it for all they have to do, and there are numbers of miserable-looking cottages between it and the hills, whose inmates are too palpably in the same condition. One of its peculiar features is the river. In passing through the village it has cut for itself a deep trench, thirty feet deep at least, in a white sandstone which, though extremely soft, stands the weather, and thus preserves on both sides a perpendicular frontage. The appearance in consequence for some distance is rather that of a canal at the masonry of the locks, than of a river,—a curious exemplification of how geological phenomena affect scenery. Where the walls are steepest, and the water blackest, there is a bridge thrown over which the public road passes.'

'Wednesday morning.

'After taking tea last night, and then filling the opposite page, I walked out along the southern bank of the Brora for about half a mile. The path led through a sort of rectilinear ravine, evidently artificial; on reaching an extensive hollow the rectilinear ravine became a rectlinear mound, and I found it terminate at an old pit mouth. Those I saw last year lie on the opposite side of the stream. Though about two hundred feet in depth, it is now filled up with water to within thirty feet of the opening. The appearance of the hollow in which it has been opened struck me as remarkable. All around, between me and the sea, I saw a sloping rampart, terminating atop in a long even height, and, in the bottom, in a level plain laid out in fields and with a wooded island rising from out of it with the same slope as the surrounding rampart. I saw at once that it could not be the result of any glacial action, as the body of the rampart is not of *débris*, but, as shown by the artificial ravine, of white sandstone. On climbing a height to survey its general appearance, I marked that the trench-like appearance of the bed of the river, to which I referred last evening, commenced exactly at the hollow,—it is a breach in the rampart,—such a breach as you may have seen in the side of a mill-pond over-flooded and broken up by a thunder-shower. The mystery lay unveiled;—the hollow had been at one period a deep and extensive lake, which the river, by wearing down the soft white stone of the lower barrier, had gradually tapped off. I could see no one able to inform me regarding the nature of the soil in the hollow, or whether freshwater shells or moss were found in the subsoil. By the character of the shells the age of the lake might be ascertained, but ancient it must be, as the bed of the river is now a considerable distance beneath its level.

'There is a pause in the rain, and the sky is clearing, so I shall e'en wrap my plaid about me, and set out for the quarries, and the seat of the glacier. I had intended visiting Loch Brora, but it is six miles away, and, having no change of dress, I am unwilling to put myself in the way of being too thoroughly drenched.'

'Wednesday evening.

'I have been at Loch Brora, and at the quarries, and the seat of the glacier, and I have traced the river for miles, and filled both my large pockets with fossils, and all this without catching a single drop of rain. Am I not a lucky fellow? Loch Brora is a long narrow sheet of water, that lies up among the hills in the gorge of a deep rocky valley. The mountains on both sides rise to a great height, wild and naked above, but with their bases lined very sweetly with dwarf birch. One immense hill in the distance, of bold outline and precipitous sides, is the acknowledged monarch of the scene,—the loch stretches out in front of him for about four miles, its undulating and winding shores tufted with birch, and here and there mottled with small green spots that, ere the poor Highlander had been driven from home, kept him in oats and bere. I saw in one little bay some of the prettiest water-lilies I have ever seen. They looked like so many bathing Venuses submersed to the neck. Some of the green deserted spots on the shore, with the purple hills rising behind, the blue loch in front, and the dwarf birch all around, were really very sweet. I doubt not that the thoughts of them live, set in sorrow, in hearts beyond the Atlantic. As you look down the loch, the whole opening of the glen seems a tempestuous sea of miniature hills, from thirty to forty feet in height;—the Highlanders call detached hillocks of this kind *Tomhans,* and believe them to be the haunts of fairies. Geologists have learned from Agassiz to term them moraines. They form the accumulations of stones and gravel which were ploughed up before the glacier. Amid a good deal of irregularity they have an evident bearing in their arrangement to the mountains around, from which the ice descended. In some places I could trace them in chains of half a mile in length quite parallel to the mountains; and about half a mile above Brora three of these chains lie parallel to one another, like the strings in a triple necklace of beads. There lie between them miniature valleys cultivated by the ejected Highlanders; and there are many miserable-looking hovels on their sides. Imagine a table covered with crumbs, and that you set yourself to rake them into lines with the edge of your hand, keeping a sort of rude parallelism to that end of the table at which you sat, and from which you raked them. Your little lines of accumulation would be *moraines* of crumbs. But there are more than moraines here to testify of the glacier. Suppose your hand to be moist, and that it left on the cleared table in the direction in which you pushed the crumbs before you, a number of minute streaks corresponding with the minute ridges of the skin. Such are the marks on what may be termed the cleared table here. Wherever the rock is laid bare we see it all furrowed and channeled with grooves and scratches, in the bearing of the moraines on the hills. These workings give a very high idea of the antiquity of the oolite, comparatively modern as the oolite is in the chronology of the geologist. When the ponderous ice slid over it, it was in exactly the same fossil state that it is now. The rock, a white sandstone, abounds in fossils, that are but hollow casts; and wherever one of these casts occurs, the stone is weaker than where there is none. Now, we find that the glacier discovered this weakness, and in many places broke out the hollow casts that are nearest the surface, leaving within the fragments that it pressed down into them.

'I did not take the beaten path to Loch Brora, but struck upon it through the moor, and passed on my way through a large peat mass, roughened at present by the winter fuel of the people of Brora. There are a few scattered moraines that rise over its dingy level, and which must have been there long ere the morass itself had any existence. And yet in this morass large fir-trees are found, the remains of one of our ancient forests. I saw in one pit several huge blocks laid bare;—they seemed originally to have been trees of that short-stemmed, bee-hive-looking class I have seen in some places in the central Highlands. They never assume that form save when standing far apart in a comparatively open country. Here are degrees of antiquity for you; the moraines are as of yesterday when compared with the oolite, and the ancient forests as of yesterday when compared with the moraines.

'I know you are heartily tired, but I must say something about the quarries. Imagine a border of comparatively flat country stretching between the hills and the sea, and out of the middle of this border, where at the broadest, imagine a wood-covered hill rising, quite a dwarf-looking thing when compared with the mountains behind, but a genuine hill notwithstanding. It stands quite independent, a hill set up for itself, and descends nearly as steeply towards the mountains as towards the sea. Now, this independent hill is a hill of the oolite,—*English*, if I may say so, to its very heart;—nay, more, it belongs to a more modern deposit of the oolite than any other in the country. The flat base on which it rests belongs to that *inferior* middle and lower part of the Oolite known as the *Oxford Clay* and the *Great Oolite*, whereas this hill belongs to that *superior* middle part of the system known as the *Coral Rag*. It is an isolated specimen of the deposit which, though somewhat too bulky for most private collections, seeing it measures a full mile in length by rather more than half a mile in breadth, is excellently fitted for furnishing specimens which, were it not there, could not be had nearer than England. Now, it is in this insulated hill that the quarries are opened. The stone is of a beautiful white, and was at one time much run upon. There are houses in Edinburgh built of it. It was found, however, to have one great defect,—though pretty and durable, it is so exceedingly porous that a house built of it formed but a kind of sponge that took in every drop that fell on it during a shower and retained it, dispensing damp for weeks thereafter. And so, greatly to the detriment of Brora, for which at one time it promised to make a kind of trade, it has fallen into disuse. It abounds in fossils, curiously preserved as casts. It is really wonderful how sand, however fine, could have taken the impressions so finely. I picked up to-day a cast of a piece of decayed wood on which not only the yearly rings but even the longitudinal fibres were visible. It has preserved even microscopic peculiarities. And as only the white sandstone remains, the fossils have all the appearance of being pieces of beautiful sculpture. One might almost imagine among the *débris* of these quarries—for there are two opened in the hill—that he had before him the rubbish of one of our old cathedrals after the *rascal multitude* had reformed all the carved work with their pickaxes and hammers,—only the workmanship is much finer than was ever yet fashioned for Scottish cathedral. I found two masons at work in one of the quarries, and showed them that by some unaccountable means or other I had learned to hew. Good-night, dearest,—it is eleven o'clock.'

'Thursday. Golspie.

'Another stage in advance. Taking an early breakfast at Brora I set out along the shore about half-past seven, and arrived here about ten. There is not a great deal to be seen by the way, but the walk was pleasant, and when I entered on the bounds of the Dunrobin Policies, very much so. Imagine a selvage of lawn with the beach on one side, and a continuous line of bank on the other,—the bank planted and presenting the appearance of a tall screen of wood. Behind this screen there is a bold undulating outline of hills, with the tall hill crested by the late Duke of Sutherland's monument, conspicuous in the midst. In the centre of the landscape rises Dunrobin, a square, white, many-windowed pile with a round tower at each corner, and a square detached tower, turreted and battlemented, projecting from one of the sides. The woods close thick around it,

—so thickly that a road that leads to it from the shore seems dark as a vaulted passage. There is much of antique simplicity if not of ducal magnificence in its whole appearance, and it says something for the taste of the family that they have not employed their vast wealth upon it in making it tawdry and fine. A genuine old castle, that had earls and barons in it five hundred years ago, is one of the things which money cannot produce.

'The Andersons mention in their *Guide Book* that there is a fine waterfall in the neighbourhood of Golspie. I have just been seeing it; but the weather of late has been dry, and so, though the woods and rocks do their part pretty well, the stream has almost struck work, and sputters over the face of the precipice much in the sort of sublime style I have seen the contents of a washing-tub assume when emptied over a window. But the dell in which it falls is not unworthy of being seen;—it is an Eathie on a larger scale, and richer in wood. It has been hollowed in the Old Red Sandstone; there are precipices of diluvium a full hundred feet in height in the lower part of it, and the cataract has been formed, as such cataracts always are, where the stream suddenly passes from a harder to a softer rock. I am too old, I think, for getting into ecstasies with scenes of but ordinary fineness. The Golspie water-fall would do quite well for young ladies and romantic gentlemen, but the mature man needs something strong to move him. I wish I had an opportunity of trying the effect of Niagara.

'On leaving the waterfall I struck up along a very steep path to the Duke of Sutherland's monument. The day was warm, and the labour considerable. As I approached the top I saw within some sixty or eighty yards of the building a small cottage, and beside it the *débris* of a quarry, out of which the stones for the erection had been taken. The hill itself is one of those Old Red Sandstone eminences which in this part of the country attain to so great a height. I at once inferred that the cottage was the deserted barracks in which the workmen engaged in erecting the monument had lived. What was my surprise, however, to find it inhabited;—one might as well expect to find a human family located in a crow's nest on a tree-top. It is elevated I know not how many feet over the level of the sea, far above the line of cultivation, and with only heath and moss around it. I tapped at the door, and sought a glass of water,—the matron, a sunburned, staid-looking woman of about thirty-five, brought me a drink very civilly. Her husband, she said, had charge of the planting below, and of the monument, but though they had been in the place for only a year, they were both thinking long to get away. Last winter the snow lay so thick around them, that for five weeks they could not descend the hill, and weeks sometimes passed even in summer without their seeing any one. It was so very cold a place, too, at all seasons when there blew any wind! I saw in front of the cottage a minute patch of garden ground, in which, as if by way of experiment, a few potatoes had been planted, and that the more vigorous among them had contrived to throw up three whole leaves above the soil. According to Cowper,

"So farewell envy of the peasant's nest."

'The monument of the Duke, as becomes a man who had four hundred thousand pounds a year, is of colossal proportions. There is a huge statue of White, and a huge pedestal of Red, Sandstone. The large dead eyes look over Sutherland and the sea, and if they do not see as profoundly into the real interests of the country as the eyes of the living Duke, they see wonderfully little considering that they are six inches across. The view from the base of the monument is very extensive. Suppose one half of the whole occupied by an immense acute angle of sea,—the Moray frith. Imagine, further, a long sharp tongue of land striking out into the middle of this angle from the apex, and with a tall tower-like building at its point. This tongue is the peninsula of Easter Ross, and the tower-like erection the lighthouse of Tarbet Ness. At the base of this tongue there is a spindle-shaped lake,—the bay of Cromarty. The hills of Sutherland and Ross make a splendid frame to the landscape on the north and west, and on the south and east stretch the light blue hills of Moray, Banff, and Inverness. Remember I am not vain enough to attempt drawing a picture;—I am merely scratching as with a burnt stick a few careless outlines.'

'Saturday morning, Cromarty.

'One of our neighbours here, *old* Mrs Forbes, has just been furnishing me with a new evidence that I am getting old myself. My father was but fifty at the time of his death, and I am now turned of forty. "Oh," she said, after shaking hands with me, "you are growing your father's very image." I have reached the same period of life to which he had attained when she knew him best, and years are bringing out resemblances to him in my face and figure, which seem to have lain latent before. I had a walk yesterday to the rocks with my hammer, and succeeded in picking up several tolerable specimens.'

'Saturday evening.

'I have been holding a *tête-à-tête* to-night with your

picture in Mrs Fraser's, and calling up old times. I have been just re-reading, too, your letter to Mrs Gray, the letter which Sir David Brewster could not believe written by a woman; and I felt both letter and picture do my heart good. Think what you may of the matter, dearest, you have no such admirer as your own husband. You would laugh at me as fond and foolish did you know how often I have been thinking of you during the past week, and how much pleasure I have had in writing this scrawl, just because, however trivial my topic, I felt that I was conversing on it with you. You were beside me, as now, and I felt the lover as strong in my heart as I did seven years ago. Do you not remember how many hours we used to sit together, and yet how very short they always seemed? Shall we not, dearest, renew these times at our meeting, ere I plunge once more into the stern turmoil of controversy, and be a man of war? We shall have one or two quiet walks together ere we cross the country to Edinburgh.

'The first person I met as I landed was poor forlorn Angus, now grown a great fellow. He had been watching the steamer by which Andrew came, but on finding that I did not accompany him he waxed indignant and high, and refused to shake hands with him. I got a double shake. He then seized on my carpet-bag with a shout, and bore it home in triumph. He is not quite pleased, however, that I have not brought "Miller's boy" with me. Angus is not a Moderate,—he has gone out, regularly attending sermons in the factory close. His head finely illustrates the phrenological doctrine that size is power. He wears my old broad bonnet, and finds it not a hair's breadth too large.'

'Sunday evening, 7 o'clock.

'I have been hearing Mr Stewart in the factory

close. The congregation was larger than used to assemble in the parish church, and apparently more attentive. It was striking enough to see the familiar faces in such a place, ranged, not in the accustomed pews which some of them have occupied in my recollection for full thirty years, but all mixed up in one group,—poor and rich together. The day was gusty and fitful; the surrounding walls of deep red, with their many windows, now glared upon us in sunshine, now lay deep in shadow,—there was the occasional rattle of a window in the upper stories as the gust shook the boards, and the measured roar of the sea mingled from the beginning till the close of the service with the tones of the psalmody and the voice of the speaker. Poor Angus was greatly engaged in carrying forms and chairs, and in looking, uncouthly enough, after the comforts of a few favourites, mostly old women. Aunt Jenny tells me that he prays every night in his own peculiar style for "Miller and Miller's boy."'

'9 o'clock.

'I have been seeing the old chapel of St Regulus and poor Liza's grave. The little mound is as well marked as it was four years ago, and it is now wrapped over with a mound of rich unbroken turf. The little head-stone bearing your and my name has whitened somewhat under the influence of the weather, and leans slightly to one side, but there is no other change. The sun was hastening to his setting, red and broad, and throwing a strong bright gleam on the upper foliage of the surrounding wood, and on the top of the ruin, while the tombs and graves lay in deep shadow. Poor Liza! The little events of her span-long life rose all before me, from the time that I first felt that I was a father, till I flung myself down in uncontrollable anguish on my bed, a father no longer. The spring in which we lost her was peculiarly a dark time;—but it is over; and Liza still lives, though not with us.'

'Monday, 3 o'clock.

'And I have heard from you. I am glad you should have fixed on so interesting a locality, but not at all glad that you should be so little able to avail yourself of its various points of interest. But you will, I trust, get stronger, and that by the time I join you you will be able to accompany me in my walks. You must take especial care and not over-exert yourself, nor walk much in the sun. I am pretty sure that by putting yourself in what boxers and pedestrians call a course of training,—extending your walks bit by bit as you felt your strength increase,—never exhausting yourself,—never urging exertion past the point at which fatigue is merely an agreeable languor, you could be brought into *condition*, and made strong enough to travel ten miles per day. My mother, never a strong woman, has travelled at your age thirty miles on a stretch. Only think of all you could do, could you but travel thirty miles! I shall, however, be well content if, at our meeting, I find you able to accomplish five. Five would bring us to Loch Lomond together, and take within our range a few of the striking points of Loch Long.

'And so Harriet and Bill enjoyed their voyage, and Harriet has become a shell and fossil collector. Tell her papa trusts she is to have a great many pretty things to show him when he comes to Arrochar, and that she is to be so good a girl as not to cry when nurse is bathing her. Tell her, also, that I send her a kiss for mamma, and that I send mamma a kiss for her, and that the best way of managing the matter will be for her to receive first the kiss from you, for which you are my proxy, and then to give you in turn the kiss for which she is my proxy.

'Write me, dearest, as long letters as you can without tasking yourself, and I in turn will devote to you every leisure hour, my walks, my visitings, and my dinings out will leave at my disposal. To say that my chief pleasure, in a ramble devoted to pleasure exclusively, arises from my conversations with you, is not saying too much. My heart is continually turning to you and the two little persons at Arrochar. My home is not a locality,—it is not a dwelling; it is you and the little ones.'

'Cromarty, 18th July, 1843, Tuesday.

'Mr Sage was preached out on Sunday last, and by dint of superhuman exertion among all the lairds a congregation of thirty were brought together to see that he was. Cromarty, Rosemarkie, Avoch, all sent their guests to swell the amount; and of the thirty, two whole individuals, a man and his wife, were stated hearers in the parish church. There could be found no one to ring the bell and no one to be precentor, though twenty shillings were offered as remuneration, and a man and gig had to be sent rattling to Cromarty an hour ere the service began, to procure both out of McKenzie's congregation. The story goes that, with the first tug the bellman gave, a swarm of angry bees, disturbed by the swing of the rope, came down about his ears with wrathful fizz, and that to avoid their stings he had to quit his hold and show them a clean pair of heels. The Moderates are in a perilous state, when every untoward incident that occurs is regarded as an omen, and interpreted to their disadvantage. Their spirit in this part of the country is bitterness itself. Servants dismissed, labourers thrown out of employment, angry interviews between landlord and tenant,—we hear of little else in this corner. The poor in Cromarty who go to —— for assistance from the parish are invariably asked as a first question, "Do you belong to the Free Church?"'

'Tuesday afternoon.

'I have been spending a few hours very pleasantly. Before leaving Edinburgh I mentioned to you my intention of digging into the floor of the large Pigeons' Cave to see whether shells or bones might not be found in it, as in some of the caves of England and the continent. I procured, through Andrew, a small boat and a crew of young lads, Andrew's contemporaries and companions, and providing ourselves with tools and candles, we set off. We went careering along the rocks at two oars' length from the shore. I saw the little rock where you first said *yes* to a certain interesting question,—the said important yes bearing reference to a log-house in the back-woods. I saw besides, the two beech-trees where we were so foolish, you know, as to spend a great many hours together, and so exclusive that the company of any third person we could not have endured. We landed at the cave, drew up our little boat, and then commenced digging about twenty feet within the opening. We found for two feet and a half pure guano,—the same umbery sort of mass you saw in the museum of the Highland Society, but without the overpowering stench, a difference which may possibly arise from the circumstance that the guano of South America is an accumulation of the dung of predacious birds, whereas the guano here was an accumulation of the dung of pigeons. We took home some of it with us, and James Ross is to try its effect on a drill of turnips. Beneath the guano we found a loose gravel, which became less loose as we went down, and then passed into a sort of breccia bound together by a brown oxide of iron. We found neither shells nor bones. We then opened another pit about fifty feet further in,—the guano was wanting, but the gravel bore the same appearance as in the other opening. There are masses of gravel that project, strongly enough, from the sides of the cave at the height of four and five feet from the floor. Some of them stick out like fragments of cornices for fully two feet, bound together by stalactites. They show that the bottom of the cave must have at one time stood that high, that the stalagmitical matter running from the sides bound into coherent masses the stones and gravel in contact with them, and that the sea then swept the loose matter away, leaving the projections to be crusted over still further by the stony cement. We broke into several of the bulkier of them and found them composed of water-rolled pebbles and very distinct fragments of shells. I procured a specimen in which there are five well-marked bits of shell rounded on the edges as we find such fragments on surf-beaten and rocky shores. How strange a record! We have evidence all around the coast in our raised beaches that the sea once stood higher on the land than it does now; but we see here that there must have been an alternation of elevations and depressions. When these shells were rolled in on the floor of the cave,—a floor standing five feet above its present level, the sea must have stood high. During the time when they were consolidating through the agency of the stalagmitical matter, the sea must have fallen, otherwise the consolidation could not have taken place. It must then have again returned and washed out the loose gravel which the matter had not reached. And then it again fell. There is another cave among the rocks of Cromarty that indicates similar changes.

'I saw on the walls and roof of the cave appearances that used at one time greatly to puzzle me. The commoner stalactites drop straight down like the stalks of tobacco pipes, or slant along the walls. Every drop that forms them obeys the gravitatory law. But here and there we find a drop-formed stalk in which this law does not seem to have been obeyed. Some of them curve up at the ends and form hooks, from one of which I suspended my watch, and some of them curve round so as to form loops,—both ends joining to the roof. How account for the formation of these? Did water lose its weight in some wonderful age of the world and set itself to drop upwards? The solution of the difficulty when at length discovered proved simple enough. There is a small spider that takes refuge in the colder months in these darker recesses, and forms as it passes along the sides and roof a multitude of loops with its hair-like threads. Along these the drop sometimes runs,—a drop running continuously along one end forms a hook, when a drop runs along each end it forms in course of time a loop,—and such is the mystery. In the outer part of the cave I saw some curious encrustations. A miniature moss bends out to the light, and the rock side to which it is attached is a roughed plane of mosses in stone, all turning to the light. The dead remained fixed in the attitude of the living.'

'Wednesday evening.

'I have been across the ferry in the little boat I had yesterday, and after spending several hours in hammering among the strata of that fine section of Old Red that leans against the Hill of Nigg, I have succeeded in making something approaching to a discovery. I have described in my geological volume one great platform of death in the Old Red, above which fossil remains are very unfrequent. I discovered to-day in this section that there are two platforms of death equally abundant in organisms, the one three hundred and eighteen feet above the other. Nor does any change seem to have taken place in species in the long lapse that intervened, as the fossils of the lower and upper platform seem identical. I disinterred out of both specimens of *Coccosteus, Osteolepis,* and *Cheiracanthus,* that could not be distinguished with reference to their localities. This discovery throws a good deal of additional light on the geology of the district, and solves for me one or two problems which lay unsolved before. I found besides, under the northern Sutor, a recent deposit that belongs, I am inclined to think, to the period when the sea stood high over the floor of the pigeon cave. It is composed of an adhesive blue clay, not very unlike some of the rich blue clays of the Lias, but the remains which it rather plentifully contains are all recent,—bits of decayed wood, among which I could distinguish pieces of common fir and alder,—the common mussel, and a not very rare Venus, fir-cones, and hazel-nuts. I was a good deal interested in this last discovery. We have on the Cromarty side a considerable quantity of the same blue clay, and until now I could never find aught to determine its age by.'

'Thursday morning.

'Nothing like perseverance! I have been out again at the fish-beds, and have got some capital specimens. It is truly wonderful how thickly the fish must have lain in some period of the *Old Red* on this bank. The herrings never lay so thickly on our coasts in the fishing season. Out of an area across which two average-sized nets would extend, I have been digging fish for the last twelve years. I have taken many hundreds away, and broken up many thousands more, and they seem innumerable still. Did our recent fish crowd together as closely, a single net would mesh enough to load a boat.'

'Thursday evening.

'What would I not give for an occasional peep into the magic mirror of the tale, that I might see what you are doing, and how you are looking, and what Ha-ha and Bill are about! But I dare say I bear your absence better without it. It would serve but to suspend the viands over the hungry man. I am counting the days till I receive your reply to my last. You would have my letter on Wednesday; yours will be posted to-day; it will be all this evening and all to-morrow on the road, and I shall have it on Saturday. Or suppose a day is lost through your distance from the Post-office, it will not arrive in that case till Sunday; but though business letters may take their Sabbath-day's rest, and welcome, my own Lydia's letter cannot be permitted to lie silent so near me. I shall have young Andrew to call for it, and if it tell me of your welfare, and that you enjoy yourself, I shall go to church with a heart all the more soft and grateful through its influence. A short absence, dearest, serves to show me how very dear you are to me; you grow large as you recede, and you are tall enough at present to fill my whole heart from top to bottom. Good-bye; it is past ten, and I intend rising early to-morrow morning, if well, to make another attempt on Shandwick.'

'Friday evening, 6 o'clock.

'Just returned after a pleasant, though rather fatiguing walk, of twelve miles or so. In going I passed along the elevated path that winds immediately above Nigg at the base of the hill, just where the cultivated ground borders on the moor. The day was one of speckled sunshine, and everything fresh and green after the recent rain. On one of the lower slopes, where the hill of Nigg descends towards the east, there is a fine tomhan roughened at its base with dwarf birch, and bearing very distinctly atop the marks of a hill-fort. Some six or eight other tomhans lie around it, according to Burns,

"Hillocks dropt in Nature's careless haste;"

according to Agassiz, hillocks not dropped in haste, but ploughed up with vast deliberation and perseverance. The tomhans here are all moraines,—the production of a moderately-sized glacier that once occupied the north-eastern slope of the hill of Nigg. The hill-fort may not be more than two thousand years old,—not so much, perhaps; two thousand years brings us beyond the days of Cæsar. And yet it gives a sort of felt antiquity to these moraines to find one of them occupied by an erection older than history. In my descent on the shore I was struck with what I must have often seen, but never so particularly remarked before,—the great beauty of the curve in which the lias strata recline against the coast. It forms such a sweep as Hogarth would have introduced into his dissertation. Betweeen the harder strata and the land there occurred beds of a softer quality which have yielded to the sea, leaving a long hollow, into which the higher strata descend, as if by steps of stairs, and the appearance presented is that of a low, long, elliptical amphitheatre, on which some two or three hundred thousand people might take their seats. I found very few specimens,—a few ammonites and belemnites, and a bit of fossil wood; but what I saw will be of use to me should I set myself to my proposed geological sketches. The recent formation which I found under the northern Sutor throws a good deal of light on all our lias deposits of the north. In both the lias and the deposit we find the same classes of remains,—wood, cones of the pine, and shells,—the bivalves among the latter presenting in most cases their two valves unseparated. I should have said that, before leaving the shore, I found the fish-beds of the *Old Red* appeared almost in contact with the lias. I found a Coccosteus last year in this place among the loose pebbles, and wondered where it could have come from.

'I hear on every side of the oppression of our landlords. In the parish of Logie there is a large gravel-pit in a fir-wood in which on sacramental occasions the out-door congregation used to assemble. At other times it is a famous resort of the gipsies. Their smoke may be seen rising over the trees six months in the year, and their rude tents pitched in a corner of the hollow. Some of the neighbouring farmers and cotters expressed a wish not very long ago that persons so dangerous and disreputable should be prevented from making it a place of resort, but they were told by the proprietor's *doer* to be kind to the gipsies, and they would find them harmless. On the Disruption the minister of Logie respectfully applied for leave to erect his preaching-tent in the hollow, in the expectation, fond man, of being permitted to rank with the gipsies. But, alas, no! Tinkers may be patronized as picturesque, but the Free Church is dangerous; and so the use of the hollow was promptly and somewhat indignantly refused.'

'Sunday morning.

'There has been a night of weighty rain; the streets have been swept clean, and the kennels show their accumulations of sand and mud high over their edges. I awoke several times during the night to hear the gush from the eaves and the furious patter on the panes; and I thought of the many poor congregations in Scotland who would have to worship to-day in the open air. But the rain is now over, and a host of ragged clouds are careering over the heavens before a strong easterly gale. I do begrudge the Moderates our snug comfortable churches. I begrudge them my fathers' pew. It bears date 1741, and has held by the family, through times of poverty and depression, a sort of memorial of better days, when we could afford getting a pew in the front gallery. But yonder it lies, empty within an empty church, a place for spiders to spin undisturbed, while all who should be occupying it take their places on stools and forms in the factory close.

'I felt, by the way, when listening to Mr Stewart last Sunday, that our out-door congregations would require a style of preaching for themselves. His discourse was in his average style, very ingenious, but not very powerful, and with more of point than breadth. I thought it would have told better in a moderately-sized church than in the open air. The cabinet picture requires to be placed in a cabinet;—I felt as if he were hanging his cabinet picture in a vast gallery. The times require that the power should be great, but the finish, perhaps, not very nice. In an hour from this the post will be in. If disposed to forget how I used to watch his arrival at Linlithgow when expecting a letter from you, my present state of solicitude would scarce fail to remind me of it.'

'6 o'clock, evening.

'I have been hearing sermon a second time in the factory close. Mr Stewart was in his happiest and most impressive vein,—clear, interesting, solemn; I have rarely, if ever, heard him more effective. My criticism on his last Sabbath's discourse would in no respect apply. Never have I heard better *out-of-door* discourses. His audience was most attentive, and at several of the more striking passages visibly impressed. The day cleared up beautifully, and except for a few minutes, when a slight shower led to a vast unfurling of umbrellas, we were quite as comfortable as we could have been in any building. There has been much rain of late, and it has been of great use and greatly needed, but scarce any of it fell during the time of Divine service on Sabbath. In his prayer Mr Stewart made appropriate mention of a goodness that could be at once favourable to exposed congregations and to the concerns of the husbandmen. I saw two servants in livery at the tent. They are, I have been told, domestics of Sir Hugh Fraser, a fierce intrusionist, and have been warned to quit his service at the term, for their adherence to the Free Church.'

'Monday morning, 7 o'clock.

'You have been my inseparable companion, dearest, since we parted. You were with me last night on the ridge of the hill, looking at all I looked at and feeling all I felt. Do you remember the exquisite evening we passed among the pines on the upper slope of the hill above the cultivated ground, where the hill looks down upon the town and bay? There was a bright red sunset;—the trees in front of us, where relieved against the sky of flame, seemed as if drawn in black, while the trees behind seemed as if dipped in blood. It was early in our acquaintance,—friendship had passed into love, though we had not yet become aware of the fact; but rarely, I suppose, do mere friends manifest the same unwillingness to part that we did that evening. We lingered on till all that was fine in the sunset had disappeared, and found the grey of sea and sky, and the blackness of field and wood, quite as agreeable as the many-tinted landscape we had so admired a little before. My own dear Lydia,—it *is* an advantage to have recollections such as these to summon up.'

'Wednesday morning.

'I had an invitation last evening for tea with Mrs Allardyce, but having previously accepted Mrs Smith's, I had to make a compromise by leaving Mrs S. early and calling on Mrs A. I spent two hours with her very agreeably in conversation of an exclusively literary cast. She showed me a very elegant translation of a German poem with which she amused her leisure last winter. The work itself is a great curiosity. The Princess Elizabeth of England, one of the daughters of George III., afterwards Landgravine of Hesse Homburg, seems, like her sister Amelia, to have had a considerable dash of genius. She amused herself long before her marriage, in making a series of allegorical drawings, descriptive of the progress of Genius, with Imagination and Fancy for his companions. Take one of the set as a specimen. The scene is a cavern crusted with stalactites, some of which, as is not uncommon, present the appearance of animals, some of men,—mere approximation, of course, but yet traceable. Genius with his two companions stands thoughtful in the midst. Fancy points out the resemblance to the boy, and Imagination hands him a chisel and mallet, that he may imitate them in the marble. The scene symbolizes the origin of sculpture. When the Princess became the Landgravine,—I should rather say when she became the Landgrave's widow, her drawings were published with a poetical description, the work of a German poetess, affixed to each; and what Mrs A. has been engaged upon is a translation in the Spenserian stanza of the description. It seems, as I have said, very elegant.'

'Inverness, Wednesday evening, 26th July, 1843.

'I trust to find you quite strong and ruddy, with the rounded outline of your earlier days, and quite able to climb such steep slopes as that which leads from the farther beech-tree to the pathway above. We shall have gallant walks together, and I shall be your squire and sweetheart.

'I left Cromarty by the steamer about eleven o'clock. Poor Angus carried my bag and stick to the pier, sturdily refusing to give them up, though much urged. He then took his place in the boat, and went out with me to the vessel's side, where he attracted some notice by the uncouthness of his gesticulations in bidding me farewell. "What sort of a strange man is that?" said one of the passengers to me, as the boat was returning shorewards, and Angus stood up swinging his arms like an ogre, and shouting out, "Hugh, boy, Hugh, good-bye, boy;" "he is the strangest looking person I ever saw." "A poor idiot," I said, "but a very attached one." The first half-hour of our voyage was fine though somewhat gloomy, and I enjoyed the bay as it receded,—the Hill, Navity Shore, and the Burn of Eathie. There then came on, somewhat suddenly, a disagreeable rain, accompanied by a strong head wind, and the latter three-fourths of the passage were barely endurable in consequence.

'I wished Mr Mackenzie to show me the house in which you were born. But he could show me only the place it had once occupied. The house itself had disappeared and a fashionable hotel rises on its site. Tristram Shandy made a pilgrimage once to the gate of a French city to drop a tear over the tomb of two faithful lovers, but on reaching the gate found no tomb to shed it over. And here was I, romantically seeking my wife's birth-place in a spick-span, fresh-coloured building with bottles of spirits in the lower windows and obsequious waiters at every door. And now, dearest, good-bye. It is past eleven, and I start for the west at five. Ere going to bed, however, I shall read your letter once more.'

'Steamboat, Thursday evening.

'I rose this morning before four, and looked out on the weather in despair. The rain fell thick and fast, and the whole heavens seemed as if cloaked in grey felt. In a wonderfully short time, however, the rain lightened, and such a slice of the newly-risen sun as you might clip off with a pair of scissors appeared through a narrow slit in the east, reddening Craig-Phadrig with a faint flush of vermilion. In an hour after it was a fine morning, and the steamer set off with the prospect of an agreeable voyage.

'Having spared you hitherto, I must inflict upon you a very few sentences of geology. The features of the very fine scenery around Inverness wear all a geological impress. The ridge of the Leys extending from the east side of the moor of Culloden to the shores of

Loch Ness, is marked atop by its rectilinear character, as if a ruler had been applied to it in the forming. The coast of Caithness, as seen from the sea, presents exactly the same appearance;—so also does the ridge which extends from the hill of Cromarty far into the parish of Resolis. In each case these ridges are composed of Old Red of the lower formation,—of sandstone, however, not low but high in this lower formation; and the rectilinear peculiarity seems to be one of its characteristics. The ridge fronting that of the Leys, that to which Craig-Phadrig belongs, is of an entirely different character;—the outline is more than usually waved and indented. It is composed of the great conglomerate, and this wavy character seems as characteristic a condition of it as the rectilinear outline is of the deposits that overlie it. Deep under the rectilinear ridge the fish-beds appear, as shown in the quarry of Inches; but to indicate their place with regard to their indented ridge opposite, one has to draw a line in the clouds. In the valley between these two ridges,—a portion of the great Caledonian valley, there lie a number of detached hills, of an appearance wholly different from the hills on either side. Some of them are detached; some of them lie in chains, like birds' eggs blown and threaded. They are composed of sand and gravel, and preserve a rude parallelism to the loftier hills behind, just as rebounding waves preserve a parallelism to the rock or mole from which they have been thrown back. And these gravel hills are, as you anticipate, moraines. They belong to a period when Loch Ness, that now never freezes in even the severest winter, must have existed as one vast mass of ice wedged into the earth, and when a huge glacier filled the great glen of Scotland from the east to the west sea.

'We had fine sailing through the lower part of Loch Ness. The hills, half rock, half wood, had their patches of sunshine that looked all the lovelier and brighter from the groundwork of shadow, and the previous rain had given to every ravine its white thread of foam. We could hear from the deck the dash of a thousand tiny cascades. I marked Aldourie as we passed. Does it not seem strange that in so thoroughly highland a place, in a Tory family, and full eighty years ago, when Jacobitism had still life in such recesses, the author of the *Vindiciæ Gallicæ* should have been born? After leaving Aldourie behind us we neared Urquhart. The steamboat made what Galt would have termed a *circumbendibus* into its beautiful little bay, and passed as near the picturesque old castle on its western promontory as safety permitted. Originally it must have been a place of great strength and importance; and as a ruin it is still very fine. Including the outworks, the space covered very considerably exceeds an acre. It was taken in the wars of Edward I. by the English, and rough marching must they have had ere they reached it; —according to the well-known couplet, they must have seen General Wade's roads "before that they were made," but whether they blessed the General or no history declareth not.

'I saw the Fall of Foyers; but by a narrow enough chance, and for a very short time. The captain was unwilling to stop, and urged that very few of the passengers were inclined to be at the trouble of visiting it; but we made a muster of nine, all stark fellows, who undertook to clear the ground to and from in three-quarters of an hour, and so he consented to lay to. I find I still rank with the young men, so far as strength and activity are concerned. Only one of the party contrived to keep abreast of me, and him I ultimately distanced. But I cannot expect to rank with the young men long. You are acquainted with the decision of Burns,—

"For ance that five an' forty's speel'd," &c.

The fall was in prime condition; the sun shone, spanning the caldron with a very fine rainbow, and the water, swollen by the late rains, covered in the precipice and boiled in the hollow in its best and grandest style. One of Byron's similes, descriptive of a waterfall, if not indeed rather of that to which the waterfall is compared, came athwart me as I looked:—

"Like the tail of the Pale Horse in the Revelation."

I do not know what the botanist might think of the plants and shrubs so immediately in the neighbourhood of the cataract that they lie under a perpetual drizzle, but to my very hasty survey the group seemed different from what we find either on wholly dry or wholly wet ground. Marsh plants draw their moisture from below, and plants of the dry land get only occasional showers, but here were plants living in never-ceasing rain. There was, however, neither space nor opportunity for examination, and so hurrying back we got aboard a few minutes within time.

'We made some stay at Fort Augustus in passing the locks, and I rambled over the village and the Fort for about three-quarters of an hour. The view from the ramparts is exceedingly fine. Loch Ness is assuredly the sublimest ditch in existence. The rectilinear sides, parallel throughout its whole extent, and the extreme steepness of the hills that rise out of it, in so many places without skirt or margin, give it a very peculiar character. The Fort and its immediate neighbourhood interested me in some degree as the scene of the most pleasing portion of the most pleasing work of Mrs Grant of Laggan. Her earlier *Letters from the Mountains*, written when she was a very young lady, and yet, though young, intellectually at her best, for she afterwards became stiff and pompous,—date from Fort Augustus. She speaks in them, pleasingly enough, of her walks, and of the young *Moderate* chaplain, whose wife she afterwards became. The Fort itself, though fast hastening to decay, will make, like all other modern places of strength, but an indifferent ruin. For the purposes of the landscape-painter one baronial castle, in the style of Castle Urquhart, would be worth a dozen of the most regular fortifications run to decay, Vauban ever erected; but when regular fortifications shall have run to decay, baronial castles shall have gone out of existence altogether; and so in the millennium, the ruins that tell of war and bloodshed will be very ungainly-looking things. The village beside the Fort consists of a scattered group of houses, a few of them exceedingly neat and genteel, with native ladies in them that look like the ladies elsewhere, but the greater part neglected-looking and poor. The village has its church, and Romance could doubtless be fine upon the clustered cottages associated with the place of devotion. It seems an argument for an establishment embodied in stone and lime. Alas for the fact, however. A mean-looking man, in shabby black, comes regularly down to every steamboat as it passes the locks, and stretching out his fiery nose over the side, looks beseechingly for a dram. And this mean-looking, red-nosed man is the minister of the village.

'As we passed through the piece of canal which stretches between Loch Ness and Loch Oich, I saw a lime-kiln burning on the hill-side and would fain have examined the stone,—a primary limestone, certainly, perhaps a marble,—but had no opportunity. In passing through Loch Oich, the centre of the canal with respect to level, for then the descent commences on both sides, we had a striking though somewhat exaggerated specimen of the very different climates of the east and west coasts. On looking through the valley towards the east we saw in a clear atmosphere a series of horizontal clouds, rising the one beyond the other, like level beams in a roof seen in perspective, and the mountains clear and well-defined, and coloured according to their distances. In looking towards the west we saw hills shorn of their tops by the descending fogs, a blue haze resting on the lower grounds, and a turbid and broken sky overhead. The hills, however, improve mightily in their vegetation. The heath well-nigh disappears, and fine rank grass covers their sides instead. But how sadly they are scarred with ravines,—the conduit pipes by which the upper fogs empty themselves in water into the lakes. This difference between the vegetation of the opposite coasts becomes very marked at Glengarry, when for the brown sterility of many of the Loch Ness hills we have mountains green to their summits, and woods of luxuriant foliage. The old ruined tower of the chieftain looks out with imposing effect on the lake from a cluster of trees. It does sound somewhat ridiculous that the late Glengarry should have squandered so much money in imitating the old highland chieftains who had none, that now neither the lands nor the tower of the ancestors of whom he was so vain belong to his son. He was so proud of his family, that he ruined it.

'Isn't it provoking that I can't see Ben Nevis? Twice have I passed him now, and each time has he veiled face and head like an eastern beauty. He presents this evening an enormous breadth of dark-blue base, blue approaching to black, and at the level of about three thousand feet there is a jagged outline of cloud that varies its form every moment as the long vapoury trail sweeps by. Hush!—there is the sun breaking out on his lower slopes,—the very appearance you describe as of such frequent occurrence on the hills of Loch Lomond and Loch Long. All is dark above and on the flat grounds below, but rock and corrie on the hill-side flare out in the light. And now the sunny patches have flitted over it, and all is again blue and sombre; and now a single round patch of half an acre's breadth sails slowly across. We shall have to quit the boat almost immediately;—the great breach in the canal which was formed last winter, and so nearly drowned the valley of the Loch, has not yet been repaired, and so the last three miles of our journey towards the sea must be by coach. For at least two of the three the canal has a dry gravelly hollow.'

'Aboard the Steamer.

'Here we are awaiting the arrival of our heavy luggage. Fort William lies across the loch, a cold, blue-looking town under the base of a cold, blue-looking hill. In half an hour I shall be there, and trust I shall fare better this year than I did the last, when I found for my friends Mrs D— and her sister, a locked door and shuttered windows. I leave all my plans regarding the Small Isles to be settled by them. Good-bye, dearest,—the bell rings, and we are casting off our moorings.'

'Friday Morning. Fort William.

'Fort William is exactly the last place I would choose to live in. It is a naked, mean-looking town at the foot of a steep marshy hill, that allows scarce selvedge enough between its base and the sea for the houses to stand upon. The highest ground in Great Britain lies within a very few miles of it, and there are bold hills all around, but the forms seem ill-combined, and the general result is not striking. I was, I find, too hasty in my conclusion regarding the green vegetation of the western hills, —I should, at least, not have generalized so promptly. The hills fronting me are as dark and barren as any on the east coast. As I write the haze thickens, and the first drops of the coming shower patter against the panes.'

Our next letter furnishes a trait which will not surprise us in Hugh Miller. He has attained a position of the highest social respectability in the capital of his country, and his acquaintance are the most eminent men of the age. Yet he calls to his side his old friend of the hewing-shed, plain John Wilson, and sits down with him at breakfast, John, I am quite sure, not finding in Hugh any change since those days when they lived together in the Niddry cottage, except that now, as the wealthier man, Hugh can exercise the privilege of friendship to hint delicately that, if John would be the better of a little money, he has only to say so. In this little incident of Miller's intercourse with John Wilson do we not find better proof of a faithful, simple, incorruptible heart than if he had endowed a hospital or built a cathedral? Here is the letter.

'Edinburgh, August, 1843.

'Very many thanks, dearest, for your punctuality. I always *sleep in* on Saturday mornings; this morning, however, I was awaked by Bill. Nurse or Mary was running up to him, and then running back, a famous joke to poor Bill; and he was greeting each exhibition with so lusty a shout of laughter that he awoke papa. When I became collected enough to remember that it was Saturday I rang the bell, and as coolly as you used to order breakfast in similar circumstances, I *ordered up* your letter. "Yes, sir," was the reply, and the letter came. And so again, dearest, many, many thanks for it, and for your punctuality.

'I had my old workfellow, John Wilson, at breakfast with me. He has been out of employment for some time, and has taken to dealing in tea and sugar. You will not be displeased with me for offering him the use of a few pounds for laying in stock if he should require them, or if necessary, to any small amount, the use of my name. I am much mistaken if John be not one of the class whom it is a privilege to be permitted to favour. He thanked me very cordially, but said that at present, at least, he stood in no need of assistance. I have some thoughts of getting up a descriptive article of my residence at *Niddry*, in which I might introduce John, and contrast his blameless life with the dissipation of the other workmen. It might excite interest, and do him good in his present way of life.

'There is an interesting exhibition of ancient armour in Edinburgh at present, which I have just been seeing. Some of the suits of mail are very ancient and very curious. One gigantic suit gives the idea of what sort of a man the Bruce must have been. Its height does not exceed six feet two, but the breadth of shoulder, strength of limb, and depth of chest, are enormous. The warrior who wore it, an old Teutonic Goth, must have been a full match, if his courage was equal to his strength, for at least four ordinary men.'

William Thom, the writer of the following letter, was a man of fine and brilliant genius, his vein of pathos as true as that of Burns. Born in Inverury, a small town of Aberdeenshire, at the confluence of the Ury and the Don, he earned a precarious livelihood by hand-loom weaving, and found consolation in the piercing and tender melody of his songs. His *Mitherless Bairn,* the subject of one of Faed's noblest pictures, must live as long as the language. The following letter, light as it is, has touches of an arch, brave, and piquant humour, which bespeak the child of genius. He sank mournfully, as the immense majority of his class have done.

'Inverury, March 4th, 1844.

'Accept my very sincere thanks for your lively and kind notice of me and mine in your widely-spread Journal. True enough, it is a significant way of requiting such favours by dragging your care and kindness into fresh work; but, believing that you *will* as well as *word* my prosperity, I take leave to hand my Prospectus, in the hope that, should you meet a friendly name, you will make my list all *that* the longer. My book will be a *little* book, which is sometimes a *great* mercy; but however lowly its claims in other respects, I assure you no page shall bear aught to disparage its patrons or me. I would fain have waited yet a little, and, in finishing what is pretty well on, have more nearly come up to that expectation which a (perhaps too partial) favouring press has taught to wait my appearance. Here is the secret;—customar weaving is *down,* fairly sunk before its leviathan rival, the big manufactory; cheapness, elegance, durability, all spring ahead of the solitary weaver. My loom, once my ship—Hope and Hardship, alternate steersmen—is now seen ducking.

The former of these mates "cuts," the other lends a hand to heave the lyre overboard, hen-coop ways, that we may float ashore, never doubting of his company even there. I think, if my project prosper, of a small shop stuffed with second-hand books—"Bought or exchanged." Such my El Dorado! The hope makes me happy; even that is something.'

A few extracts may be taken from the letters sent to Mrs Miller during the English tour, but as he made large use of those letters in preparing the *First Impressions*, we must glean sparingly.

'Olney, 9th September, 1845.

'Here I am in a quiet old inn kept by a quiet old man who remembers Squire Cowper and Mrs Unwin; and in the early part of the day I walked the walk described by the Squire in the *Task* with an old woman of 71 for my guide, for whose schooling Mrs Unwin had paid. She knew the Lady Hesketh, too, and, when a little thing, used to get coppers from her. A kindly lady was the Lady Hesketh,—there are no such ladies now-a-days,—that is, at Weston Underwood, I suppose. She used to put coppers into her little silk bag every time she went out, in order to make the children whom she met happy. She and Mrs Unwin, too, were remarkably good to the poor. I walked with the old woman, much entertained with her gossip, through the stately colonnade of limes whose "obsolete prolixity of shade" the poet has celebrated, and which is in sober truth a very notable thing, on to the "alcove," and from thence to the "rustic bridge," and then on through the field with the chasm in the centre of it, into which the sheep of the fable proposed throwing themselves, to "Yardley Oak." Then returning by another road, I passed by the "peasant's nest," and after making the old woman happy with half-a-crown, parted from her and struck down to the Ouse, a sluggish, sullen stream fringed with reeds and rushes, that winds through flat dank meadows, on which a rich country looks down on either side. I saw the broad leaves of the water-lilies bobbing up and down in the current, but the lilies themselves were gone. By the way, my old guide knew not only the squire and the two ladies, Mrs Unwin and the Lady Hesketh, but also the little dog Beau, and a pretty little dog he was, with a good deal of red about him.

'This part of the country lies on the Oolite, and we find fragments of Oolitic fossils in almost every heap of rubbish by the wayside. Directly opposite Cowper's house in Weston Underwood I picked up a fossil pecten and terebratula, and bethought me of his denunciations of the geologists,—who, to be sure, in his days were a sad infidel pack. His Olney house, a tall brick building not very perpendicular in the walls, is now an infant school. I entered what had been his parlour, and was almost stunned by the gabble of infant voices. There have been alterations made in the interior of the building to suit its altered circumstances, but the small port-hole in the partition through which his tame hares used to come bounding out to their evening play on the carpet still exists *in statu quo.* I saw, too, in the garden an apple-tree of his planting,—a Ribstone pippin, if you wish to be particular,—and the little lath and plaster summer-house in which he wrote so many of his poems and letters. The latter has been preserved with good taste, which one would wish to see more general, in exactly its original condition,—nothing has been changed except that, like the book in the Revelations, it is now written within and without with several thousand names. I saw among the others a name not particularly classical, that of John Tawell the Quaker, who was hung some time since for poisoning his mistress. The characters are written in a firm bold hand, and immediately beside them there is a vignette evidently of after production,—a gallows with an unfortunate wight hanging on it,—

"With his last gasp his gab doth gape."

Immediately behind the garden is the snug parsonage-house,—the home in succession of John Newton and Thomas Scott,—and the parish church in which they both preached, a fine solid structure with a tall handsome spire, closes the vista in this direction.

'So much for Olney;—the greater part of yesterday I spent in Stratford-on-Avon, where I saw both the birth-place and grave of "William Shakspear, Gentleman,"—have you ever heard of such a person? The birth-place,—a low-browed room, under the beams of which one can barely walk with one's hat on, is not half a mile removed from the burial-place. The humbly-born boy was a purpose-like fellow, and returned to his native town a gentleman, and to get himself a grave among its magnates in the chancel of the church. By the way, in utter defiance of fine taste and fine art, I pronounce the humble stone bust, his monument, incomparably superior to all the idealized likenesses of him, whether done on canvas or on marble, that men of genius have yet produced. The men of genius make him a wonderfully pretty fellow, with poetry oozing out of every feature,—but their Shakspear would never have been "William Shakspear, *Gentleman;*" neither, in the times of Elizabeth and James, when money was of such value, would he have returned to his native village a man of five hundred a year. The Shakspear of the stone bust is the true Shakspear;—the head, a powerful mass of brain, would require all Chalmers's hat,—the forehead is as broad,—more erect and of much more general capacity; and the whole countenance is that of a shrewd sagacious man, who could, of course, be poetical when he willed it,—rather more so than anybody else, but who mingled wondrous little poetry in his every-day business. The man whom the stone bust represents could have been Chancellor of the Exchequer, and in opening the budget his speech would embody many of the figures of Cocker, judiciously arranged, but not one poetical figure.'

'Birmingham, Ravenhurst Street, Oct. 5th, 1845.

'There is still a good deal to interest me within half a day's ride of Birmingham; I must revisit at least once more the ancient formation in the neighbourhood of Dudley, and see, what are rather famous in this quarter, the Clent-hills, a group of eminences from which Thomson is said to have drawn some of the noblest descriptions in the *Seasons*. It is, however, no mere love of sight-seeing that detains me in England. Have you ever noticed on the shore what fishermen call "the turn of the tide"? Often, for the greater part of an hour, it is impossible to say whether the sea is rising or falling along the beach. But then all at once there comes a change; if it be flood that has commenced, the little waves come running upwards, bearing on their tops a slight crust of dried sand, and pebble after pebble disappears; if it be ebb, the waters are drawn off as if by suction, and strip after strip of the damp strand is laid bare. Now, for the last month I have felt, with regard to my health, as if the "tide was on the turn; " I have been rallying slowly but not decisively; and until I feel the flood-stream of health setting fairly in, I hold it would scarce be justice to you, myself, or the bairns, to return to Archibald Place, and commence my labours with but the prospect of sinking under them.

'I walked yesterday considerably more than ten miles along very uneven ways, and some six or eight miles to-day. It would have been interesting could we have traced together in the Leasowes the marks which still remain of the artistic skill of Shenstone. I never yet saw any place in which a few steps so completely alter a scene; in the space of half a mile one might fill a whole portfolio with sketches, all fine and all differing from each other. There is not in all Shenstone's works a finer poem than the Leasowes.

'It grieves me to hear that there is still something radically wrong with your constitution. You must lose no time in getting out to Gifford; if you have this bracing weather in Scotland, and if it continue for but a fortnight, it may do you a world of good. Mrs Fraser describes the country in her neighbourhood as fine, the house as comfortable, and the society as good. I suppose you will take Harriet with you;—is she still looking about her as when she first set out on her travels, and exhausted her mamma's knowledge of the plants and grasses?

'My present mode of life is not very suggestive of topics for discussing in the *Witness*, and I am afraid my last two articles will betray the fact. My evenings in London hung heavy on my hands; I bought a cheap two-shilling edition of Eugene Sue's last work, "*The Wandering Jew*," to while away the time in my lodgings, and the perusal suggested an article on the State of Opinion on the Continent. It is no wonder that Popery, in its contest with mere liberalism, should have the better. Bad as it is, it is not so bad as the antagonist system of morals which the writings of Eugene Sue serve both to illustrate and to spread.

'You speak, dearest, of temperament, and the difficulty of bearing up against it by any mere effort of the will when it is adverse to small but not unimportant everyday duties. I know somewhat of that difficulty from experience in myself; willing may do much, but it will not change nature, or convert uphill work into downhill. But I trust we shall both get on, bearing and forbearing, with a solid stratum of affection at bottom. I have been conscious since my late attack of an irritability of temper, which is, I hope, not natural to me, and which, when better health comes, will, I trust, disappear. I keep it down so that it gives no external sign; since I entered England it has found no expression whatever; but I am very sensible of it, especially after passing a rather sleepless night. To-day I am in a very genial humour, the entire secret of which is in the excellence of last night's rest, induced, I think, by the fatigue of the previous day. I mention the thing merely in corroboration of your remark;—we cannot be independent of the animal part of us. I am a good boy to-day because I slept well last night, but I was not so good a boy a week since, for my nerves were out of order, and my sleep had been bad.'

'Dudley, Oct. 16th, 1845.

'Fine weather at last,—clear, bracing, warm, but not too warm, exactly the sort of weather for getting well in; and here I am, you see, enjoying it. Had such weather come six weeks ago, I would have been well, I think, for the last month. When nervousness mingles with one's other complaints, one is exceedingly dependent on what sort of a sky chances to be overhead; it is well to be out-of-doors getting strong and all that, but when Nature's face is as gloomy as one's own, as if she too had nerves, the mere sense of duty lacks strength to drag one out. It cost me no effort, however, to get out yesterday morning. I took coach for the Leasowes, where I spent some hours in re-exploring, and then, passing through Hales Owen, walked on to the Clent Hills, through a picturesque country, rich in historic associations. Then descending on Hagley, I walked on to Stourbridge, a considerable town, and as the evening was darkening, took coach for Dudley, a ride of a peculiar sort of interest. The coal-field to which this part of the country owes its prosperity, and which has made it the workshop of the Empire in whatever is wrought in iron, is of no great extent, but of astonishing richness. One of its seams, known as the Ten-yard coal, is actually thirty feet in thickness, thrice that of the next best seam in Britain, and the tract of country over it is studded—as seen last night, I should rather say, spangled—with furnaces. The view on both sides, as seen from the coach-top, had, if I may venture on such a combination, an infernal beauty; it seemed, at least, a bit of the scenery described by Milton. The darkness was sprinkled thick with roaring, flickering, comet-like fires, and the heavens above glowed in the reflected light a blood red.

'To-day I have spent very agreeably in exploring the caves and the ruins of the Castle Hill of Dudley, and in geologizing at the *Wren's-nest*, a very singular hill rich in the Silurian fossils and honey-combed to a vast depth by lime workings. I spent some time, too, in examining the Dudley museum,—a well-arranged collection, considerably richer in the more ancient organisms than the British one. I saw in it not a few of the originals figured by Murchison, and found it of use to acquaint myself with them in their actual forms. But I will be unable, I find, to add materially to my collection here. It is rare to find a well-preserved trilobite,—so rare that the fossil dealers charge for them from 10*s*. to £5, and I cannot afford to collect specimens at such a price. I had no little pleasure, however, in hammering among the rocks, though I found but little;—I was in a region which I had not hitherto explored, and all I did find in it was new to me,—new, at least, as fossils, though Murchison had brought me acquainted with their forms.

'Dudley Castle is a fine specimen of the very ancient and very extensive English castle,—consisting of keep, chapel, dungeon, great hall, servants' hall, ladies' rooms, &c., &c., with a vast court, in which I found a company of soldiers on parade, and surrounded by a deep moat. The keep, a picturesque pile of great strength, bears marks of the iron hand of Cromwell. It was garrisoned during the civil wars by the Royalists, and held out until battered down on one of its sides almost to the ground. Balls of thirty-two pound weight have been found among the ruins,—some of the Lord Protector's arguments, of whom it may be said as Barbour said of the Bruce, that

> "Where he strook wi' even straik
> Nothing mocht against him stand."

The castle, though now thoroughly a ruin, was inhabited so recently, that the grandmother of Mrs Sherwood (an authoress of some celebrity) spent some time in it in the capacity of lady's-maid; and the grand-daughter gives some amusing gossip in one of her works of her ancestor's recollections of its splendour. The English have less of family pride in their composition than we. I question whether a Scotch authoress would choose to introduce her grandmother to the public as a waiting-maid, even though her story should be a very good one.

'The caverns under the Castle Hill are very extraordinary, and the work of excavation is still going on. It is one of three hills of the Silurian system,—the Wren's-nest is another,—that rise in the middle of the coal-field, and the lime which they abundantly furnish is in great use as a flux for fusing the dry ironstone, besides for building and agricultural purposes. And so they have been wrought to their very centres, and perforated by railways and canals. One of the caverns I visited is a most extraordinary place, a full half-mile in length, with a deep sluggish canal winding through it, and supported with vast columns of rock. Had it not been excavated by human labour I would term it sublime, but the idea that it was all picked out by barrowfuls militates against its respectability. Dr Buckland, however, makes it the subject of a really fine description, and lectured on it to the members of the British Association in 1839.

'This is a letter somewhat like a bit of a guide-book; but you must bear with it. I am telling you of what I have come to this part of the country to see, and have had much pleasure in seeing. Though not yet very strong, I have borne the fatigue of my yesterday and to-day's explorations tolerably well,—greatly better than my sight-seeing in London, but the clear sky and pure bracing air have mightily assisted. I lodge here in a Temperance Coffee House, kept by very quiet people, and am on particularly good terms with their youngest boy, a little fellow of eight years, who makes speeches at Temperance meetings, and divides all society into Teetotallers and Drunkards. He is beside me at present writing a speech, which begins—"Ladies and Gentlemen, I don't mind saying a few words; I have been a teetotaller eight years." He is a clever big-headed little fellow, but somewhat spoiled by the notice which has been taken of him. The circumstance of one's self having children wonderfully softens the heart towards the children of others.

'There were some four or five points during my journeyings of yesterday and to-day in which I wished you were with me,—at the Leasowes,—at the Clent hills, at Hagley,—on the coach-top as we drove through the region of furnaces,—on the Dudley Castle Hill,—and in the caverns of the Wren's-nest. But though not strong myself, you are less strong still, and could not have borne half the fatigue. By the way, if this reach you ere you have left the north, I would suggest to you coming south by the omnibus that plies from Fort William to Loch Lomond,—passing through Glencoe and the Deer Forest of Breadalbane. You would see by this means some of the wildest scenery in the kingdom,—scenery that ere the establishment of the present conveyance was scarcely accessible to ladies at all, and to men only at a considerable expense of money and exertion. It would be well for you to secure an inside seat, with the stipulation that you might if you liked ride outside. Glencoe is often drenched by deluging rains, and if you took merely an outside seat,—by far the most advantageous for sight-seeing,—you might suffer as much as you did in the storm of Loch Katrine. If you take this route there are one or two things to which it might be well to direct your attention. After leaving Fort William the scenery for the first eight or ten miles is tame, but this part of the ride would be passed over in the twilight darkness of the morning. You would then get into a very fine tract of mingled hill, island, and mountain,—the formation here is the *mica-schist*, and the scene finely represents its characteristic picturesqueness. You then enter Glencoe, a porphyritic region, and you at once find the picturesque giving place to the sublime. The forms of the hills are different, every summit is a pyramidal peak, and the vast precipices are barred by imperfectly-formed columns. At the top of the glen you enter a granitic region; there are vast barren plains, and the hills are rounded, not peaked.

'Where the porphyry ends and the granite begins you see a hill of each, set up in close juxtaposition as if for specimens. You then enter on a dreary gneiss district; the hills are vast, but truncated—mere lower stories of hills; the valleys, too, are on a large scale, but not picturesque. You then a second time enter a mica-schist district, and all is picturesqueness and beauty. And then you arrive at Loch Lomond, and all is known ground. I know not better illustrations anywhere of the dependence of scenery on geologic formation than are to be found in the line from Fort William to Dumbarton, so interesting on many other accounts, and wish you would take it.

'I have not yet said anything of London, and yet I passed my week there—nearly a week, at least—most agreeably. The greater part of two days I devoted to the Museum, but I did not derive such advantage from the study of its fossils as I had expected. They are numerous, but in most cases the arrangement is not good. I had some conversation on the subject with the curator of this department, and found he had enough of geology to discover that I, maugre my "leather clothes," had more. By the way, in the shelves devoted to fossil fish I saw a very considerable number of specimens of my own gathering, many of them still bearing young Andrew's handwriting. I think they must be those I sent to a young Englishman, in exchange for fossils of the English Oolite and Lias. I think I remember being hurried when I sent them, and getting Andrew to name them for me. Over one of the specimens, in a conspicuous corner, I saw printed large, "Miller's Winged Fish." I may mention, but I am afraid you will not care very much for the information, that the large fragments of icthyolites found in Russia, which, from the description of Murchison, I concluded to be identical with those I have got from Thurso, are altogether different from them; so both my specimens and the Russian ones are alike unique.

'The time which I could not devote to the Museum,—for it is open only on alternate days,—I spent in general sight-seeing. I contrived to get to the Golden Gallery of St Paul's, from which I saw London in the form of a whole county covered with brick and smoke. Its greatness, however, though it contains many fine things, is merely the result of aggregation; by adding brick to brick and beam to beam London has been made great. If we pit cities against each other, as the English pit their boxers, weight for weight, Edinburgh, for its weight and inches, is by much the greater city of the two.

'By the way, you said nothing in your last of poor Harriet. Tell her that I have put a kiss in the heart of this round O, which she must try to bring out of it. This is a fine country for nuts, and I must get some for her Halloween, brought to Archibald Place. I did not get to the Liverpool Meeting. . . . From the tone of the dissenting papers here regarding it, I have lost all hope of its producing aught except bad speeches.

Voluntaryism has eaten the very pith out of Dissent; like a goodly tree eaten by white ants, it will yield to the first shock of the tempest.'

The scene now changes again to the North of Scotland.

'Cromarty, Dec. 15th, 1846.

'The snow in these northern regions lies very deep, the mail-gig for the last three days has come in at midnight, and at Inverness on Saturday last four-and-twenty hours passed during which no post, South or North, entered the town. I had a short walk at noon to-day with Mr Stewart; we attempted going up as far as Newton in the expectation of finding a beaten track along what is usually so well frequented a road, but in some places we found the snow-wreaths lying across from side to side level with the fences, and so, after some wading, were glad to turn back. Bill, like Yorick's starling, "can't get out," but he requires to be well watched, and, despite of his Granny's painful tending, has already made several desperate bolts out into the snow. His history since the storm came on has been wholly barren of incident. He builds houses with boards, and asks when there will be another Martinmas market? But he merely shares in the general stagnation;—the snow-blanket seems to deaden the very surface of society here; and so I cannot give you one item of Cromarty news.

'In all likelihood I will pass both my Christmas and New-Year here. My sleeping-closet is by no means so cold and comfortless a place as you seem to think it. The snow, by drifting against the window, has caulked up the chinks, and I get quite snug every night among the blankets. Would that our poor Highlanders at this dreary season of cold and starvation were but half as snug! By the way, how comes it that it is for the *Irish* the Edinburgh ladies are exerting themselves? The Irish are, to be sure, human creatures, but they are human creatures of a considerably lower order than the Scotch Highlanders, who are at least suffering as much; and they have surely not such a claim on the Scotch. They will be better cared for, too. They are buying guns, and will be by-and-by shooting magistrates and clergymen by the score; and Parliament will in consequence do a great deal for them. But the poor Highlanders will shoot no one, not even a site-refusing laird or a brutal factor, and so they will be left to perish unregarded in their hovels. I see more and more every day of the philosophy of Cobbett's advice to the chopsticks of Kent, "If you wish to have your wrongs redressed, go out and burn ricks; government will yield nothing to justice, but a great deal to fear!"'

'Thurso, 25th August, 1849.

'I have been greatly more fortunate in my voyage than you were in yours. You, a good sailor, were sick; and I, a very bad one, had not a qualm. I see, however, that I am not the very worst of sailors. We had several sea-sick ladies on board, and one or two sick gentlemen, with a considerable batch of pale ones. We were at the quay-head of Wick exactly twenty-four hours after leaving that of Granton.

'Miss T— was regretting the absurdities of ——, which are, I see, very widely blown. Absurdity sticks to some people by a kind of fate. I saw Dr McCrie immediately before leaving Edinburgh; he had just returned from the South country and had brought with him an exceedingly droll account of ——'s marriage. Tall Mr —— dropped as if from the clouds upon the quiet of a Southland Free-Church Manse, accompanied by a muffled lady—as broad as she was long, and considerably advanced in life. To the broad lady Mr ——— was, it would seem, to be married; and the broad lady had a friend with her,—a grand person,—captain, major, or colonel, with huge whiskers and an air. The friend had never seen a Presbyterian marriage; and when the preliminary prayer was over he thought all was done, and began to speak in a joking vein very loud. "Marriage is honourable," said the minister, beginning the proper business of the transaction. "Yes it is, my boy," said the military man, clapping him on the shoulder. The young people looked unutterable things, and the minister was so much put out, that he was in danger of sticking the whole concern.

'I found a sad accident which had just happened at Wick the general subject of conversation on my arrival. The Sheriff-substitute, a young Edinburgh gentleman, has been drowned in boating on a long narrow lake, the Loch of Watton, which skirts for three miles the Thurso road. He was courting, it is said, a daughter of Sir John Sinclair, and was sailing up the loch in a small skiff to meet her, accompanied by a single boatman, when a sudden squall caught the sail, and the skiff first capsized and then went down, and boatman and sheriff were both drowned. I saw men engaged as I passed the loch to-day in dragging for the body of the boatman,—that of the sheriff was found the day before. They perished in a quiet inland scene hardly a gunshot distant from a grassy shore rising into green fields.'

Our two next letters, the first from Sir Roderick Murchison, the second from Lord Ellesmere, referring, the one to the *First Impressions*, the other to the *Footprints of the Creator*, require no introduction.

'Athenæum Club, London, April 19th, 1847.

'MY DEAR SIR,

'Many thanks for the acceptable present of your *First Impressions of England.* I hope they may not be the last, and that during your next visit you will endeavour to find your Ross-shire contemporary, who is located in Belgrave Square. Now, however, I am on the wing for a year's Continental touring, and this time it is the Mediterranean, Sicilian, and Spanish type which allures me. I shall, however, be back for the Oxford Meeting of the British Association, where I shall give up the chair to my successor, and shall doubtless have to undergo the annual persecution of the able but bigoted *Times*.

'I feel deeply indebted to you, on public as well as private grounds, for your vigorous defence of the noble cause of science, and for the enlightened manner in which you entwine it with Christianity. Such reasoning ought to put to shame the efforts of those who would put down science as antagonistic to religion.

'I am also much flattered by your passing allusions to my Silurian labours. On two points, however, I have to offer critical remarks. The first of these is purely personal, and may be considered as a little symptom of Silurian (not Scotch) *pride*.

'In your excellent work you speak of Buckland haranguing the philosophers of the British Association in the Dudley Caves. Now, the fact is, that in virtue of my labours on that ground, I was named generalissimo of the expedition, and I executed my duties, I assure you, like an autocrat. When all the barges, laden with geologists, had entered the caverns, which were previously filled with "Ladies and Gentlemen" of the neighbourhood, perched all around the lighted vaults, I commanded silence, and gave them a regular half-hour's harangue on the structure of the Wren's-nest and the surrounding tract. This address has been graphically alluded to by Schönbein, the chemist (now so famous for his gun-cotton), in a lively German work, descriptive of his visit to England, and he specially adverts to my *military air* and voice, the relics of the old peninsular soldier. I have not the slightest jealousy towards my dear and good friend Buckland, and if the den had been tenanted by hyænas I should have ceded my place to him; but being in the very heart of my Silurian, I was necessarily the commandant. He, it is true, got together some portion of the party, and *after* my harangue he amused that detachment, whilst I led the chief forces to the top of the Wren's-nest, and descanted on the exterior face of the surrounding country. The press got hold of his name, and forgot mine, so my operations had no recording poet. I give you the hint in case of a second edition, of which I doubt not.

'And now to a more material point. I am at this moment very sensitive on everything *Silurian*, as my old ally Sedgwick, after waiting 12 YEARS (i. e. since I first proposed the Silurian classification), has made a great onslaught on old Caractacus from the recesses of Cambria. In other words, he now wishes to substitute Cambrian for Lower Silurian. This is IMPOSSIBLE, and no geologist will assent to it, for the plainest reason,—that, as the fossils now known all over the world as Lower Silurian are identical with those subsequently found in Cambria, so all fairness and equity speak for the conservation of the original name. I doubt not that you also will be of this opinion when you have thrown your eye over my last publication in the May number of the Quarterly Journal of the Geological Society. I will send you a copy of this Memoir as well as of my last words on Sweden, and also on the Silurian rocks in Cornwall, if you will tell me to what London bookseller I may address them ;* and I beg to call your special attention to my conclusions on the paper on the Silurian Rocks of Sweden. Pray read them first, and then my reply to Sedgwick's new proposal. I have letters from Russia, Scandinavia, Germany, France, and America, all declaratory of firm adhesion to the Silurian faith.

'Alas! for one of my generalizations founded on negative evidence (of a very extensive kind, however) on which you built. The Lower Silurian is no longer to be viewed as an invertebrate period; for the Onchus † (species not yet decided) *has been* fixed in Llandeilo Flags, and in the Lower Silurian Rocks of Bala. In one respect I am comforted by the discovery, for the form is so very like that of the *Onchus Murchisoni* of the Ludlow Rocks, that it is clear the Silurian system is one great Natural History series, as proved indeed by all its other organic remains. Professor E. Forbes and every palæontologist is of this opinion, although the system is divided into upper and lower groups. It must be stated that ichthyolites are most rare, and very peculiar in these protozoic rocks.

'P.S. I should like you to have seen the Cornish ichthyolites, which are unlike anything known in the

* 'I will send them to your London publisher in Paternoster Row whose name is on your title-page.'

† This so-called discovery misled both Sir Roderick and Hugh Miller for the time. It led to the subsequent intention of cancelling one of the chapters in 'Footprints of the Creator.'

Devonian or Old Red, and which occupy a "quasi" Upper Silurian place. (*Vide* my Memoirs.)

'I shall be in Bohemia all July clearing out a grand Silurian Basin.'

FROM LORD ELLESMERE.

'Worsley, Manchester, November 27th, 1849.

'I have on two grounds to apologize for the liberty I take in addressing this letter to the author of *Footprints of the Creator*. In the first place, I have not the privilege of your acquaintance. In the second, I am one of a numerous class of readers justly stigmatized as "dilettanti," who are reduced to take upon trust statements of scientific facts, and whose opinion upon questions of inference from such facts is of no value or authority to others. I believe, however, I am scarcely wrong in supposing that though your work is addressed to readers of the highest class, its aim and intention is not confined to such, and that a large portion of it at least is intended for the use, instruction, and protection of persons like myself. If so, in labouring for the benefit of such persons to neutralize the emanations of a poisonous book, you have, as seems to me, done that which demands the gratitude of the Christian world, and performed with signal felicity and success the office of a sanitary commissioner of science. I cannot speak of myself as one whom your work has rescued from positive danger. I read the Vestiges with disbelief and detestation, but it has been a deep satisfaction to me to find such impressions justified and buttressed by arguments which I could not supply, and to feel assured that my own incredulity and aversion are not based merely on my own ignorance and incapacity. The extensive diffusion of the work in question forbids its being treated with silence and contempt, but it is a positive and great misfortune, when the task of answering such works is taken up by zealous but incompetent men. This happened to a great extent in the memorable instance of Gibbon's History, and he gained an easy and pernicious victory over several inferior scholars and feeble reasoners who hurried into the fray. It is true, as you have remarked, that both in England and Scotland the Church is adorned by men competent to deal with the scientific questions at issue, but he must be a bad churchman, whether in England or Scotland, who is not thankful for the appearance in the field of such a lay auxiliary as yourself. I have no right further to occupy your time with my own acknowledgments for pleasure and instruction received.

'P.S. A fortnight since, after finishing your volume, I left it in the hands of an old and valued friend, who, now in his 83rd year, at the close of a life of public service and activity, is much addicted to such studies as prepare him for another. He had read with pleasure a well-meant but somewhat superficial work of Dr Wiseman's, and I thought I was supplying him with something better. The following is an extract I have just received from him:—"I wish I were knowing in geology, but unknowing as I am, it is not often I meet with books so intensely interesting as I have found Hugh Miller. It is a pleasure reading him, and scarcely a less pleasure to think and reflect upon what he has written. His work has caused me to reflect more than almost any work I have ever read. It warms one's heart to find such a true believer. It astonishes me how a man of his antecedents can have taught himself to write so beautifully, and to be so clear in all his expressions, and how he can have been so extensive a reader. He amazes me throughout, and I am not surprised that Murchison and Agassiz should esteem and consider him so highly as they evidently do. I long to fly off to the north to see and hear him talk, and if he talks as well as he writes, what must be the charm of living with him!"'

This note to his mother-in-law, on occasion of the birth of his son Hugh, is worth printing on account of the autobiographic touch respecting his own birth, and the characteristically mournful reflection which follows.

'2, Stuart Street, Edinburgh, 4 o'clock, Thursday afternoon.

'The doctor has just been with us, and he is well pleased with the appearance of both mother and child. Baby, in his introduction into the world, had a sore struggle for life, and in pugilists' phrase, but with a deeper meaning than theirs, was for about five minutes "deaf to time." Accidents can scarce be hereditary; but my mother has told me that, when making my début, I refused to breathe for a still longer period. Were all the future known to the little *entrants* such refusals would, I dare say, be more common than they are, and more doggedly persisted in.'

Miller was particularly gratified by the terms in which the gift of the *Footprints* was acknowledged by Professor Owen, and quotes, in the following letter to a lady, the passage in which the work was mentioned by that eminent philosopher.

'Unpopular as I supposed my little book was to prove, the first thousand has gone off bravely, and I am passing a second through the press. Some of the letters regarding it are of a very gratifying character. One, in especial, from the first comparative anatomist in the world (Richard Owen) is singularly warm-hearted and cheering; and as you will not set it down to the score of vulgar vanity, I must just give you an extract, partly in order to show you, *as one of my pupils*, that I was not mistaken in supposing that the least popular portions of my book would be exactly those to which a certain class of students would attach most interest. "I have just received and, setting all other things aside, devoured your Asterolepis of Stromness! I find it not so hard and indigestible as it may prove to the Vestigesians! I have been instructed and delighted by it. I have also derived a heartfelt encouragement from it. It is almost the first contemporary work in which I have found some favourite ideas of my own weighed out and pronounced upon. This cheers one up after the despondency that will, in spite of reason, creep over one through the blank silence in which one's favourite works are received by those in whose especial behoof they have been cogitated and printed. I allude to the host of estimable anatomists, anthropotomists, and zoologists that we live and move amongst in our scientific coteries of London." This, you will agree with me, is worth whole volumes of ignorant criticism; a newspaper reviewer, very favourable in the main, speaks of my "rather tedious introduction;" it was, however, not for newspaper reviewers but for men such as Professor Owen that that introduction was written; and the Professor, you see, does not deem it tedious.'

In the autumn of 1850 the annual meeting of the British Association for the advancement of science was held in Edinburgh. In the beginning of June the following note from Sir Roderick Murchison reached Miller.

'16, Belgrave Square, 31st May, 1850.

'At a meeting of the Council of the British Association held yesterday I was named President of section C (or Geology and Geography), with yourself and James Nicol as secretaries for our special science, and with A. Keith Johnston for geography.

'As I moved that you be placed in this office, and as my motion passed unanimously (and indeed with acclamation in a full meeting, Professor Kelland from Edinburgh being present), I trust you will not allow anything to prevent your accepting it. I honestly confess that no honour could be more gratifying to me than to occupy the geological chair in my native country, and if I know that the author of the *Old Red Sandstone* will be one of the secretaries, I shall be still more proud, for I consider that *we* come from the same nook of land, the Black Isle and Cromarty being inseparable.'

He replied as follows:

'Accept my best thanks for the great and unexpected honour you have done me in proposing me as one of the secretaries for the Geological Section of the British Association at its coming meeting. I am afraid my qualifications for the office are not of the highest kind, my newspaper, too, will engage much of my time, and my health for the last few years has not been very strong; but I feel that I could scarce decline the appointment without, at least, appearing to fail in respect to the Council of the Association and in gratitude to you. I shall, therefore, do my best to fulfil its duties, trusting that the good nature of the Association may excuse my shortcomings, and the activity and business habits of my coadjutor, Professor Nicol, more than compensate for them. I have, I think, some curious things to set before the English geologists illustrative of the first ages of ganoidal existence,—fossils either without duplicate, or in a better state of keeping than elsewhere,—which may serve to show them that at least one of the pages of the geological record (that which you were the first to open) is more fully and clearly written in the rocks of our native country than in perhaps those of any other. My set of the remains of the *Asterolepis* is, in particular, very curious and, I believe, unique; and though I do not know that I can say much regarding them in addition to what I have already said in my little work *The Footprints*, it may be something to verify and illustrate by fossils so rare and little known what in the work I had to illustrate by but a set of greatly reduced woodcuts.

'I had the pleasure of hearing, a few days ago, from Professor Agassiz. He is doing me the honour of editing, at the request of a book publisher, an American edition of my *Footprints*, and I am at present engaged in making him a set of casts from some of my less fragile specimens of *Asterolepis*, on which he may possibly found some of his notes. At least two of the fossils, that of the interior and exterior surface of the cranial buckler of this huge ganoid, come out with beautiful effect in the plaster, and if I thought you would attach any interest to them, I would have much pleasure in sending you a pair. The interior surface, when viewed in a slant light, is quite sculpturesque in appearance. From a curious combination of plates, an angel robed and winged seems to stand in the centre, and the effect is at once singular, and, to at least the geological eye, beautiful;—so much so that I propose getting a pair of them framed.'

Hugh Miller was a tenderly affectionate parent, and never, until the last year of his life, when the serenity of his temper was shaken by reiterated and agonizing attacks of inflammation of the lungs, did he display the least severity to any of his children. One or two specimens may be given of the letters which, when absent on his geological tours, he used to address to them.

TO HIS DAUGHTER HARRIET.

'Cromarty, 20th September, 1850.

'You tell me you were "considering whether I wrote anything in the Album at the John o' Groats' Inn, and, if I did, what I wrote." Well, I did write in it; without adding my name, however; and what I did write was, though not poetry, a kind of verse. In one respect, papa resembles a great many other grown-up people; he occasionally writes bad verses; but then, unlike many of the others, he knows the verses to be bad, and does not make roundabout speeches in order to create an opportunity of repeating them in company. In all verse-writing a sort of marriage should take place,—a marriage between the lady *Rhyme* and the gentleman *Reason;* but in many verses the parties do not come together at all, and in many more the union is far from being a happy one. It is only those Heaven-made marriages that are happy, in which Genius enacts the part of the priest. That in which I took a part in the Inn at Huna was at best only a kind of humdrum fisher-wedding;—the bridegroom, though not quite a fool, was decidedly commonplace, and the bride, if not a *fright,* was at least plain.

"John o' Groats is a shapeless mound,
John o' Groat is dust;
To shapeless mound and wormy ground,
Man, and man's dwelling, must;—
Rottenness waits on the pomp of kings,
On the sword of the warrior canker and rust.

"John o' Groat lives still,
Lives in another sphere;
Evil his fare if his life was ill;
Happy, if righteous his course when here.
Traveller, ponder on these things,
And depart in God's fear."

'It is possible that the curious creature which Miss Allardyce found in the salt-water pool may be the Amphioxus or lancelet,—the lowest of all the fishes. There is a figure of this creature in a book on Natural History by Agassiz and Goold, which you will find, I think, on one of the tables in my store-room. Like many of our own species, but in a more literal sense, it wants brains.

'When across the ferry two days ago, I explored that sandy spit or point which runs into the sea, and is known to the Cromarty fishers as "Majock's (i.e. Marjorie's) Point." It seems to owe its formation to the tides which run out and into the bay of Nigg, acted upon by the high winds from the west, which are so frequent in this part of the country; and as it forms in lagoons or salt-water lakes, it is a good place for marking how the vegetation of the land encroaches on that of the sea, and for determining what the plants are that bear to be covered by every tide. I found a salt-wort to be the most abundant and the most encroaching,—some of the plants were thriving outside the castings of the sandworm.

'I have seen most of our friends here, and found them, as usual, very kind; but I have fenced off all the invitations I could, as I prefer my mother's potatoes and fish, with full liberty to go about with my hammer at all hours, to much better dinners, and the necessity of being punctual to an hour, and then sitting up quite proper in my chair for some four or five hours at a time. I must apologize to you for this letter, which is not at all so amusing as it might be; but the day is rainy and dull, and I am dull too.'

TO HIS SON WILLIAM.*

'Cromarty, August 24th, 1851.

'On the morning of Saturday I rose at five o'clock, and set out on foot (from Assynt) on my way to the Low Country, with the purpose of examining the marble quarries of Ledbey ere the mail-gig, with which I was to travel, should come up. The morning was lovely beyond description. While all was in deep shade in the valleys, the tops of the tall mountains gleamed like fire to the rising sun; and Loch Assynt, that seemed as black and as polished as a jet brooch, had here and there its patches of reflected flame. I passed one other very pretty loch, not of great extent, but speckled with green islands waving with birch and hazel, and abounding in fish, that as I went by were dimpling it with a thousand rings, and leaping for a few inches into the air. I here met an English manufacturer in kilt and hose, surrounded by half-a-dozen Highland gillies all tightly breeched. I found the marble, white and grey, rising amid the heath and long grass, like old snow on a mountain-top in midsummer, and detached several specimens. But I was fairly beaten off from examining the deposit as thoroughly as I could have wished, by armies of midges, that rose in clouds about my face every time I bent down to strike a blow, and made it feel as if it had been bathed in boiling water. There is perhaps no place in the world in which these little creatures are more troublesome than in the western parts of Sutherland. I travelled on by the gig to Lairg, which I

* Now Lieutenant in the 37th Grenadiers in India.

reached about two o'clock, and took in one meal breakfast and dinner. When I was a boy a few years older than you I used to spend some of my holidays in the neighbourhood of this place; and being desirous to revisit the localities with which I was so well acquainted of old, I determined on passing a day or two in the Lairg Inn. But I was told by the landlord that his house was quite full, and that he could not accommodate me. "I have a plaid of my own; could you not give me the use of a sofa?" I asked. The landlord looked at me, and then beckoned me into a corner. "Mr M., you are a man of sense," he said. "All my bed-rooms have been engaged for months by sporting gentlemen from the South, and my public room is occupied chiefly by their servants. The engaged rooms I cannot give you; and the servants are no company for you. Even the very bed I can give you is in a double-bedded room, occupied in part by Lord Grosvenor's valet. I state to you the real case, while you are yet in time to ride away by the mail to Golspie." I thanked the landlord, who is really a very decent man, and determined that, as I have prosecuted my researches during the last thirty years under great difficulties, the difficulty of the servants in general, and of the valet in particular, should not turn me back now. And so I took my share of the double-bedded room. Papa thinks that his status, such as it is, is that of the man of science, and that it is so dependent on what he achieves for himself, that it could not be improved by sleeping in the same bed with a lord, nor yet depressed by going to bed in the same room with a lord's valet. I had calculated on attending church at Lairg;—the Free Church is only a few hundred yards from the inn, and though a rapid river lies between, the late Mrs Mackay of Rockfield had given money to make a suspension-bridge over it, just in order that the people on the Lairg side might get easily to church. But on this Sabbath there was no preaching, the minister being from home; and though the *men* of the district officiated, their addresses were all in Gaelic. The father and grandfather of Mrs Mackay's husband were, in succession, ministers in this parish, and there is an interesting monument in the churchyard to their memory, and to that of two of the sons of the *son*, the one a captain in the army, who "fell," says the epitaph, "in the moment of victory at the muzzle of the enemies' cannon, at the memorable battle of Assaye, fought between General Wellesley and the Mahrattas;" the other a commander in the navy, of whom it is said that "his narratives of the shipwreck of the *Juno*, and of his exertions in the Red Sea, where, under God, he saved part of the 86th Regt., will commemorate his talents, fortitude, and humanity." Now, regarding the narrative of the shipwreck of the *Juno*, something curious can be told, in which you may take an interest when you grow older. There is a very famous description of a shipwreck in Byron's poems, into which there are introduced many circumstances new to poetry. And it was in this narrative that Byron found almost all of them. Indeed, his description may be in great part regarded as but a metrical rendering of Commander Mackay's narrative.'

TO THE SAME.

'Assynt, August 20th, 1852.

'Harriet, you, and Bessie are the public for which I write; while poor little Hugh, who is, I suspect, not intelligent enough to feel any interest in papa's adventures, must be regarded as that ignorant, but not uncared-for portion of the community which education has not yet reached. I broke off late on Saturday night in the little inn of Huna;—the Sabbath morning rose clear and beautiful;—I never saw the Orkney Islands look *so near* from the main land; their little fields, at the distance of many miles, gleamed yellow in the sun; and the tall Old Red Sandstone cliffs of Hoy, cliffs nearly a thousand feet in height, were sharply relieved against the sky, and bore a blood-hued flush of deep red; while the Pentland Frith, roughened by a light breeze, was intensely blue. I walked on after breakfast to the Free Church, and heard from Mr Macgregor two solid, doctrinal discourses. The congregation, however, was very thin; but I ought not to judge of it, I am told, by present appearances, as many of the men connected with it are at the herring-fishery. Still, however, it is only half a congregation at best, the other half congregation of the parish being in the Established Church. Papa and his friends could not help being Free Churchmen, as you will learn when you get older; we could not avoid the Disruption; but papa does sometimes regret that the Disruption should in so many parishes have, as it were, made two bites of a cherry; that is, broken up into two congregations a moiety of people that would have made one good one, but no more. After sermon, and after dinner, I walked on in the cool of the evening towards Wick, which I reached about nine o'clock. The little inn at Huna is far, far from any butcher's shop; and so a poor hen had to die every day papa was there, in order that papa should dine. There is an album kept at it, as I told Harriet two years ago, in which all visitors write their names, and whatever else may come into their heads; and so it contains, as you may think, a great deal of nonsense;

with here and there a witty thought, and here and there a just sentiment. Papa wrote some not very good rhymes in it about two years ago, and some not very good rhymes on the present occasion; though in the space between the two visits he wrote only solid prose. But those who pass evenings in the inn at Huna have usually not much to do; and so many of them, in imitation of those who have rhymed in the album before them, set themselves to break the Queen's English into very clumsy lines, with not much sense in them, that chime at their terminations. The following formed papa's last effort.

"Right curious volume! two long years have pass'd
Since I glanced o'er thy chequer'd pages last;
And leaves then blank in thee are fill'd, rare book,
E'en in a way on which 'tis sad to look;
So scant the wit, the wisdom seldom seen,
While dreary wastes of folly spread between.
Yes, sad, full sad, for, curious book, I see
Too true a picture of my life in thee;
Folly in both prevails, in both how few
The points t' instruct the heart, or glad the view;
In both the past admits of no recall,
E'en as the thing was done we stand or fall;
The ill express'd we cannot well suppress,
Nor can we make the by-past nonsense less.
Yet must I hold, poor curious book! that thine
Is e'en by much a harder fate than mine;
For thee no new editions wait,—no *Mind*
Shall add the wisdom that it does not find;
Raise into worth the grovelling and the low,
And mental force and moral weight bestow.
E'en faulty as thou art, thou still must be:—
A nobler dest'ny waits, I trust, on me;
The Great Corrector shall revise my past,
And in a fairer mould its lines re-cast;
For folly, wisdom; strength for weakness lend,
And all my blots erase, and all my faults amend."

'So much for papa's rhymes. There are some people who rhyme not much better than papa who believe themselves to be poets; but that is a mistake, and it usually does those who entertain it some harm. They rhyme, and rhyme, and rhyme, and deem themselves neglected because the world does not buy and praise their rhymes. But for rhymes such as theirs and papa's the world has no use whatever; the article bears no money value in the market; though to devote twenty minutes or so, in two years, to the manufacture of it, as papa does, is productive of no manner of harm.

'But I must pursue my narrative. I took the mail-coach from Wick very early on Monday morning, and travelled on, for the greater part of the way, under the cloud of night to Helmsdale, where I passed a day. On the morning of Wednesday I walked on along the shore from Brora to Golspie, and saw at a place called Strathstever a cave high up in a sandstone rock, to which there ascended a flight of steps. A high wall fenced round the bottom of the steps, but I contrived to climb over it; and, ascending to the cave, found it to be in part the work of nature, but also indebted to art. The space within was about the size of a rather large room, and there was a range of seats cut in the live rock that ran all around it. I was told that in the days of the late Duchess it used to be the scene of many good *pic-nics;* and as it occurs in a sequestered corner of the old coast line, with wild shrubs hanging from the rocks all around, and with a green level strip of land and the wide sea in front, one could, I dare say, take one's dinner in it very sentimentally. I passed Dunrobin Castle on my way;—it is an immense pile, on which there has been expended money enough to purchase a large estate. But though a part of it,—the part papa remembered of old,—be very ancient, and though the additions be in the newest style, the general effect is that of a very large fine *upstart* modern building. The new has swallowed up and overborne the old. I had to rise at a little after four o'clock next morning, to take the mail-gig from Lairg to Assynt, and reached the former place at breakfast-time, passing through a fine valley, that which opens into the sea, at the Little Ferry, for about sixteen or eighteen miles. Along the flat bottom of this valley the sea must once have flowed, and the precipices along its sides are very steep and abrupt, as if they still retained the forms given them by the waves; but the bottom is occupied by rich corn-fields; and cottages and trees appear where once canoes may have been.

'At Lairg I used to be well-acquainted, but it was long, long ago; and the people whom I knew were all away or dead. On my way to Assynt I saw a gentleman in a kilt, and, before the gig came up to him, set him down from his dress *as an Englishman*. And an Englishman he was, who had come out to the road to get his letters. As a general rule Scotchmen—save pipers, who are paid for it, and soldiers, who can't help it—don't wear the kilt. Assynt is a great marble district, and it was to examine the marble deposits with the rocks that lie over and under them that papa has paid it the present visit. He thinks he will be able to render it probable that the thick, stratified beds to which the marbles belong represent those flagstones of Caithness which abound in fossil fishes, and that the quartz-rock above and red sandstone below were once the Old Red Sandstone of the great conglomerate, and the deposits that overlie the fish-beds. But you will be able to understand all this a few years hence, should you live and attend to your lessons. Assynt is remarkable for its fine springs. They burst out of the limestone beds in volume enough, some of them, to turn a mill. Their water, too, from its great purity and clearness, forms a contrast to that of the Highlands generally, which is usually tinged with peat. One of those (the largest spring papa ever saw) is really a fine object. It comes bursting up out of the earth,—a little river, very clear, and in summer very cool,—though in winter it feels warm, and during hard frosts smokes as if it had been heated over a fire; and rank grass and rushes spring up along its edges. In the neighbourhood of Edinburgh it would be of great value, and would more than supply all the city. The day on which I date my letter (Friday, the 20th) I have spent very agreeably in tracing out the relations of the rocks here to each other; and I now think I understand the structure of Assynt, and have succeeded in reading appearances which some of the elder geologists misunderstand. But you would not much care to hear about these. There is not an inhabited dwelling-house in sight; and the two shattered houses, and the wasted brown hills, and the solitary country, where for hours together not a human being may be seen, gives one the idea of a world whose inhabitants have become extinct. The old castle was the stronghold of the McLeods of Assynt, and is famous as the place in which Montrose was confined previous to his being sent off to Edinburgh to be hanged. He was apprehended and delivered up by one of the McLeods,—a transaction base in itself, and which must have done the family no good when the Restoration came on, for the lands of Assynt then passed into another family, named Mackenzie. It was the Mackenzies who built the comparatively modern house; but, selling the property to the old Earls of Sutherland, the Mackenzies of Ross-shire were so angry at them, that they came and burnt their house, and houghed all their cattle. Those old times must have been very rough and uncomfortable; and, I dare say, it is better to be living now.'

FROM THE DUKE OF ARGYLL.

'Inverary, October 16th, 1852.

'May I ask you whether you have examined in detail the brecciated strata on the shore near Helmsdale—on both sides; and if so, whether you have formed any opinion as to the position they occupy relatively to the Brora Oolites, and the more Liassic-looking strata along the shore-line from Dunrobin upwards?

'That breccia appears to me to include no fragments which are clearly referable to the *Brora* strata, although it contains many which belong to the beds immediately behind and *below*—on the line towards Dunrobin. Very many, also, belong to the Caithness Flagstones—a portion of the Old Red of which, I think, there is no vestige remaining *in situ* now in Sutherland. I found in the breccia a fragment containing a ganoid scale, apparently that of the Dipterus like those at Banniskirk.

'The paste which cements this breccia seems to be a calcareous mass of pounded shells, belemnites, &c.: and this part of the breccia seems to me more referable to a break-up of the Brora strata than any of the included fragments.

'The remarkable continuity and uniformity of dip in the shore strata from Dunrobin to the Ord, and their total *unconformability* with the Brora beds, which generally terminate in the cliffy banks of the old sea beach—some distance from the present one—and are much more horizontal, suggested to me that the latter is a subsequent and superior formation, and that the Helmsdale breccia represents the close of a period embracing the Lias and some of the lower Oolites.

'I did not observe any of the lignites and fossil woods *in situ* in the breccia; but as all the other pebbles of the beach seem to be derived from that breccia, I conclude that they also do. But it strikes me that both the woods and the corals must have been already fossilized when they were pasted up in the breccia. One or two specimens of the latter I found firmly embedded amongst the other fragments.'

HUGH MILLER'S REPLY.

'I regret that I had not the good fortune of meeting your Grace this autumn at Helmsdale, as for the last eight or ten years I have been in the habit of spending a day in that neighbourhood every season, and could wish to have some of my findings regarding the rocks which form its framework tested by an eye so discerning as that of your Grace. The locality, though of much geological interest, is one that opens its stores but slowly. I visited it several seasons in succession ere I felt satisfied that I really understood it; and I still feel a desire to go back upon my conclusions respecting it, and have them put to the question.

'The conglomerate which at first sight seems so puzzling, consists of numerous beds which decidedly alternate with beds of argillaceous * strata, Liassic in their character, but unequivocally, as your Grace suggests, of the age of the Lower Oolite. The shale finely and regularly laminated, and blackened by carboniferous matter and abounding in fossils, must have been slowly deposited on an undisturbed sea bottom,—whereas the alternating beds of conglomerate, composed almost

* The word in Miller's manuscript is 'arenaceous,' but, on the suggestion of Mr Carruthers of the British Museum, who has kindly looked over the scientific chapters, I substitute 'argillaceous' as alone compatible with the argument.

exclusively of Old Red Sandstone materials bound together by an Oolite cement, seem to represent periods of great disturbance, during which the broken ruins of some neighbouring coast composed mainly of the Caithness Flagstones were spread to the depth of several feet, and for great distances, over the previously quiet bottom. And then another period of tranquillity took place, and finely laminated shales were deposited over the rough breccia until another paroxysm of violence overspread them also with a breccia as rough as the underlying one. And thus during a very protracted period the process of alternate formations—quiet and disturbed by turns—went on in what is now the Helmsdale Oolitic basin; the entire basin from Port Gower to the Ord consisting of successive beds of conglomerate and shale. Should your Grace revisit the locality, it might be well to walk out along the shore to the north-east of Helmsdale for about three quarters of a mile, till you reach a little cottage standing on a point. The shore below consists of a very rough conglomerate bed—the uppermost in the basin,—and you will have no difficulty in convincing yourself, as you retrace the way backwards through the space laid bare by the ebb, that that bed overlies a bed of shale rich in characteristic fossils of the deposit—such as Terebratula, Avicula, Serpula, Belemnites, &c., and which overlies in turn another conglomerate bed, that overlies yet another shale bed. And thus in regular succession till you reach the river will you find bed succeeding bed—the remains and evidence of many periods of tranquillity and of many alternating paroxysms of violence; and that some of the paroxysms must have been strangely violent indeed, you will find proof in the great size of some of the blocks which the conglomerate contains. Within five minutes' walk from Helmsdale, immediately below a bulwark that protects the road, there is a block of pale grey Old Red Flagstone enclosed in the shelly paste whose upper surface measures twenty-seven feet in length by twenty in breadth. The same alternations of shale and conglomerate may be seen in passing eastwards from the little cottage towards Navidale, where a few of the same beds are laid bare in the descending order as are uncovered in the Helmsdale direction, but the shore here is much roughened by detached blocks, and the section not so good. They may be seen, also, in the face of that portion of the old coast-line under which the road runs for rather more than a mile between Helmsdale and Port Gower—the shale beds usually presenting a hollowed and weakened appearance, while the conglomerate stands out in bold and very rough cliffs, in some places overlying, in others underlying, the shale beds.

'I am rather inclined to hold that the fragments in the Helmsdale breccia, which your Grace regards as Oōlitic and of a quality similar to that of the Oolite strata near Dunrobin, must be in reality Old Red, as I have found in many of the exposed masses fossils—such as Fucoids, jaws of Diplopterus, scales of Asterolepis, &c., but never a single Oolitic organism, the Oolitic organisms being restricted to the shelly paste which connects them together. I am further of opinion that the conglomerate does not represent one great break-up of the system after its formation, but, as I have already intimated, as many paroxysms of violence during the deposition as it consists of alternating beds.

'On passing towards the south-west after quitting Port Gower we again come upon the same beds of shale and conglomerate as appear at the little cottage to the east, and then, rising in the scale towards the middle of the south basin, we find beds of a later age than any of those in the Helmsdale one. I have been unable to trace them continuously, as the shore is much covered towards the centre and south-western side of this passage by sand and shale. The north-eastern side of the Brora basin is equally obscured, and so I cannot regard myself as acquainted with the deposits in succession from Helmsdale to Brora. I think, however, there is evidence that in the Helmsdale basin we have the base of the system,—that there is evidence that, in the south basin we have the same base overlaid to a considerable thickness by higher and newer deposits, overlaid by deposits still higher and newer till we ultimately reach, in the bed of white stone of which the additions to Dunrobin Castle have been built, the uppermost deposit of the Oolite on the east coast of Scotland. But though the upper bed occupies a curiously insulated position,—the effect apparently of extensive denudation,—it lies, so far as I can judge, conformably to the other rocks of the basin. Sir Roderick Murchison, with whose elaborate and valuable paper on the Sutherland Oolite your Grace must be familiar, holds that some of the Navidale conglomerates were formed by the upheaval of the granite of the Ord in a period posterior to the formation of all the Oolites, but it must be a very local conglomerate that has been so formed,—and not the conglomerate of Old Red Sandstone materials that alternates with the shale beds.

'I have detected fossil wood in the conglomerate, though not so frequently as coral. Three years ago I saw a piece of wood projecting from a conglomerate cliff immediately above the harbour at Navidale, and it may be there still. Fossil trees were at one time very abundant in the shale beds on both sides of the little cottage, but almost all the exposed ones were dug out about ten or twelve years ago, and burned into lime. I saw, however, this season, the stump of a very large tree projecting from the wasted surface of a shale bed that, in the year 1842, had of itself furnished a whole kiln-full. As to the question of the period of the petrifaction of these woods, it must, I fear, be deemed a hard one. They were, however, evidently petrified in the tranquilly deposited shale; and if so, why not in both the underlying and overlying conglomerates, whose cementing paste is so much more calcareous? But this special question, like not a few of the others raised, will be found to hinge on the pressing one, Do the conglomerates in reality alternate with the tranquilly deposited, fossil-bearing shales? I trust that, should your Grace revisit Helmsdale, you will find materials for its solution in the section exposed between that place and the little cottage.'

During the year 1853 a large proportion of the chapters of Hugh Miller's autobiographical work appeared in the *Witness*. In the beginning of 1854 the book was published under the now well-known title, *My Schools and Schoolmasters*. Its success was immediate and decisive. Literary men were glad to meet Miller in a field from which controversy of every kind was excluded, and to mark the skill with which he wielded the instruments of pure literary art. Their disposition was to pronounce it his most valuable work. Readers will with interest peruse the two following letters on the subject.

FROM MR ROBERT CHAMBERS.

'1, Doune Terrace, March 1, 1854.

'I cannot think of confining my thanks for your

volume to the few hurried words I had an opportunity of saying last week. Not that I am excitedly grateful for the kind reference you have made to myself, but because I have read your first 350 pages with a heart full of sympathy for your early hardships and efforts, and an *intense* admiration of the observant and intelligent mind which I see working in that village boy on the shore of the Cromarty Frith. I cannot refrain from congratulating you on the publication of this book, which I consider as yet your best, and the one that will prove most enduringly useful, interesting, and popular—simply because yourself have been the best phenomenon you have ever had to describe. I cannot refrain from congratulating you on the triumphs you have achieved over the great difficulties of your early position, which now appear to me far beyond anything I had previously imagined. And believe me, I am most cordially sincere when I offer you my best wishes for the remainder of a career, the early part of which has been so creditable to you. Be assured that Scotland has few dearer living names at present than Hugh Miller, and must henceforth feel the deepest interest and concern in everything you do.

Your autobiography has set me a thinking of my own youthful days, which were like yours in point of hardship and humiliation, though different in many important circumstances. My being of the same age with you, to exactly a quarter of a year, brings the idea of a certain parity more forcibly upon me. The differences are as curious to me as the resemblances. Notwithstanding your wonderful success as a writer, I think my literary tendency must have been a deeper and more absorbing peculiarity than yours, seeing that I took to Latin and to books both keenly and exclusively, while you broke down in your classical course, and had fully as great a passion for rough sport and enterprise as for reading—that being, again, a passion of which I never had one particle. This has, however, resulted in making you, what I never was inclined to be, a close observer of external nature—an immense advantage in your case. Still I think I could present against your hardy field observations by frith and fell, and cave and cliff, some striking analogies in the finding out and devouring of books, making my way, for instance, through a whole chestful of the *Encyclopædia Britannica*, which I found in a lumber garret! I must also say, that an unfortunate tenderness of feet, scarcely yet got over, had much to do in making me mainly a fireside student. As to domestic connections and conditions, mine being of the middle classes were superior to yours for the first twelve years. After that, my father being unfortunate in business, we were reduced to poverty, and came down to even humbler things than you experienced. I passed through some years of the direst hardship, not the least evil being a state of feeling quite unnatural in youth, a stern and burning defiance of a social world in which we were harshly and coldly treated by former friends, differing only in external respects from ourselves. In your life there is one crisis where I think your experiences must have been somewhat like mine; it is the brief period at Inverness. Some of your expressions there bring all my own early feelings again to life. A disparity between the internal consciousness of powers and accomplishments and the external ostensible aspect led in me to the very same wrong methods of setting myself forward as in you. There, of course, I meet you in warm sympathy. I have sometimes thought of describing my bitter, painful youth to the world, as something in which it might read a lesson; but the retrospect is still too distressing. I screen it from the mental eye. The one grand fact it has impressed is the very small amount of brotherly assistance there is for the unfortunate in this world. I remember hearing the widow of Mr —— tell how she had never been asked by any relation of either herself or her husband to accept of a five-pound note, though many of them were very well off. The rule is to leave these stricken deer to weep themselves away unsuccoured. I have the same experience to relate. Till I proved that I could help myself no friend came to me. Uncles, cousins, &c., in good positions in life—some of them stoops of kirks, by-the-by—not one offered, or seemed inclined to give, the smallest assistance. The consequent defying, self-relying spirit in which, at sixteen, I set out as a bookseller with only my own small collection of books as a stock—not worth more than two pounds, I believe—led to my being quickly independent of all aid; but it has not been all a gain, for I am now sensible that my spirit of self-reliance too often manifested itself in an unsocial, unamiable light, while my recollections of "honest poverty" may have made me too eager to attain and secure worldly prosperity. Had I possessed uncles such as yours I might have been much the better of it through life.

'Pray accept with lenity these hurried and imperfect remarks, into which I have been led by a sort of sympathetic spirit, almost against my own sense of propriety.'

FROM MR CARLYLE.

'5, Cheyne Row, Chelsea, London, 9th March, 1854.

'I am surely much vour debtor for that fine Book

you sent me last week, which was welcome in two ways, the extrinsic first, and now the intrinsic; for I have now read it to the end (not a common thing at all in such cases), and found it right pleasant company for the evenings I stole in behalf of it! Truly I am very glad to condense the bright but indistinct rumour labelled to me by your Name, for years past, into the ruddy-visaged, strong-boned, glowing Figure of a Man which I have got,—and bid good speed to, with all my heart!

'You have, as you undertook to do, painted many things to us; scenes of life, scenes of Nature, which rarely come upon the canvas;—and I will add, such *Draughtsmen*, too, are extremely uncommon, in that and in other walks of painting. There is a right genial fire in the Book, everywhere nobly tempered down into peaceful radical heat, which is very beautiful to see. Luminous, memorable; all wholesome, strong, fresh and breezy, like the "Old Red Sandstone Mountains" in a sunny summer day:—it is really a long while since I have read a Book worthy of so much recognition from me, or likely to be so interesting to sound-hearted men of every degree. I might have my objections and exceptions here and there (not to the matter, I think, however, if sometimes to the form); but this is really the summary of my judgment on the business. And so, once more, I return you many thanks, as for a Gift that was very kind, and has been very pleasant to me.'

In the spring of 1854 Miller lectured in Exeter Hall, London, to the Young Men's Christian Association. The following notes of his visit to the metropolis occur in a letter to Mrs Miller of February 9th:—

'I was safely delivered of my Address between the hours of eight and ten on Tuesday evening (skilfully assisted by Mr Allon), and am now as well as could be expected. Mr Allon is a clergyman, but an Independent one.

'I had a noble audience of, as the *Morning Advertiser* says, five thousand persons, and I carried them with me throughout. During the earlier stages of the Address they were attentive, which I hardly expected, as my preliminary matter was somewhat scientific and dry, and throughout the concluding part rapturous in their applause. On the whole, a more successful address was never delivered in the great Hall. Allon's pronunciation was beautiful, and showed me better than a hundred lessons the faults of my own. He read, too, with great vigour, and gave to my style the proper classical effect.

'I dined on the Tuesday at Dr James Hamilton's, and met several members of his Session,—among the rest our old friend, James Robertson, who was making kind inquiries after you. I began to-day with a wild-goose chase after Mr Searles Wood, the geological conchologist; but, after driving to Kentish Town, found only a vacant house, and that he had removed to Egham. I have, however, had my shells determined for me by a still higher authority, whom I fortunately encountered at the British Museum,—*Deshayes*, the greatest conchologist in the world. I find I would soon get very tired of London.'

In June of the same year he was again in Cromarty, and we may glean a jotting or two from his letters. The town was now sinking into decay, and the light of the past touched the prospect for his eye with melancholy hues.

'Cromarty, 18th June, 1854.

'Cromarty is fast becoming a second "Deserted

Village," and seems chiefly remarkable, at present, for its fallen houses and its vacant streets. The laird was calling upon me, and we had a long conversation together. He purposes having some memorial over his mother's grave in the old chapel, and I gave him a rude sketch of the remains of the old Gothic inclosure around the spot, as this existed thirty years ago. Two buttresses occupied each corner of a rectangular erection which stood on a broad sloping base, and showed, a little higher up, a portion of a massive moulding. He kindly invited me to dine at Cromarty House, but the invitation I of course declined. Till I get stronger I must take care and avoid dinners, save very quiet ones with my mother.

'Among the other things changed for the worse in Cromarty is the old captain's house. —— seems to have dealt by it as he usually does by other good texts. The shrubs and trees on the green in front are all cut down, the hillock in the centre levelled, and a stiff, bare stone wall built in front. The place, in short, just looks the thing which it has become, — a commonplace Moderate manse. But the change is merely representative of what is taking place, under the name of improvement, in the district generally. The fine old woods and picturesque hedges are away, and bare cornland with raw dry-stone fences occupy their place.

'The bairns will, I trust, have good collections of Portobello shells ere my return, and will be able to stand an examination on their names. I had a brief walk two days ago along the shore, to the west of the factory, and found the group of shells strewed along the beach essentially different from the Frith of Forth one. Astarte sulcata was the prevailing shell, Lucina borealis was also very common, and I found in tolerable abundance Nuclea nucleus and some specimens of Saxicava borealis. I must remember to bring to the children a few shells which, though common here, are rare at Portobello.

'I have just returned from a walk along the hill-side. I have been in the old chapel and seen poor Eliza's little tomb-stone, half buried in long grass; I have looked out on the sea from the "Broad Bank," and down upon the Dripping Cave and the Lover's Leap; and in looking back on more than twenty years when we used to meet, evening after evening, among the trees, I felt how surely life is passing. It is now more than fifteen years since we buried Eliza. The hill is changed, like all else about Cromarty, and our haunt, right over the Doocot Cave, with its scattered beech-trees and its thick screen of tall firs, is now a bare, heathy slope, without shrub or tree. The view, however, from the corner part of the hill down upon the town and bay is still fine as ever,—the forest trees, which are still spared, are in fresh leaf, though Ben Wyvis continues to retain his patches of snow, and the winding outline of the Frith is so much beyond the reach of improvement that it cannot be spoiled.'

It probably tended to deepen the pensiveness of his mood at this time that he found one of his dearest and oldest friends battling stiffly with the *res angusta domi*. How, thought Miller, can I contrive to assist him? He well knew that a friendship of forty years did not give him, in dealing with a proud spirit, the right to offer a pecuniary gift. He would try, he wrote Mrs Miller, whether a loan might not be accepted. Mrs Miller, hastily reading his letter, mistook his meaning, and thought, while cordially sympathizing, that the loan had been applied for. He wrote as follows in reply.

'Cromarty, 15th July, St Swithin's day (mingled sunshine and shadow, with showers in the distance), 1854.

'I fear I must have sadly misled you by what I said regarding "*trying* to lend —— a score of pounds or so." From your remark I infer that you think that he wished to borrow. No such thing. I have, since I wrote you, offered him the use of the sum proposed, and he has point-blank refused receiving it. Trusting to be able to work himself out of his present difficulties, he believes that the spur of necessity might come to be very considerably blunted were I his creditor, and so he chooses rather for the present to be indebted to others. We must not give things wrong names. There is not only *honesty* but high honour in such a determination; —nay, Christian principle of a considerably more genuine kind than that which leads so many vain Christians of the common type to come under unnecessary obligation to their neighbours. Possessed of the necessaries of life, they become beggars for the sake of its gentilities; —beggars for the sake of unexceptionable bonnets, and supernumerary frills, and the ability of playing Italian music on the piano; and fashionable saints charitably minister to their fashionable wants by enabling them in pure charity to enjoy "the vanities of life." Depend on't there is an abyss of humbug in this direction.'

Some years previously to the period at which we have now arrived, he received a letter from certain students of Marischal College and University, Aberdeen, asking him to permit himself to be nominated as candidate for the Lord Rectorship. Here is his reply.

'I am deeply sensible of the great honour you do me in entertaining the purpose you intimate of proposing me as a candidate for the Lord Rectorship of Marischal College and University. In present circumstances, however, the honour is one which I beg respectfully to decline. I am, perhaps, not quite without apology for having done but little for literature and science compared with what I had once hoped, and still wish to do;—but I am conscious that what I have yet accomplished is but little,—and you will, I trust, attribute to the right feeling my determination of accepting no place to which my claim on the score of merit might with justice be challenged. The last ten years of my life have been exceedingly busy ones, nor were the harassing occupations in which they were spent of a nature very favourable to acquirement of the more solid or thought of the profounder kind. In that period, however, I was enabled to give to the public two little works,—one descriptive of the second period of vertebrate existence on our planet, and one an examination of those evidences which connect the first beginnings of life in the remote past with the *fiat* of a Creator,—that have been favourably received by men of science on both sides of the Atlantic. And should there be some ten or twelve years of active life, or a greater time still, before me, I may, I trust, succeed in doing for the geology of Scotland what may render me at least more worthy than now of an honour in connection with some of our Scottish Universities such as that which, in your too partial kindness, you at present propose. Trusting that you will sustain my reasons for declining your very gratifying proposal both as valid themselves and as proffered in good faith, I am, gentlemen, &c.'

In 1853 the chair of Natural History in the University of Edinburgh fell vacant. It was the place which, of all others, Hugh Miller would have been gratified to fill. The crisis of the Church controversy had long been past, and the position of the Free Church was thoroughly established. He felt that he had earned a right to comparative repose, and that mind and body required it. A large proportion of his countrymen agreed with Lord Dalhousie in his Lordship's brief and emphatic opinion on the subject of the chair:—'I have written to Lord Aberdeen as strongly as I can in favour of Hugh Miller's claims, and I shall consider that the best man amongst us for this chair is passed over should he not be chosen.' With all respect to the memory of the eminent naturalist who received the appointment, we may be permitted to hold that, in a Scottish University, no man could have so deeply stirred the enthusiasm of geological students as the author of the *Old Red Sandstone*. Professor Edward Forbes, however, had the advantage of being thirteen years the younger man, and this advantage may have turned the scale. The disappointment was deeply felt by Miller.

In 1855 Lord Breadalbane offered him, through Dr Guthrie, the office of Distributor of Stamps and Collector of the Property Tax for Perthshire. Income 'about £800 per annum—may be more,' the holder being required to disburse from £200 to £300 yearly in office expenses. 'It is not an onerous duty,' wrote his Lordship; 'residence at Perth will be necessary, but occasional absence is permitted, and five or six weeks' holidays during the year. All I should ask of Mr Miller would be to come and see me at Taymouth as often as possible, for I am very anxious to know him, and to cultivate his acquaintance.' Dr Guthrie handed Lord Breadalbane's note to Miller, and, after considerable hesitation, he replied to it in these terms:—

'It is with heartfelt gratitude that I tender your Lordship my sincere thanks for the kindness you have shown me in proposing to recommend me to an office so important and respectable as that of Distributor of Stamps for Perthshire. My utter lack of any claim, personal or political, on your Lordship, makes me, I trust, all the more sensible of such a spontaneous exertion of goodness on my behalf. I find, however, that, partly from a too facile disposition, and partly from that rigidity of mind and habit which, making itself felt in most cases after the period of middle life has been passed, usually renders men turned of fifty unfit to take up a new profession to any good effect, I would be but ill-qualified for the efficient performance of the duties of an office of such responsibility and care. I would, I have too much reason to fear, break down under the difficulties and responsibilities of a course of exertion which might seem to many even less laborious and exciting than the one in which I am at present engaged, and thus fail to do justice to your Lordship's kind recommendation. And so, with a sense of your goodness which will, I trust, remain with me through life, I respectfully and gratefully decline availing myself of that offer of exerting your influence on my behalf which you have so generously made me through my kind friend, Dr Guthrie.'

At this point it will be convenient to take in those recollections of Hugh Miller for which my acknowledgments are due to Dr McCosh.

DR M'COSH'S RECOLLECTIONS.

'I HAVE been requested to write out some recollections of my old friend, and I have a melancholy pleasure in complying.

'The name of Hugh Miller first became known to me, as it did to Scotchmen generally, by his *Letter from*

One of the Scotch People to Lord Brougham. Multitudes rejoiced when they heard soon after that the friends of Non-Intrusion and Spiritual Independence had made arrangements for his conducting a newspaper in Edinburgh. They felt that from that day the cause received an accession of strength. We were particularly pleased to find that from the first, and to the end, he maintained an attitude of independence, not only of politicians, but of Church leaders. The movement had hitherto assumed too much of an *ecclesiastical* aspect. No doubt it had in it all along the popular element of Non-Intrusion; but the Spiritual Independence side was the one mainly dwelt on in Church courts; the Scottish people did not always understand what was meant by overtures to the General Assembly about independence of the civil courts; and not a few entertained the suspicion that the ministers were seeking for power to themselves which would set them above the law of the land, in the settlement of ministers. Certain it is that the cry of the clergy did not always carry with it the popular sympathy; and there were times and places in which public meetings called to support it were put down by mob-opposition. It was most important that at that crisis the grand old cause of Scotland should have one of the people to support it, and a shout of joy burst out all over Scotland when Hugh Miller came forth as the champion of popular rights. The newspaper was published twice a week, and Wednesdays and Saturdays, the publishing days, were looked forward to by many, because they brought the *Witness*. The paper was read not so much for its news, nor even its reports of Church courts, as for its leading articles. Many a retired country minister living, perhaps, in the midst of a body of farmers, who had no appreciation of the questions at issue, had his spirit aroused as by a trumpet clang by the powerful appeal made to him, and he paused in the midst of writing his sermon on the Saturday to read the leading article. The day labourer or weaver could not afford to take the paper, but he was grateful when some one, minister or elder, lent it to him. The literary men in our Universities had, as a whole, a deep suspicion of the movement, and did not like to be thought readers of the stone-mason's paper, but were glad when they could get furtive glances at it, and were obliged to acknowledge its literary superiority. From a very early date Hugh Miller's name became a household one in the best families of Scotland. The common people never called him Mr Miller—they would no more have done this than they would have called Robert Burns by the name of Mr Burns; they identified themselves with him and identified him with themselves by calling him *Hugh Miller*. They felt as if he still carried his chisel and his hammer, and as if he were now forming and fashioning, by firm and manly stroke, a nobler edifice than ever his mason's tools had constructed. I was in circumstances to know the feeling of Scotland at the time, and I am convinced that the old national cause which was defeated in 1843, but which gained the victory in its defeat—as, with reverence be it spoken, the cause of Christ did when He was crucified—was indebted, among the great mass of the people, to Hugh Miller as much as to any other man.

'I read his paper from the first, but shy as I have ever been to court the great men whom I admired, I was not personally acquainted with him till 1850. That year I published my first work, *The Method of Divine Government.* I had made up my mind to meet failure as well as success; and I believe that if the work had met only with popular neglect I would have clung to it the more resolutely, as the father will dote the more lovingly on that daughter in whom the silly beaux see no merits. I had fortified myself against both praise and blame; but there were two men in Edinburgh about whose good opinion I felt somewhat sensitive: these were Sir William Hamilton and Hugh Miller. I asked my publishers to send a copy of my work to each. Meanwhile the public scarcely knew what to make of my big, and, as some deemed, pretentious book. The *Athenæum* noticed me contemptuously, evidently without reading me—it praised some of my works afterwards, when I had earned a reputation. My very friends shook their heads, and were expecting a failure. Now it was at this time, when readers were hesitating which side to take, that the two Edinburgh giants spoke out, and spoke out courageously; not uttering ambiguous oracles, which would make them right whatever the way in which the public might ultimately decide. I was not known at the time beyond a limited district in the north of Forfarshire, and the south of Kincardineshire; I believe neither of the eminent men referred to had ever heard of me before; and this led them to exaggerate my merits. I feel in this distant land a deep gratitude to the many kind Scottish friends who helped to bring my book into notice; but I feel most to the great Scottish metaphysician; and with him, and above him, to the man who spoke first, to Hugh Miller.

'I felt now as if I ought to seek the acquaintanceship of Hugh Miller. This was brought about by our mutual friend, the Rev. Dr Guthrie, who invited him to meet me at dinner. And this may be the fittest place for describing his outward appearance as it first came under my notice. In dress he neither affected a slovenly carelessness nor a prim gentility. It was very much the dress worn on Sundays by the better class of tradesmen and upland farmers. It must be confessed that at first sight he had somewhat of a shaggy appearance, relieved, however, by a look of high independence and an air of indomitable energy and perseverance. He was a *man* of the highest type throughout; but if he had a counterpart in any of the lower animals it was in the noblest of the dog tribe, such as Landseer loves to paint, as in his Dignity as contrasted with Impudence. Yet he was withal wonderfully shy, and unwilling to seem to be seeking the favour of any man. A little incident may show what I mean. Dr Guthrie and I had been walking together on the day on which he had asked him to meet me at dinner; and when we were at some distance we saw him approaching the door. "Let us run," says Dr Guthrie; "for if he goes to my house and finds me not in he will set off;" and we did run to catch him. I remember another circumstance illustrating the same point. I was talking to him of the *litterateurs* of Edinburgh, great and small. He did not seem to have much intimacy with them. We talked of Lord Jeffrey, of whom he spoke kindly and respectfully. "He expressed," he said, "a wish to make my acquaintance when I came to Edinburgh; but as he did not call on me I did not see my way to call on him, and we have not had much intercourse." I happened to be in his house one forenoon when he was expecting a call from Sir Roderick Murchison in the afternoon. He was too proud a man to make any boastings about it; but it was evident that he was highly gratified by the proffered visit; and he had his splendid museum in the highest possible order to show it to the distinguished geologist. He had a heart to cherish a sense of favours when bestowed without any condescension, or endeavour to interfere with his independence, but certainly with no stomach to seek favours even from those who would have been most willing to grant them.

'We met at dinner on the day I have referred to. Beside Dr Guthrie, Mr Miller, and myself there were only one or two others present. Dr Guthrie restrained his usual flow of mingled manly sense, humour, and pathos to allow his friend to speak freely; and he had soon to go out to a congregational meeting. So I had the great man to myself for the evening, and he under no restraint. I have observed that in large promiscuous companies he was apt to feel awkward and restrained, and to retire into himself, and sit silent. But when there were only a few persons present, and these of congenial tastes, his conversation was of the most brilliant description. You saw the thoughts labouring in his brain as distinctly as you see the machinery in a clock when the clock work is in a glass case. That evening we talked of subjects that were familiar to him, and which I was at that time studying, such as the typical forms which Professor Owen was detecting in the vertebrate skeleton, and the possibility of reconciling them with the doctrine of final cause and the mutual adaptation of parts. He was not sure about some points, and my delight was to set his mind a working. He afterwards brought out his matured views in a very brilliant article which he wrote, reviewing a paper of mine in the *North British Review*. But that night his thoughts came out tumbling with a freshness, an originality, and a power, which somewhat disappeared when he came to write them out in elegant English.

'From that date he expected me to go out to Portobello or Musselburgh and see him when I went to

Edinburgh, which I commonly did once or twice a year. I took care not to intrude upon him the night before the bi-weekly issue of his paper; but I always found him welcoming me on the Wednesday or Saturday night when the hard work of composition was off his mind. On these occasions he showed me his museum with the feeling of a boy showing his toys to his companion. We sometimes, but not often, talked of Church matters; but it was evident that having to write and speak so much about them he rather kept off them with me—except, indeed, that he was ever ready to speak of the great religious cause of the freedom of the Church, as imbedded deep in the hearts, even as it was in the history, of Scotchmen, and certain in the end to triumph. Sometimes we talked of geology and religion and the difficult problems which they started. At times I introduced a topic new to him, as on one occasion Comte's Classification of the Sciences and the Positive Philosophy, not so well-known then as now. It was extremely interesting to watch his mind grasping the new ideas, apprehending but not yet fully comprehending them. In next paper he had an article on the subject, but it is evident that his mind had not yet settled down into clearness, and the written composition had not the full expanse of the conversation. Had he lived he would certainly have grappled with the Positive Philosophy as he did with the *Vestiges of Creation.*

'Having great confidence in his singleness of purpose and his far-sighted wisdom, I consulted him when I was about to be called to the Chair of Logic and Metaphysics in Belfast, stating to him my difficulty about giving up my pastoral work. He gave me a clear and unequivocal advice. "If a man," says he, "has decidedly a high heaven-bestowed gift,—even if it should be that of a mason or mechanic, he should exercise it to the glory of God." "You have," he was pleased to add, "such a gift; go and use it, and God will open spheres of usefulness to you."

'The last conversation I had with him was in the early autumn of the year before his decease. He was completing his last work, *The Testimony of the Rocks*, and we went over the topics discussed in it. I was struck then, as I ever was, with his powerful memory and his special acquaintance with the English literature of last century—I suspect it was the literature most accessible to him in his younger years. He could quote verbatim long passages from the poets of that epoch, illustrating points casting up in the conversation. As we took the usual walk through his museum he freely allowed that the apparent breaks between the various geological epochs and animals were being fast filled up by new geological discoveries, and showed me some examples as we went along. It was evident to me that he was setting himself to a thorough grappling with these facts, and to a consideration of their relation to the great truths of natural and revealed religion. Often do some of us wish that he had been spared to take his place in the more formidable conflicts of these times.

'In common with not a few others, I looked on Hugh Miller as the greatest Scotchman left after Thomas Chalmers fell. These two men differed in many points, but they were essentially kindred spirits: they were alike in their high aims; in their lofty genius; in the moving power of their writings; in their partiality for the study of the works of God; in their deep reverence for the word of God; in their desire to unite science and religion, and attachment to the principles of the Church of Scotland. What Chalmers did for the older sister, astronomy, Miller has done for the younger, geology, in wedding her to religion. Both lived for the purpose of elevating their countrymen and their race; and in order to effect this end both laboured to promote the Church's independence and the freedom of its members. Each had his own field of influence; each had a class of minds on whom he exercised a burning and enduring power for good. Most appropriately, now that their day's work is done, do they sleep side by side in the same grave-yard.

'I am tempted to compare Miller, and when I compare him, to contrast him with another eminent Scotchman, Robert Burns. Both were sprung from the nobler order of the Scottish peasantry; neither was originally educated as a scholar; both rose to the highest eminence in the midst of difficulties which led the one as well as the other seriously to propose emigrating to America; both had a deep love for Scotland and her common people; both will go down through all coming ages as household words, and as representatives of the intelligence of the sons of toil in their native land; and both were characterized by a noble modesty and a manly independence of nature,—"Owre blate to seek, owre proud to snool." But with the resemblances there were differences. In respect of native genius they rank in my view equally high; but the complexion and bent of that genius differed in the two individuals. No one would compare any poetry published by Hugh Miller with the poetry of Robert Burns; though there are passages of very high poetic power in all the prose works of Miller. Let us only look at that bold sketch of a proposed Epic towards the close of the sixth Lecture of his *Testimony of the Rocks*, in which he represents Lucifer, son of the morning, cast down on the

Pre-Adamite earth, while yet a half-extinguished volcano: then, as ages roll on, moving amidst tangled foliage and ravenous creatures, horrid with trenchant tooth and barbed sting, and enveloped in armour of plate and scale, marking all the while and wondering at the progressive work of God, as animal follows plant, and man succeeds animal, and seeking to frustrate it by diabolical wiles, which, however, only fulfil the eternal counsels of Heaven, and issue in the crowning work, the descent and death and ascension of the Incarnate Son of God. Greater as a poet, Robert Burns cannot be placed as a thinker, or a man of science, or a writer of prose, on the level of Hugh Miller. Dugald Stewart expressed his surprise to find Burns form so correct an idea of the then prevailing theory in Edinburgh which referred all beauty to the association of ideas, a theory which seems to have gained the momentary assent of Burns in spite of his own better sense and truer feeling:—"That the martial clangour," he says, "of a trumpet had something in it vastly more grand, heroic, and sublime, than the twingle-twangle of a Jew's-harp; that the delicate flexure of a rose-twig, when the half-blown flower is heavy with the tears of dawn, was infinitely more beautiful and elegant than the upright stalk of the burdock, and *that* from something innate and independent of all association of ideas,—these I had set down as irrefragable orthodox truths, until perusing your book shook my faith." No one who had conversed with Hugh Miller would have expressed surprise that he was capable of understanding such truths. He must have been in the clouds himself, who, in talking with him, did not confess that Hugh Miller could soar as high as he, and keep well-balanced pinions all the while. He had no difficulty in understanding the Association

Theory, and many other theories of Hume and Brown as well; and in brushing away them and some other meagre or misty explanations by a few brief but cogent facts and arguments. In his works he has combined, as no working man ever did before, lofty speculation with rigid science, and irradiated the whole with the corruscations of poetry. Nor is it to be omitted that there is a far more important point of difference between the Ayrshire ploughman and the Cromarty mason. Both were men of naturally strong passions; but where the one yielded to the temptations that assailed him,

> "And thoughtless follies laid him low,
> And stained his name,"

the other resisted with all his might. As I conversed with Hugh Miller, or after parting with him, with his words of power still ringing in my ears, often have I felt, and said too—but not in his hearing—"What an amount of mischief would that mighty man have done had he, say on his being tempted by his brother masons at Niddry, given way where he stood firm; had he, like Burns, joined the foes of evangelical religion, instead of becoming its defender; or had he, when at one time tempted to scepticism, abandoned the religion of the Bible, with 'its grand central doctrine of the true humanity and true Divinity of the adorable Saviour' (to use Miller's language), and gone after some plausible form of nature-worship or man-worship." I feel as if his country and his Church had not, when he was yet alive, been sufficiently grateful to God for raising such a man to guide aright so large a portion of the thinking mind of Scotland at a most critical era in its history.

'Every man in Scotland had heard of Hugh Miller. Noble Lords were in the way of pointing their sentences and securing a plaudit by an allusion to him. The artizan and peasant felt that he was one of themselves, and one who (unlike some others sprung from their ranks) never felt ashamed of them, or his connection with them. The infidels knew him well, for many a hard blow had he dealt them; and they were obliged to respect while they feared him. The religious community recognized him as in certain departments the ablest, as he was the most disinterested, defender of the faith. Scientific men recognized in him one who could cope with them in their own department, who knew the facts as well as they, and could reason them out with greater power. Literary men acknowledged in him a brother who could mould a sentence or turn a period with the best of them. The ablest and boldest man in the country would have felt his knees shaking at the thought of engaging in a controversy with the stone-mason. Even those who had no learning relished him; and some have earnestly wished to be better scholars that they might understand him; and some have made themselves scholars by spelling their way through his writings. Thinkers in no way inclined to agree with him in his ecclesiastical or political opinions, took the *Witness* because they liked to have thoughts awakened within them; and even those who were not particularly disposed to think read his writings for the sake of their pictorial power and noble sentiment. One of the most distinguished assemblies I ever looked on met in the City Hall of Glasgow in September, 1855, to hear the opening Address of the President, the Duke of Argyll, to the British Association for the Promotion of Science. There were present a very large number of the *savans* of the age, and mingling with them a number of others quite sufficient to make the audience a singularly promiscuous one,—shrewd merchants who traded with the ends of the earth, and other not less excellent merchants who were not particularly shrewd, and who were conversant with little other literature than the *Glasgow Herald*, and along with them their wives and daughters, some of them *blue stockings*, but others—quite as useful members of the family—who knew cookery vastly better than geology. In addressing the assembly, the noble Duke gave us a panoramic view of a number of the most distinguished scientific men of the day, and an epitome of their discoveries. Many of them were cheered as their names passed in brief review; but there were two whose names called forth the loudest and most repeated shouts. The one of these was a prince in rank, even as he is a prince in science. Prince Lucien Bonaparte, the cousin of the then ally of England in the Russian war, received a cheer worthy of Glasgow. There was just one other who was acclaimed by so loud a burst; and some of us observed with interest that in his case the cheer came from every heart, and from a greater depth in the heart: that cheer was in honour of one—we need not name him, but when his name was pronounced it moved the vast assembly simultaneously like an electric shock—of one of Nature's nobles, made noble not by the hand of man, but by the evident mark of God upon him.

'But Fame was not the idol before which this great man bowed. The love of reputation was but an undercurrent in his soul. He lived to do a work, but it was for the glory of God and the good of mankind. I watched him with interest, as many did, in the meetings of the Geological Section of the British Association; and I observed that he sat, and stood, and spoke, and moved with the most perfect simplicity. There was no bravado on the one hand nor mock humility on the other; there was no courting of popularity, no tricks to draw attention; no looking round to see if men and women were gazing at him. He received the advances of distinguished individuals with deference, and was gratified by them; but there was no fawning or flattery on his part, and he received in precisely the same manner (as some of us can testify) the most obscure of his old friends, and assumed towards them no airs of superiority or of patronage. Whatever might be the situation in which he was placed, one felt in regard to him that the fiddlers struck up the right tune, when after his health was drunk at a parting dinner at Cromarty they played, "A man's a man for a' that." But I would not be exhibiting his full character if I did not add, that, bending before no man, he ever bowed in lowliest reverence before his God; that seeking no patron, climbing by no dirty arts, and determined to be dependent on no man, he ever felt and acknowledged his dependence on a Higher Power.

'Princeton, New Jersey, U. S., Jan. 1870.'

CHAPTER IV.

CLOSING SCENES.

THE last time I saw Hugh Miller in his own house he mentioned to me that his capacity for work was not what it once had been. He used, he said, to write an article at a sitting; he now liked to do it in two, relieving himself by a walk in the interval. This was in the summer of 1855, and the weakness which even then was stealing over him continued, month by month, to increase. The mason's disease—the presence of particles of stone in the lungs—augmented the torturing irritation of repeated inflammatory attacks in this most sensitive organ. The tendency to brood—to live in a world of thought, and meditation, and phantasy, apart from that of living men—which he had manifested from childhood, grew upon him as his physical energies decayed. That imaginative timidity, also, which had made a man who, if confronted by a lion, would have looked it down, arm himself with pistols against the assassin who might lurk in the recesses of a wooded glen, or haunt a lonely road at midnight, fed itself on the accounts of garotte robberies, house-breakings, outrages by ticket-of-leave men, of which, in the autumn of 1856, the newspapers had more perhaps than the dismal average. Hugh

Miller sympathized with Mr Carlyle in his views as to the necessity of subjecting to a rigorous discipline our professional criminals, and the folly of leaving men whom repeated conviction has proved to be incorrigible to prey upon the community. He wrote upon the subject in the *Witness*, and it was much in his mind. A slight circumstance was sufficient to give the matter a personal turn, and to convince him that he was himself exposed to danger. In a corner of the grounds at Shrub Mount, Portobello, where he now resided, a building had been reared under his direction for the accommodation of his beloved specimens, and he was impressed with the idea that it might be broken into and robbed by the prowling miscreants who were never absent from his imagination. One evening his eldest boy, having been in the garden after dark, returned with the news that he had seen a lantern moving among the trees, and had heard whispered voices. Miller went out to survey the ground; and though nothing appeared to be amiss, the attention of the household was awakened, and night after night the children and the servants had tales to tell of mysterious sounds having been heard and strange sights having been seen. All this influenced his imagination, and pistol and sword were ever in readiness to repel attack.

At this time Mrs Miller and her husband occupied different sleeping apartments. A severe illness had almost deprived Mrs Miller of the use of her limbs, and it was not without great pain that she could go upstairs. It was therefore necessary for her to have her bed-room on the ground floor. For him, on the other hand, the air of an upper apartment was considered best, and he slept in a small room adjoining his study, on the second floor. Every morning and evening, during one of his illnesses, Mrs Miller ascended to his room, ' at the cost of an hour's severe pain,' to minister to his wants. At other times, when his strength permitted, and her limbs were powerless, he would carry her in his arms to her sofa, ' in the kindest and tenderest manner.'

One night Mrs Miller was startled by her husband bursting into her room at midnight, with fire-arms, she thinks, in his hand, and asking in a loud voice whether she had heard unusual noises in the house. She answered composedly that she had heard nothing. He went into his eldest daughter's room, and made the same inquiry. Soothed apparently by the result, he retired to his own room.

An incident occurred at this time, which had the effect of partly setting his mind at rest. Lord and Lady Kinnaird resided during the winter of 1856 in a villa by the sea-side at Portobello, and a cordial intimacy had sprung up between them and Mr and Mrs Miller. Lord Kinnaird, hearing of Miller's apprehensions for his collection, presented him with a man-trap which some sagacious inventor had recently offered to the public. The combination of gentlest philanthropy with chronic dread of burglars appears to have struck this ingenious person as not infrequent, and the man-trap which he devised had the engaging property of holding the robber fast without hurting him. Lord Kinnaird and Mr Miller set up this trap at the porch of the Museum, and the fears of the latter were considerably allayed.

There were other matters, however, besides this imaginative excitement on the subject of robbery, which, as the months of autumn were succeeded by those of winter, occasioned deep anxiety to Mrs Miller. The time was approaching when the *Testimony of the Rocks* was to see the light, and her husband was working at it with indomitable resolution. His activity had always been high-strung, but there was now a feverish intensity in his application which amazed and saddened Mrs Miller. He had been on the whole a calm and regular worker, had loved the morning air and devoted the hours of night to slumber; he now moved restlessly about during the day, as if unable to concentrate his thoughts, and only as the darkness fell aroused his intellectual energies and compelled them to their task. Night after night, in spite of entreaties, he commenced his toil when the rest of the family retired to rest. Through the long silent hours his tired and throbbing brain was forced by his iron will to forge link after link in the argument he was drawing out. Sometimes, when Mrs Miller awoke in the morning, she heard, as she thought, the servants beginning their work, but found that it was her husband leaving his. The slightest noise distressed him. That his nerves were in a state of disorder Mrs Miller could not doubt, but the dread which tormented her was that of apoplexy. Of insanity she never thought until the appearance of other symptoms, but the vision of Hugh Miller struck down by apoplexy and carried into the house constantly haunted her. At night, before bidding him farewell, she would linger, on one pretence or another, trying to find an opportunity to remonstrate against his vigils, but she saw that he was nervously irritable, and she often feared to speak, lest the evil she wished to abate might be aggravated.

Although any one who was constantly and closely with him could not but remark the change which had taken place, his manner with friends who saw him in occasional interviews remained unaltered. Perhaps a deeper tone of earnestness mingled in the genial flow of conversation with which he entertained every visitor, and the reverence and godly fear which lay at the very roots of his being became more than usually conspicuous. On Thursday, the 18th of December, a friend who had a long conversation with him, 'never enjoyed an interview more, or remembered him in a more genial mood.' On the Saturday following, another friend from Edinburgh found him in the same state. True to a habit which had characterized him from his youth, of leading the conversation to some book or topic which was occupying his mind, he repeated with deep feeling a prayer of John Knox's, which, he said, 'it had been his frequent custom to repeat privately during the days of the Disruption.' There was no name which represented more for Hugh Miller than that of John Knox. The Scotland of Knox and the Puritans was the Scotland which he loved; the Church of Knox was the Church for which he had toiled when his strength was in its meridian and when his dawning fame first thrilled him with rapture; the faith of Knox was the faith to which, after Hume and Voltaire and Lamarck had done their worst, he still anchored his soul. This is the prayer which was passing through the mind of Hugh Miller on that Saturday:—'O Lord God Almighty and Father most merciful, there is none like Thee in heaven nor in earth, which workest all things for the glory of Thy name and the comfort of Thine elect. Thou didst once make man ruler over all Thy creatures, and placed him in the garden of all pleasures; but how soon, alas! did he in Thy felicity forget Thy goodness? Thy people Israel, also, in their wealth did evermore run astray, abusing Thy manifold mercies; like as all flesh continually rageth when it hath gotten liberty and external prosperity. But such is Thy wisdom adjoined to Thy mercies, dear Father, that Thou seekest all means possible to bring Thy children to the sure sense and lively feeling of Thy fatherly favour. And, therefore, when prosperity will not serve, then sendest Thou adversity, graciously correcting all Thy children whom Thou receivest into Thy household. Wherefore we, wretched and miserable sinners, render unto Thee most humble and hearty thanks that it hath pleased Thee to call us home to Thy fold by fatherly correction at this present, whereas in our prosperity and liberty we did neglect Thy graces offered unto us. For the which negligence, and many other grievous sins whereof we now accuse ourselves before Thee, Thou mightest most justly have given us up to reprobate minds and induration of our hearts, as Thou hast done others. But such is Thy goodness, O Lord, that Thou seemest to forget all our offences, and hast called us of Thy good pleasure from all idolatries into this city most Christianly reformed, to profess Thy name, and to suffer some cross amongst Thy people for Thy truth and Gospel's sake; and so to be Thy witnesses with Thy prophets and apostles, yea, with Thy dearly beloved Son Jesus Christ our Head, to whom Thou dost begin here to fashion us like, that in His glory we may also be like Him when He shall appear. O Lord God, what are we upon whom Thou shouldst show this great mercy? O most loving Lord, forgive us our unthankfulness and all our sins, for Jesus Christ's sake. O heavenly Father, increase Thy Holy Spirit in us, to teach our hearts to cry Abba, dear Father; to assure us of our eternal election in Christ; to reveal Thy will more and more towards us; to confirm us so in Thy truth that we may live and die therein; and that, by the power of the same Spirit, we may boldly give an account of our faith to all men with humbleness and meekness, that whereas they backbite and slander us as evil-doers, they may be ashamed and once stop their mouths seeing our good conversation in Christ Jesus, for whose sake we beseech Thee, O Lord God, to guide, govern, and prosper this our enterprise in assembling our brethren to praise Thy Holy name. And not only to be here present with us Thy children according to Thy promise, but also mercifully to assist Thy like persecuted people, our brethren, gathered in all other places, that they and we, consenting together in one spirit and truth, may (all worldly respects set apart) seek Thy only honour and glory in all their and our assemblies. So be it!'

Next day Mr and Mrs Miller went to church in the forenoon, and, on the way home, he remarked that the wind was cold, and that he did not feel well, and asked whether she would remain at home with him in the afternoon. She consented, adding that she was very tired, and that one of her limbs pained her. Mrs Miller usually went to church in a basket phaeton, but did not use it that day, and her husband observed affectionately that he wished he could carry her. In a lane opening on the main road, a few yards from the gate of Shrub Mount, there was a poor woman who, some days previously, had met with an accident, and Mrs Miller now said that she would go and inquire for her, remaining not more than a few minutes. An expression of pain crossed his face, as if he disliked the momentary separation.

The time of the afternoon service, the rest of the household being in church, was passed by Hugh Miller and his wife in solemn, thoughtful converse. The reader knows what she was to him. She had been his friend, respected for her intellect, honoured for her character, before he loved her; and when he did love her, it was with the intense and passionate devotion of a strong man, who never loved woman but one. Had he never loved her, he might never have laid down the mallet and the chisel, or quitted Cromarty. They had known fair and foul weather in their wedded life, but on the whole it has been a fitting sequel to those old days when his 'gentle blue eyes' would 'melt with benevolence and a chastened tenderness,' looking into hers, while the green leaves softened the sunlight above, and the summer wave threw its shattering crystal at their feet. His trust and pride in her had never changed, and she held to the last the throne of his affections, supreme and apart from any other human being. On this Sunday afternoon he was in his most tender and confidential, which was always also his religious, mood. No secular matters were spoken of, but in what he said on spiritual things Mrs Miller observed, what she had recently noticed in his prayers at family worship, 'an increasing earnestness, a child-like humility, a more entire reliance upon the merits of the Saviour to blot out all sin, a more awful sense of God's immediate presence.' His affection was so ardent that Mrs Miller regarded it with something of surprise. He suddenly seized her hand, and kissed it with a manner she had never seen before. 'There was in it a great deal more than affection,—an air of *courtliness*, so to speak, indescribable.' In pondering on this action, Mrs Miller has asked whether it could possibly have had the meaning of a farewell. Comparing it with the strange and painful expression which flitted across his countenance as they came from church, she is persuaded that he was haunted by the dread of some prostrating stroke, and that there were sensations in his brain which gave him the idea that it might be near. He was certainly haunted by some great terror; but it was not the fear of apoplexy, it was the fear of an overmastering paroxysm of insanity. He spent the evening quietly, reading a little book on a religious subject, and writing a brief notice of it for the paper.

On Monday morning Mrs Miller made her way upstairs before breakfast, and met him on the top of the stairs. He said that he had passed a bad, restless night. At breakfast, which only Mrs Miller and his eldest daughter partook with him, his conversation was animated and copious. He ate nothing, however, merely swallowing a cup of tea, and his mind was evidently occupied with his sensations in the night. He spoke of sleep-walking, and told an anecdote of a student who had left his room, clambered on the roof, entered an adjoining house, divested himself, night after night, of his shirt, and hidden the garments, to the number of half-a-dozen, in a cask of feathers. Breakfast over, he recurred to the subject which had never been from his thoughts. 'It was a strange night,' he said; 'there was something I didn't like. I shall just throw on my plaid, and step out to see Dr Balfour.' Dr Balfour lived in Portobello, and was in customary attendance upon Mr Miller and his family. The proposal of her husband astonished Mrs Miller. During his whole life he had shown the utmost reluctance to take medical advice, and this was the first time she had ever known him speak of going voluntarily in quest of a doctor. She cordially approved of his determination, and, at about ten o'clock, he presented himself to Dr Balfour.

Of the consultation which followed Dr Balfour has given a full report. 'On my asking,' says the Doctor, 'what was the matter with him, he replied, "My brain is giving way. I cannot put two thoughts together to-day: I have had a dreadful night of it: I cannot face another such: I was impressed with the idea that my museum was attacked by robbers, and that I had got up, put on my clothes, and gone out with a loaded pistol to shoot them. Immediately after that I became unconscious. How long that continued I cannot say; but when I awoke in the morning I was trembling all over, and quite confused in my brain. On rising I felt as if a stiletto was suddenly, and as quickly as an electric shock, passed through my brain from front to back, and left a burning sensation on the top of the brain, just below the bone. So thoroughly convinced was I that I must have been out through the night, that I examined my trousers, to see if they were wet or covered with mud, but could find none. I was somewhat similarly affected through the night twice last week, and I examined my trousers in the morning, to see if I had been out. Still, the terrible sensations were not nearly so bad as they were last night. Towards the end of last week, while passing through the Exchange in Edinburgh, I was seized with such a giddiness that I staggered, and would, I think, have fallen, had I not gone into an entry, where I leaned against the wall, and became quite unconscious for some seconds."' Dr Balfour informed him that he had been overworking his brain, and agreed to call at Shrub Mount on the following day to make a fuller examination.

Mr Miller had no sooner left the house to seek Dr Balfour, than Mrs Miller, turning to her daughter, said:—'There is something unusual the matter with your papa. I cannot be satisfied without more advice. I am quite certain he would make some difficulty about it; therefore you and I will go up by the 12 o'clock coach, and make an appointment with Professor Miller to come down here.' When Mr Miller returned, he mentioned that he had a funeral to attend at two o'clock, and would go to Edinburgh by the 12 o'clock coach. She told him that she and Miss Miller, having something to do in town, were to go at the same time. He inquired particularly what was their errand, and Mrs Miller, 'acting under a stern and inexorable necessity,' put him off with the statement that they wished to see a picture then being exhibited in Prince's Street. The picture was there, they wished to see it, and they actually did so on this occasion; but the principal object of their trip to town remained unseen in the background. Professor Miller, 'with his usual quick decisive kindness,' said, 'I'll be down to-morrow at three o'clock.' Mrs Miller arranged that Dr Balfour should come at the same hour.

When she returned home, she found her husband already there, resting on the sofa in the dining-room. He told her what he had been doing since they parted, dwelling especially on the precautions which he was using to avoid taking cold. In return, he wanted to know exactly what his wife and daughter had been about in town. 'We saw the picture, as we intended.' 'Nothing else? Was that really your chief business?' Mrs Miller saw that a suspicion of the true state of affairs had crossed his mind. She feared his serious displeasure, and thought it possible that, if not treated with frankness, he might keep out of the way and defeat her main aim. With hesitation, therefore, and placing her hand on his forehead, 'Don't,' she said, 'be displeased. I went likewise to Professor Miller to ask him to come and see you. He is to come at three to-morrow.

You won't object—you won't throw any obstacle in the way—if it were only to relieve *my* mind?' He made no answer, and remained silent for a considerable time. At the moment he was, Mrs Miller thinks, displeased; but the shadow soon passed from his face, and during the evening he was in his gentlest, kindest mood.

Next morning Mrs Miller again met him at the top of the stairs before breakfast, and was relieved to find that he had passed a better night. Immediately after breakfast he began to correct the last proofs of the *Testimony of the Rocks*. About midday he became restless, and she feared that he might make some movement which would prevent the consultation. The day was bitterly cold, with drizzling rain. She made pretences to be near him, watchful lest he should slip away unnoticed. At last he proclaimed his intention of going up to town to anticipate Professor Miller's visit. 'I cannot bear,' he said, 'taking him down in this way. You know his generosity, and he has so much to do—' 'Well, believe me,' replied Mrs Miller, 'there's not much he has to do he would put in competition with coming to see you when you need it. Just look at the day. You know that if you go out, you will bring on another inflammatory attack in your chest, and then I shall have done more harm than good. Do stay now, and let things go on, if you never do again.' 'Well,' he answered, 'I will.' Mrs Miller was passing his chair at the moment, and putting her hand into the shaggy hair which he used to wear on the top of his head, she gave it a slight tug, 'half a caress, half a playful rebuke for his contumacy,' while she thanked him for complying. 'Don't,' he said mildly, 'it hurts me.'

The medical gentlemen arrived and the interview commenced. 'We examined his chest,'—such is Professor Miller's report,—'and found that unusually well; but soon we discovered that it was head-symptoms that made him uneasy. He acknowledged having been night after night up till very late in the morning, working hard and continuously at his new book, "which," with much satisfaction he said, "I have finished this day." He was sensible that his head had suffered in consequence, as evidenced in two ways: first, occasionally he felt as if a very fine poniard had been suddenly passed through and through his brain. The pain was intense, and momentarily followed by confusion and giddiness, and the sense of being very drunk, unable to stand or walk. He thought that a period of unconsciousness must have followed this,—a kind of swoon, but he had never fallen. Second, what annoyed him most, however, was a kind of nightmare, which for some nights past had rendered sleep most miserable. It was no dream, he said: he saw no distinct vision, and could remember nothing of what had passed accurately. It was a sense of vague and yet intense horror, with a conviction of being abroad in the night wind, and dragged through places as if by some invisible power. "Last night," he said, "I felt as if I had been ridden by a witch for fifty miles, and rose far more wearied in mind and body than when I lay down." So strong was his conviction of having been out, that he had difficulty in persuading himself to the contrary, by carefully examining his clothes in the morning to see if they were not wet or dirty; and he looked inquiringly and anxiously to his wife, asking if she was sure he had not been out last night, and walking in this disturbed trance or dream. His pulse was quiet, but tongue foul. The head was not hot, but he could not say he was free from pain. But I need not enter into professional details. Suffice it to say, that we came to the conclusion that he was suffering from an overworked mind, disordering his digestive organs, enervating his whole frame, and threatening serious head affection. We told him this, and enjoined absolute discontinuance of work, —bed at eleven, light supper (he had all his life made that a principal meal), thinning the hair of the head, a warm sponging-bath at bed-time, &c. To all our commands he readily promised obedience. For fully an hour we talked together on these and other subjects, and I left him with no apprehension of impending evil, and little doubting but that a short time of rest and regimen would restore him to his wonted vigour.'

It may occur to many to ask how it could happen that medical men so circumspect, so vigilant, so able as Professor Miller and Dr Balfour, having become acquainted with these symptoms of insanity, did not suggest that precautions should be taken, to the extent at least of removing fire-arms from the person and presence of their patient. Miller's fear of robbery had returned in all its force. A revolver lay nightly within his reach. A broad-bladed dagger was ready to his hand. At his bed-head lay a naked sword. Why were not these taken away? The reply is that, though paroxysms of madness had already visited Miller, he had revealed no trace of suicidal mania, and the circumstances under which his mental disorder had originated were not such as to suggest alarm. His brain had been overworked; but the labour which had shaken his nerves and sapped his strength, had been congenial to him; and his conversation was that of one who looked with hope and with interest to the future. His intellect, besides, apart from the maniacal belief which at moments oppressed it, that he was made the sport of demons in the night, was strong and clear. That powerful action of the mind may take place even when it is under the spell of overmastering mania is demonstrated by the fact that Cowper wrote in a single day, while suffering 'the most appalling mental depression,' the *Castaway*, a poem which, in the masculine brevity of its narrative, and the concentration of its tragic power, is altogether masterly. Listening to the conversation of Hugh Miller, which, even while he described the agonies that tortured him, attested the vigour of his faculties, the medical gentlemen never thought it possible that intensity of mental horror might suddenly paralyze his will, and deprive reason of control over his actions. It is so easy to be wise after the event! so difficult to forecast the steps of destiny! The riddle of the Sphynx, once we know its solution, looks childishly simple; and yet it took an Œdipus to read it.

Professor Miller and Dr Balfour having left Shrub Mount, it was now time for dinner. The servant entered the dining-room to spread the table, and found Mr Miller alone. The expression which once or twice already had been observed on his features was again there. His face was so distorted with pain, that she shrank back appalled. He lay down upon the sofa, and pressed his head, as if in agony, upon the cushion.

The paroxysm flitted by, and when Mrs Miller returned to the dining-room he was in apparent health. She naturally shared the hopeful anticipations which Professor Miller had expressed, and her mind would dwell with comfort on the regimen and rest which were now, by medical authority, to be brought to bear upon his case. After dinner the conversation turned upon poetry. Miss Miller, 'then just blooming into womanhood, between sixteen and seventeen years of age,' was at the time attending classes, and among other tasks, had to produce verses upon given themes. She consulted her father upon her performances, and he would take the opportunity of delivering a chatty little lecture upon poetry, taking down from the shelf the works of some well-known author, to serve for the illustration of his remarks. This evening the subject of lecture was Cowper, one of his supreme favourites. He ranked the bard of Olney, both in poetry and in prose, among the great masters of the English language. The verses on Yardley Oak he would often refer to, as evincing the wonderful power with which Cowper could bend the roughest words to suit his purposes of delineation and of melody. The lines are perhaps the fittest which a critic could select to illustrate the genius of Cowper. They have that brief, decisive force by which, at his best, he recalls the mighty touch of Dryden, with a vivid, eye-to-eye truth to nature which reminds us that Cowper, if the poetical child of Dryden, was the poetical sire of Wordsworth.

> 'Thou wast a bauble once ; a cup and ball,
> Which babes might play with ; and the thievish jay,
> Seeking her food, with ease might have purloined
> The auburn nut that held thee, swallowing down
> Thy yet close-folded latitude of boughs,
> And all thine embryo vastness, at a gulp.
> But fate thy growth decreed ; autumnal rains
> Beneath thy parent-tree mellowed the soil
> Designed thy cradle ; and a skipping deer,
> With pointed hoof dibbling the glebe, prepared
> The soft receptacle, in which, secure,
> Thy rudiments should sleep the winter through.'

The vein of reflection, too, obscured by no mysticism yet touching on deep things, which runs through the piece, would please Miller.

> ' While thus through all the stages thou hast pushed
> Of treeship—first a seedling, hid in grass ;

Then twig; then sapling; and, as century rolled
Slow after century, a giant bulk
Of girth enormous, with moss-cushioned root
Upheaved above the soil, and sides embossed
With prominent wens globose—till at the last
The rottenness, which Time is charged to inflict
On other mighty ones, found also thee.
What exhibitions various hath the world
Witnessed, of mutability in all
That we account most durable below!
Change is the diet on which all subsist,
Created changeable, and change at last
Destroys them. Skies uncertain now the heat
Transmitting cloudless, and the solar beam
Now quenching in a boundless sea of clouds;
Calm and alternate storm, moisture and drought,
Invigorate by turns the springs of life
In all that live, plant, animal, and man,
And in conclusion mar them. Nature's threads,
Fine passing thought, even in her coarsest works,
Delight in agitation, yet sustain
The force that agitates, not unimpaired;
But worn by frequent impulse, to the cause
Of their best tone their dissolution owe.'

Mrs Miller's little Christmas volume, *Cats and Dogs*, which has since been highly popular, was then passing through the press. Mr Miller took a lively interest in it, and gave a warmly favourable opinion of its qualities. He now asked the children if they knew Cowper's lines *To a Retired Cat*, and, on their answering in the negative, read them with sprightly appreciation. The cat, it may be remembered, finding an open drawer, lined with the softest linen, concludes that it has been prepared expressly for her accommodation, falls fast asleep in it, is immured by the chambermaid, and is in danger of being starved.

'That night, by chance, the poet watching,
Heard an inexplicable scratching;
His noble heart went pit-a-pat,
And to himself he said—"What's that?"
He drew the curtain at his side,
And forth he peep'd, but nothing spied;

Yet, by his ear directed, guessed
Something imprisoned in the chest,
And, doubtful what, with prudent care,
Resolved it should continue there.
At length, a voice which well he knew,
A long and melancholy mew,
Saluting his poetic ears,
Consoled him and dispelled his fears;
He left his bed, he trod the floor,
He 'gan in haste the drawers explore,
The lowest first, and without stop
The rest in order to the top.
For 'tis a truth well known to most,
That whatsoever thing is lost,
We seek it, ere it come to light,
In every cranny but the right.
Forth skipped the cat, not now replete
As erst with airy self-conceit,
Nor, in her own fond apprehension,
A theme for all the world's attention,
But modest, sober, cured of all
Her notions hyperbolical,
And wishing, for a place of rest,
Anything rather than a chest.
Then stepped the poet into bed
With this reflection in his head:
Beware of too sublime a sense
Of your own worth and consequence.
The man who dreams himself so great,
And his importance of such weight,
That all around in all that's done
Must move and act for him alone,
Will learn in school of tribulation
The folly of his expectation.'

The father reading and remarking—the children with happy faces and merry trills of laughter clustered round his knee—the mother tranquillized and hopeful after her terrible anxieties of the preceding days—such is the spectacle which, on this Tuesday evening, two days before Christmas, 1856, we behold in the home of Hugh Miller. Mrs Miller was making tea, when she heard his voice 'in tones of anguish' reading *The Castaway*. Here are a few of the verses.

'Obscurest night involved the sky,
The Atlantic billows rolled,
When such a destined wretch as I,
Washed headlong from on board,
Of friends, of hope, of all bereft,
His floating home for ever left.

'Not long beneath the whelming brine,
Expert to swim, he lay;
Nor soon he felt his strength decline,
Or courage die away;
But waged with death a lasting strife,
Supported by despair of life.

'He long survives who lives an hour
In ocean, self-upheld:
And so long he, with unspent power,
His destiny repelled;
And ever, as the minutes flew,
Entreated help, or cried—"Adieu!"

'At length, his transient respite past,
His comrades, who before
Had heard his voice in every blast,
Could catch the sound no more.
For then, by toil subdued, he drank
The stifling wave, and then he sank.

'No poet wept him: but the page
Of narrative sincere,
That tells his name, his worth, his age,
Is wet with Anson's tear;
And tears by bards or heroes shed
Alike immortalize the dead.

'I therefore purpose not, or dream,
Descanting on his fate,
To give the melancholy theme
A more enduring date:
But misery still delights to trace
Its semblance in another's case.

'No voice divine the storm allayed,
No light propitious shone,
When, snatched from all effectual aid,
We perished, each alone:
But I beneath a rougher sea,
And whelmed in deeper gulfs than he.'

Mrs Miller, however, was not alarmed. She felt that her husband, in capacity of critic and reader, was merely bringing out by sudden and skilful contrast the range of Cowper's power, now archly droll, now sternly tragical. He turned, last of all, to the *Lines to Mary*, and, at certain of the verses, she could perceive half-stolen glances at her over the page.

'Thy silver locks, once auburn bright,
Are still more lovely in my sight
Than golden beams of orient light,
My Mary!

'For could I view nor them nor thee,
What sight worth seeing could I see?
The sun would rise in vain for me,
My Mary!

'Partakers of thy sad decline,
Thy hands their little force resign;
But, gently pressed, press gently mine,
My Mary!

'Such feebleness of limb thou prov'st,
That now, at every step thou mov'st
Upheld by two, yet still thou lov'st,
My Mary!

'And still to love, though pressed with ill,
In wintry age to feel no chill,
With me is to be lovely still,
My Mary!'

Early in the evening, as might have been expected after that agitating day, Mrs Miller began to feel weary. She went to the kitchen and gave particular orders about his bath; then, returning to the sitting-room, she remarked to him that she was 'very, very tired' and would be forced to leave him sooner than usual. 'Now,' he said, 'that I am forbidden ale or porter, don't you think that in future I might have a cup of coffee before going to bed?' With some hesitation she assented, and promised that the coffee should be brought him. The good-night kiss followed, and she retired to slumbers which were probably the deeper on account of the excitement and fatigue, the anxieties and consolations, of the preceding day. Miller went up-stairs to his study. At the appointed hour he took the bath, but, alas! his intense repugnance to physic prevailed over him, and the dose of prescribed medicine was left untouched. From his study he went into his sleeping-room and lay down upon his bed.

At what hour can never be ascertained, but either in the dead of night or in the grey dawn of morning, he arose from the bed and half dressed himself. Then the trance of paroxysmal horror again came over him, and the maniacal persuasion which had for days been haunting him drove him mad. He rushed to the table, and, on a folio sheet of paper, on the centre of the page, traced the following lines.

'DEAREST LYDIA,

'My brain burns. I *must* have *walked;* and a fearful dream rises upon me. I cannot bear the horrible thought. God and Father of the Lord Jesus Christ, have mercy upon me. Dearest Lydia, dear children, farewell. My brain burns as the recollection grows. My dear, dear wife, farewell.

'HUGH MILLER.'

The iron resolution and courage of the man appeared even in the maniac. He wore a thick woven seaman's jacket over his chest. This he raised on the left above the heart, and, applying the muzzle of his revolver, fired. The ball perforated the left lung, grazed the heart, cut through the pulmonary artery at its root, and lodged in the rib on the right side. The pistol slipped from his hand into the bath which stood close by, and he fell dead instantaneously. The body was found lying on the floor, the feet upon the study rug.

A *post-mortem* examination having been made, the following report was the result:—

'Edinburgh, December 26, 1856.

'We hereby certify on soul and conscience, that we have this day examined the body of Mr Hugh Miller, at Shrub Mount, Portobello.

'The cause of death we found to be a pistol-shot through the left side of the chest; and this, we are satisfied, was inflicted by his own hand.

'From the diseased appearances found in the brain, taken in connection with the history of the case, we have no doubt that the act was suicidal under the impulse of insanity.

'JAMES MILLER,
'A. H. BALFOUR,
'W. T. GAIRDNER,
'A. M. EDWARDS.'

It is a melancholy satisfaction to reflect that, in no case of suicide which ever took place, can the evidence of insanity have been more express or conclusive. Had no trace of disease been found in the brain—had no word written by Hugh Miller at the last attested madness—the overwork to which he had subjected himself, the excitement to which he had been a prey, would have afforded adequate grounds for believing him insane. But the actual mania which was gaining the mastery over him had been defined by himself some days before his death; and this mania, namely, that he was driven by witches or demons in the darkness, is specified beyond possibility of mistake or doubt, in the thrilling words, 'I *must* have *walked*.' That even when he was the victim of mania, the tenderness of his nature survived, that he could still discriminate the supremacy of his affection for the wife of his youth, that the cry of his heart, when reason was eclipsed in madness and the shadow of death fell on the reeling brain, rose clear to God the Father and the Lord Jesus Christ, will be dwelt on with sad interest by those whom Hugh Miller taught to love him with inexpressible love.

'The body,' writes Dr Hanna, who was one of the first to see it, 'was lifted and laid upon the bed. We saw it there a few hours afterwards. The head lay back, side-ways on the pillow. There was the massive brow, the firm-set, manly features, we had so often looked upon admiringly, just as we had lately seen them,—no touch nor trace upon them of disease,—nothing but that overspread pallor of death to distinguish them from what they had been. But the expression of that countenance in death will live in our memory for ever. Death by gun-shot wounds is said to leave no trace of suffering behind; and never was there a face of the dead freer from all shadow of pain, or grief, or conflict, than that of our dear departed friend. And as we bent over it, and remembered the troubled look it sometimes had in life, and thought what must have been the sublimely-terrific expression that it wore at the moment when the fatal deed was done, we could not help thinking that it lay there to tell us, in that expression of unruffled majestic repose that sat upon every feature, what we so assuredly believe, that the spirit had passed through a terrible tornado, in which reason had been broken down; but that it had made the great passage in safety, and stood looking back to us, in humble, grateful triumph, from the other side.'

The excitement occasioned by the event throughout Scotland was tremendous, and no such funeral had taken place in Edinburgh since that of Chalmers. He was laid in the Grange cemetery, near the spot where Chalmers rests.

From a large number of letters received by Mrs Miller on the occasion of her husband's death, the three which follow are selected for publication.

'Tavistock House, London, Thursday, April 16th, 1857.

'DEAR MADAM,

'Allow me to assure you that I have received the last work of your late much-lamented husband with feelings of mournful respect for his memory and of heartfelt sympathy with you. It touches me very sensibly to know, from the inscription appended to the volume, that he wished it to be given to me. Believe me, it will fill no neglected place on my book-shelves, but will always be precious to me, in remembrance of a delightful writer, an accomplished follower of science, and an upright and good man.

'I hope I may, dear Madam, without obtrusion on your great bereavement, venture to offer you my thanks and condolences, and to add that, before I was brought into this personal association with your late husband's final labour, I was one of the many thousands whose thoughts had been much with you.

'Yours faithfully and obliged,

'CHARLES DICKENS.'

MRS HUGH MILLER.

'Chelsea, April 15th, 1857.

'MY DEAR MADAM,

'Last night I received a Gift of your sending, which is at once very precious and very mournful to me.

'There is for ever connected with the very title of this Book the fact that, in writing it, the cordage of a strong heart cracked in pieces; that the ink of it is a brave man's life-blood! The Book itself, I already see, is full of grave, manly talent, clearness, eloquence, faithful conviction, inquiry, knowledge; and will teach me and others much in reading it: but that is already an extrinsic fact, which will give it a double significance to us all. For myself, a voice of friendly recognition from such a man, coming to me thus out of the still kingdoms, has something in it of religion; and is strange and solemn in these profane, empty times.

'In common with everybody, I mourned over the late tragic catastrophe; the world's great loss, especially your irreparable and ever-lamentable one: but as for *him*, I confess there was always present, after the first shock, the thought that at least he was out of bondage, into freedom and rest. I perceived that, for such a man there was no rest appointed except in the countries where he now is!

'Dear Madam, what can we say? The ways of God are high and dark, and yet there is mercy hidden in them. Surely, if we know anything, it is that "His *goodness* endureth for ever." I will not insult your grief by pretending to lighten it. You and your little ones, yes, you have cause, as few have had, to mourn; but you have also such assuagements as not many have.

'With respectful sympathy, with many true thanks and regards, I remain,

'Sincerely yours,

'T. CARLYLE.'

'Denmark Hill, April 9th, 1857.

'My dear Madam,

'I received yesterday evening the book which I owe to the kindness of your late husband, and which I receive as from his hand; with mingled feelings, not altogether to be set down in a letter, even if I could tell you them without giving some new power of hurt, if that be possible, to your own sorrow. But there are one or two things which I *want* to say to you. Humanly speaking, I cannot imagine a greater grief than yours, or one which a stranger should more reverently or more hopelessly leave unspoken of, attempting no word of consolation; and yet I can fancy that there is one point in which you may not yet have enough regarded it. To all of us, who knew your late husband's genius at all,—to you, above all, who knew it best,—it seems to me that the bitterest cruelty of the trial must lie in the sense of his work being so unaccomplished, of all that he might have done, had he lived; and of the *littleness* of the thing that brought about his illness and death. It seems so hard that a little overwork, a few more commas to be put into a page of type, a paragraph to be shortened or added, in the last moment, should make the difference between life and death. Perhaps your friends have dwelt too much—if they have attempted to help you at all—on ordinary beaten topics of religious consolation, not, it seems to me, applying to the worst part of this sorrow, and they may not have dwelt enough on what does fully bear upon it, namely, the general law of Providence in God's "*strange* work." We rarely *see* how small the things are which bring about what He has appointed, nor do we see, generally, the strange loss, which takes place continually, of the powers He gives.

If you *could* see this, you would not feel that He had set you up as a mark, and spared no arrows. That which has befallen you, though you do not think it, is yet the common lot of man. The earth is full of *lost* powers; no human soul perishes, but, if you could only read its true history, you would find that not the thousandth part of its possible work had been done; that even when the result seemed greatest the man either was or ought to have been conscious of irreparable failure and shortcoming; that, in the plurality of cases, the whole end and use of life had been more or less lost, and, in *many* cases, in the cruelest way, by accident or adversity. And in like manner, if you could only see the *origin* of all diseases, you would see that what we called a natural disease, and received as an inevitable dispensation, did in reality depend on some pettiest of petty *chances* (I speak humanly); on the man's having untied his neckerchief near a window, when he should not; on his having stopped at the street-corner in an east wind to talk to a friend for half a minute; on his having worried himself uselessly about an overcharge in a bill; nothing is so trivial, but it may be the Appointed Death-Angel to the man. And when once you feel this fully (my own work has taught me this more than most men's, for no wreck is so frequent, no waste so wild, as the wreck and waste of the minds of men devoted to the arts), when once you feel it, and understand that this waste, which seems so wonderful to us, is intended by the Deity to be a part of his dealing with men (just as the rivers are poured out to run into their swallowing Death-sea, only a lip here and there tasting them), and that this law of chance, which seems so trivial to us, is as entirely in His hand as the lightning and the plague-spot; then, while to all of us who are still counting the hours, the truth is a solemn one; to those who mourn for their dead, it ought not to be a distressing one. It is only to our narrow human view that anything is lost or wasted. God gave the mind to do a certain work, and withdrew it when that work was done; we, poor innocents, may fancy that something else should have been done; so assuredly, in all cases, it should; but in no special and separate instance can we say,—here is a destiny peculiarly broken, here a work peculiarly unfulfilled. I read that God will say to his good servants, "Well done," but not, "*Enough* done." It is only He who judges of and appoints that "enough."

'Pardon me if I pain you by dwelling on this, but I know that many persons do not feel this *generalness* in human shortcoming; we are all too apt to think everything has been right if a man lives to be old, and everything lost if he dies young.

'I have not been able to look much at the book yet, but it seems a noble bequest to us.

'Believe me, my dear Madam, always respectfully and faithfully yours,

'J. RUSKIN.'

I was personally acquainted with Hugh Miller only in the few last years of his life. I knew that my relatives in the Highlands were on terms of intimate friendship with him, and that I could be introduced to him whenever I chose; but I had a vague feeling that I could know him most advantageously by thoroughly studying one or more of his books. I used to attend the lectures delivered in connection with the Edinburgh Philosophical Institution, and it was on its platform that I first had full view and hearing of

Hugh Miller. The spell of his personality fell upon me at once. The large shaggy head, the massive bust, the modest grandeur of demeanour, the unmistakable impress of power on calm lip and in steadfast eye, the look of originality and rugged strength about the whole man, arrested and enchained me. There breathed around him, also, a certain indefinable influence of gentleness and noble affection. You felt that he was honest and good. The mood of admiring indifference with which I had regarded him gave place to ardent interest. His language, clear, idiomatic, melodious, derived zest from his provincial accent, and I shall never forget the moment when he concluded a classically picturesque and beautiful passage, descriptive of an ancient Scottish forest, with the words, 'the exe hed been buzzy in its gleds,' i. e. the axe had been busy in its glades. I now lost no time in becoming personally acquainted with Hugh Miller. My welcome was cordial; I soon felt myself at home in his society and household; and would look in not unfrequently of an evening, taking part in his favourite meal, supper, and engaging with him in free and varied conversation.

One of the first things I noted in the domestic circle to which I had been introduced was the harmonious and seemly relation subsisting between husband and wife. There was no trace of intellectual dictatorship on his part; he carefully considered and highly valued every remark of Mrs Miller's. On her side there was reverence for his mental power, without the slightest sacrifice of independent judgment, and with an obvious and lively delight in argumentative conversation. At times, in the friendliest tone, she would hint to him that some part of his dress might be improved, and pleasant little banterings—'netted sunbeams' on the surface of the stream, showing the depth of the flow of conjugal happiness beneath,—would occur upon the subject. I remember, for instance, that his hat was once pronounced exceptionable, and that, by way of providing himself with an ally against Mrs Miller on the point, he had trained his son Hugh, just beginning to toddle and to lisp, to say,

> 'Papa has got a very bad hat,
> And many a word he hears about that.'

The impression formed in my mind was that Mr and Mrs Miller were on exactly those terms on which it was desirable and beautiful that a man eminent in the intellectual world and his wife should be.

By Hugh Miller's manner to myself I was fascinated, and not a little surprised. He talked little and heard much. It seemed to please him to observe the lights cast on subjects by a younger mind; the line of my culture, also, was somewhat different from his own, for I then cared comparatively little for science, and was in a perpetual glow of enthusiasm about Shelley, Keats, Tennyson, Carlyle, and Ruskin; and my opinions, quietly allowed to gush forth in the artless gabble of youth, may have interested him as natural curiosities. Happily I have forgotten my own remarks, unhappily also most of his. He never spoke on the impulse of the moment, though he never hesitated. The opinion had been considered long before, and put into its place in his memory; and it was brought out with a calm precision which suggested that he had ceased to discuss the subject. His gentleness, his willingness to listen, his intellectual tolerance and fine kind sympathy, took captive my heart. This man, I felt, whatever else he may or may not be, is a born gentleman. The idea got hold of my mind, and in a sketch of Hugh Miller contributed to an Edinburgh magazine at the time, I dwelt on it at some length. The passage, slightly abbreviated, may as well be printed here. It conveys, in spite of juvenility of expression, a more vivid notion of what then struck me in his deportment than any words I could now use.

'Perhaps the highest compliment, all things considered, which we can pay to Hugh Miller is this,—that he is in the best sense a gentleman, that he is truly and strictly polite. True politeness is one of the rarest things; it may be met with in the hut of the Arab, in the courtyard of the Turk, in the cottage of the Irishman; and it is excessively scarce in ball-rooms. It is independent of accent and of form; it is one of the constant and universal noble attributes of man, wherever and howsoever developed. We venture to define it thus: Politeness is natural, genial, manly *deference*, with delicacy in dealing with the feelings of others, and without hypocrisy, sycophancy, or obtrusion. This definition excludes a great many people. We cannot agree that Johnson was polite; that is, if politeness is to be distinguished from nobleness, courage, and even kindness of heart; in a word, from everything but itself. Burns was polite, when jewelled duchesses were charmed with his ways; Arnold was polite, when a poor woman declared that he treated her like a lady; Chalmers was polite, when every old woman in Morningside was elated and delighted with his courteous salute. But Johnson, who shut a civil man's mouth with 'Sir, I perceive you are a vile Whig,' who ate like an Esquimaux, who deferred so far to his friends that they dared to differ with him only in a round-robin, was not polite. Politeness is the last touch, the finishing perfection, of a noble character. It is the gold on the spire, the sunlight on the corn-field, the smile on the lip of the noble knight lowering his sword-point to his ladye-love. It results only from the truest balance and harmony of soul. We assert Hugh Miller to possess it. A duke in speaking to him would know he was speaking to a man as independent as himself; a boy, in expressing to him an opinion, would feel unabashed and easy, from his genial and unostentatious deference. He has been accused of egotism. Let it be fairly admitted that he knows his name is Hugh Miller, and that he has a colossal head, and that he once was a mason; his foible is probably that which caused Napoleon, in a company of kings, to commence an anecdote with "When I was a lieutenant in the regiment of La Fere." But we cannot think it more than a very slight foible; a manly self-consciousness somewhat in excess. Years in the quarry have not dimmed in Hugh Miller that finishing gleam of genial light which plays over the framework of character, and is politeness. Not only did he require honest manliness for this; gentleness was also necessary. He had both, and has retained them; and therefore merits fairly

"The grand old name of gentleman."'

It was impossible to be long in Miller's company without perceiving the ardour of his devotion to science. He considered literature inferior to science as a gymnastic of the mind. For the facile culture of the age he had great contempt, and ranked both religion and labour as stimulating, training agencies for mind and character, higher than what is commonly called education. 'As for the dream,' he says in one of his books, 'that there is to be some extraordinary elevation of the general platform of the race achieved by means of education, it is simply the hallucination of the age,—the world's present alchemical expedient for converting farthings into guineas, sheerly by dint of scouring.' All that he had won had been won by stern effort, and he had no faith in royal roads to any kind of attainment.

Supremely devoted to science, he naturally regarded men eminent in science as standing higher than men eminent in literature. I recollect that once when the claims of Scotland to honour in the world of intellect were on the carpet, I remarked that, at all events, the first man of mind for the time being was a Scotchman. 'Who is that, Mr Bayne?' he asked. My reply was prompt and decisive, 'Thomas Carlyle!' 'Ah, no!' he said with great deliberation, 'Carlyle is not the greatest living man.' 'Who, then?' I inquired. He would not name any one, but repeated 'Ah, no! Mr Bayne, Carlyle is not the greatest.' From something in his manner at the moment the impression was conveyed to me that he would not have been surprised to hear himself named as the greatest of living Scotchmen. I gathered, also, from some of his remarks which I cannot precisely recall, that, if forced to say who was, in his opinion, the most remarkable man of the day, he would have sought him among the servants of science rather than among the cultivators of literature. He recognized Carlyle's great powers, however, and sympathized with many of the views of Rhadamanthus as to the superficiality and sentimentalism of the time. He more than half approved of Carlyle's stern views as to the treatment of criminals, and would have liked to see our grim disciplinarian try his hand upon a few drafts from the 'devil's regiments of the line.'

Once or twice Mr Gilfillan was mentioned. Miller had a friendly feeling towards him and expressed admiration of his brilliancy, but thought it overdone.

Mr G.'s style, he said, had an amount of ornament disproportioned to his thought. It suggested to him a lady's dress so stiffened with jewellery that it could stand without anybody in it. We once spoke of Professor Aytoun's satire on Mr Gilfillan. Miller had not seen it, and was interested by my account of the production. I mentioned that the satirized critic, Apollodorus, is represented as writing to eminent men, and commenting thus upon the manner of their reply :

> 'What have they answered? Marry, only this,
> Who, in the name of Zernebock, are you?'

He laughed heartily. There were, I added, far harder things in the poem. Apollodorus, for example, was introduced soliloquizing :—

> 'Why do men call me a presumptuous cur,
> A vapouring blockhead, and a turgid fool,
> A common nuisance and a charlatan?'

This sharply displeased him. 'That is too severe,' he said. He became silent, showed no disposition to continue the conversation, and seemed to brood indignantly on the cruelty of the words. I am not sure that it would have been safe at that moment for Mr Aytoun to have come into the presence of Hugh Miller. 'There was,' says Professor Masson, in his masterly sketch of his friend, 'a tremendous element of ferocity in Hugh Miller. It amounted to a disposition to kill. He was a grave, gentle, kindly, fatherly, church-going man, who would not have hurt a fly, would have lifted a child tenderly out of harm's way in the street, and would have risked his life to save even a dumb creature's; but woe betide the enemy that came athwart him when his blood was up! In this there was more of the Scandinavian than of the Celt. It appeared even in his newspaper-articles. At various times he got into personal controversies, and I know no instance in which he did not leave his adversary not only slain, but battered, bruised, and beaten out of shape. It seemed to be a principle with him—the only principle on which he could fight—that a battle must always be *à l'outrance,* that there could be no victory short of the utter extermination of the opposed organism. Hence, in the course of his editorial career, not a few immense, unseemly exaggerations of the polemical spirit,—much sledge-hammering where a tap or two would have sufficed. A duel of opinions was apt to become with him a duel of reputations and of persons.'

This is strongly put. I have not found in Hugh Miller anything amounting to a disposition to kill. But he possessed a huge capacity of wrath, and that love of adventure, of danger, of battle, which distinguishes his nation. It has often occurred to me that he had in him the making of a great general. His eye would have embraced in a moment the points of a position, and his spirit would have risen into clearness and stern joy in proportion as difficulty and peril increased. He would have inspired his men with absolute faith in him, as a rock to stand, as an avalanche to charge.

It was frequently late at night when I took leave, and he would insist on giving me convoy from Stuart Street, near Portobello, where he lived at the time, to the outskirts of Edinburgh. We used to part at the Calton Hill. On one occasion he pulled out a pistol, and remarked that he was prepared if any one should attack us. Having no thought of danger, I was much surprised. By way of explanation, he gave me an account of his adventure when walking by night between Wolverton and Newport Pagnell, as detailed in the *First Impressions*. The district was at the time infested with ruffians and blackguards, who had come to see a prize-fight between Caunt and Bendigo for the championship of England. 'About half-way on,' says Miller, 'where the road runs between tall hedges, two fellows started out towards me, one from each side of the way. "Is this the road," asked one, "to Newport Pagnell?" "Quite a stranger here," I replied, without slackening my pace; "don't belong to the kingdom, even." "No?" said the same fellow, increasing his speed as if to overtake me; "to what kingdom, then?" "Scotland," I said, turning suddenly round, somewhat afraid of being taken behind by a bludgeon. The two fellows sheered off in double-quick time, the one who had already addressed me muttering, "More like an Irishman, I think;" and I saw no more of them. I had luckily a brace of loaded pistols about me, and had at the moment a trigger under each fore-finger.' He first carried pistols when conveying the Bank's money between Cromarty and Tain. Soon after coming to Edinburgh, he commenced a series of investigations in the Lothian coal-field, which brought him into lonely places where he believed that suspicious characters might lurk. He took with him his fire-arms, therefore, and I should suppose that, in subsequent years, he was seldom abroad in the night without them.

In conversation, as in his books, he was sensitively orthodox. I once spoke with enthusiastic admiration of that famed vision of Jean Paul's, in which the author, with a view to symbolizing the horror of atheism, introduces Jesus Christ looking up into a blank universe, one vast hollow eye-socket, emptied of its eye, and wailing for His Father. Miller would see in the piece nothing beyond the poetical expression of a lofty Unitarianism, and maintained that Jean Paul intended to deny the Divinity of Christ. His Unitarianism might be more spiritual than that common in England, but Unitarianism it was. Mrs Miller and I took the opposite view, arguing that it was legitimate in the imaginative dreamer to introduce Christ as the representative of created being, and to illustrate the ghastliness of atheism by letting us see Him, a homeless orphan, filling with moans the black hollow of the universal night; but Miller held to his point.

We did not speak often of the Church controversy or the Disruption; but once, in his most earnest mood, he expressed that opinion as to the Jehu-like driving of the Evangelical leaders which I formerly mentioned. It was easy to perceive that he transferred to the Edinburgh Radicals that dislike with which he had regarded those of Cromarty. He was a steadfast Whig, but his sympathies were more with Conservatism than Radicalism. One of his profoundest enthusiasms was for the poetry of Burns, and I have heard him declaim one of Burns's grandest passages,—that about Scotland in the Vision,—with great fervour and deep rhythmic intonation. His memory seemed to be full of English poetry, and, at sauntering times, in waiting for omnibuses or the like, he would pour out stanza after stanza with astonishing profusion. The last conversation I ever had with him was at Shrub Mount in 1855, when we walked about the garden in earnest talk. There was a great noise at the time about the antiquity of the human race, and the question whether mankind have proceeded from one local centre, or from more than one. He agreed with me in considering the question of man's antiquity of slight importance compared with that of the unity of the race; and cordially assented to the proposition that the relationship of men to Christ is a matter of infinitely more consequence than their relationship to Adam.

I had not much intercourse with Hugh Miller, but, such as it was, I treasure the recollection of it with reverent affection. Here was a man to inspire infinite trust, a sterling, incorruptible man, with whom one would like to stand shoulder to shoulder in the eddies of a stubborn fray. Incapable of malignity, his sympathetic sensibility vibrating in fine response to every touch of right sentiment, his intellectual interest playing over the whole field of human life and every province of nature, he had only to be known in order to be loved. He was very gentle, very kind, but you felt that, in the calm background of his mind, like a lion slumbering by a rock, slumbering but not asleep, lay a grand sense of moral rectitude, which, if wrong was threatened or baseness done, could spring forth in crashing ire.

I can now see that my impressions of him, not incorrect so far as they went, were essentially first impressions. With this biography I have been fitfully engaged for ten years, often interrupted in the work, and forced to go again and again over the same ground. The delay may have been in some respects inconvenient, but it has had the advantage of making me live all these years in constant converse with Hugh Miller. Nothing, I believe, of considerable importance, which he did, said, thought, or felt, has escaped me, although, of course, it was only a portion of those materials which could be used in this book. The better I have known, the more deeply have I revered, the more entirely have I loved him. A man of priceless worth; fine gold, purified sevenfold; delicate splendour of honour, sensitive and proud; perfect sincerity and faithfulness in heart and mind. He never failed a friend. His comrade of the hewing-shed sits down at his table when he has become one of the most distinguished men of his time; another friend is discovered to be at hand-grips with fortune, and he applies himself, with cunning delicacy, to solve the problem of inducing him to accept assistance. This was the manner and habit of the man.

Of his power of brain,—of his genius and originality,—his books, viewed in connection with the circumstances of his career, are the living witnesses. To their testimony must be added the fact of the great influence he exerted upon his contemporaries, the personal weight, the intellectual mass and magnitude, he was felt to possess. Professor Masson, commenting on the curious notion of a Fleet Street oracle, that he was devoid of genius, has declared his conviction that, 'if the word was applicable to the description of any mind, it was to the description of Hugh Miller's.' If we estimate the amount of obstruction which lay between the mason lad of Gairloch and Niddry, and the Hugh Miller of Edinburgh, whom Murchison, Lyell, Agassiz hailed as a brother, we shall admit that the opinion is not, *primâ facie*, unreasonable. I take liberty to add that, if genius means an indefinable something, conferred by nature, inimitable, incommunicable, never given twice in exactly the same form and colour,—a power of enchantment which all men feel, but no man can quite describe,—then the critic who denies genius to Hugh Miller does not understand his craft. He owed, without question, much to culture. Twenty years of study and practice, assiduous reading, careful self-correction, were required to perfect his prose style and to give him the complete command of it which he ultimately obtained. But all this only brought into clearness and use the gift born with him, a gift traceable in his earliest letters, a gift of tempered mental strength, of brightly keen perception and broad imaginative vision, a rare gift of expression, in subtly-modulated sentence, and exquisitely-felicitous image, and solemn harmony of sense and sound, and tenderly-brilliant colour lighting up the whole. Mr Robert Chambers proved himself to have the true critical eye when he referred, at a very early period of their intercourse, to the singular *interest* attaching to all he wrote. The omniscient little critics who deny him genius may imitate that if they can; it secured Miller a large and intelligent audience throughout the civilized world, an audience whose ear he caught so soon as his power was revealed, and which, thirty years after the publication of the *Old Red Sandstone*, continues to extend.

So far as I can penetrate the charm of his composition, it lies mainly in the fine continuity of it, in the absence of all jerking, jolting movement, in the *callida junctura* not of word to word merely, but of sentence to sentence, thought to thought, illustration to illustration. An author's peculiar excellence, if we have rightly discriminated it, will give us a hint as to where we should look for his besetting fault, and in reading Miller long at one time, we may find in his billowy regularity and smoothness of movement a sense of monotony. Yet on the whole there is a marvellous enchantment in his books; the breath of the hills is in them, the freshness of the west wind and the sea. Shall we not now venture to decide—the question was left open when it last came before us—that it was advantageous for him to break away from school, and betake himself to the caves and the wood? Nature is the only safe. nurse of genius; education is indispensable, but even the education must be suggested by nature, and come at her prompting. Shakspeare was an educated man; he had a large knowledge of the books of his time; but all generations would have been poorer if his brain had been drugged in boyhood with the trite erudition of Universities. Books came to Miller at the right moment; when he had already so filled his mind with Nature's imagery that they could do no more than genially assist him to use it. To read him is like taking a walk with him; we are never far from the crags and the waters, the dewy branch and the purple heather. Compare this with the prim urban elegance of Jeffrey, or the hard vehemence, like rainless thunder, or the full gallop of cavalry, of the style of Sir William Hamilton, and you will begin to realize how much Hugh Miller owed to the circumstance that his *Alma Mater* was his mother earth. As a naturalist, also, and as a geologist, his power came essentially from the same source. The hours on the ebb shore with Uncle Sandy, the tracing of the subtleties of life among the minute denizens of the crystal pools, the watching of the race of the waves when the tide turned, first slowly, tentatively, listlessly, each timorous wavelet with its lifted handful of dusty sand, then in hurrying clamorous advance of leaping foam and marshalled surge along the reaches of the shore,—the long years of toil in the quarry, and of wandering among the hills, to mark the fellowship of the rocks, and learn the joints and curvature of the bones of the world,—these gave him his intuitive sympathy with Nature's ways, his geological eye to discern how THE ARCHITECT had put together this and that bit of the planet. 'The thing was done so,' he could say, 'and not in the way you mention; you can fold up your theoretical demonstration when you please.'

May we not call his life, first and last, beautiful, august, heroic? From his father, whose very image he in his later years became, he derived the ground-work of his character, and for the education of conscience he was primarily indebted, though he little knew it at the time, to his Uncle James. In early manhood he was encompassed with hardship, with coarseness, with manifold temptation. His soul took no taint. He rose superior to every form of vulgarity, the vulgar ambition of wealth, the vulgar ambition of notoriety, the vulgar baseness of sensuality and licence. He aspired to fame, but it was to fame which should be the ratification of his own severe judgment. 'I have myself,' he said, 'for my critic;' and while the decision of this sternest censor was even moderately favourable, no sneers could depress, no applause elate him. His course, thenceforward, was a steadfast pursuit of truth and of knowledge, an unwearied dedication of himself to all that he believed to be true, and honest, and lovely, and of good report. In the meridian of his years, he threw himself into the noblest religious movement of his time, impelling and directing it, a movement which he largely contributed to carry to a triumphant issue. As his science had begun in converse with nature, so he carried on the study day by day and year by year, traversing thousands of miles for express purposes of observation, and at all moments, at home or in the field, he was awake with keenest vigilance to the powers of nature at work around him. Modern in all his habits of study, he did not, in tracing the surface light of science, forget that Divine mystery which nature shadows forth. God in the universe was for him a reality as forceful and present as for those Hebrew psalmists whom it inspired with the ublimest strains in existence. The autumn blast raised his mind to One that walketh on the wings of the wind: at *His* look the earth trembled in the throes of earthquake, and at *His* touch the volcano smoked.

Hugh Miller stands alone, so far as I am aware, among self-educated men of recent times, first, in the thoroughness of his education, the technically disciplined and ordered thinking to which he attained, secondly, in the absence from his books and letters of all extravagance, histrionism, paradox, of all trace of that furious, teeth-gnashing humour which has been so much in vogue in our century. Great instincts of order and of common sense, inherited from his father, allied him to what was stable in the institutions of his country. Religion, integrity, continence, moderation, obedience,—all those virtues against which the waves of modern anarchism beat wild,—saw him fighting behind their bulwark. They are shallow critics who recognize genius only, as Uriel recognized Satan, by the violence of its gestures and the devilishness of its scowl; in healthful times men of genius have neither affected a perverse singularity, nor taken as their dialect an everlasting snarl. That Hugh Miller was a man of genius would never have been called in question had his works not been so free from the distempers of genius.

THE END.

www.ingramcontent.com/pod-product-compliance
Ingram Content Group UK Ltd.
Pitfield, Milton Keynes, MK11 3LW, UK
UKHW040700180125
453697UK00010B/300